Günter Burkard,
Michael Fremerey (Eds.)

A Matter of Mutual Survival

Southeast Asian Modernities

edited by

Christoph Antweiler, Claudia Derichs,
Rüdiger Korff, Frauke Kraas, Boike Rehbein,
Jürgen Rüland, Judith Schlehe, Susanne Schröter,
Hermann Schwengel

Volume 10

LIT

A Matter of Mutual Survival

Social Organization of Forest Management in Central Sulawesi, Indonesia

edited by

Günter Burkard and Michael Fremerey

LIT

Cover Picture: Michael Fremerey

Layout: Susanne Bierwirth

Bibliographic information published by the Deutsche Nationalbibliothek
The Deutsche Nationalbibliothek lists this publication in the Deutsche Nationalbibliografie; detailed bibliographic data are available in the Internet at http://dnb.d-nb.de.

ISBN 978-3-8258-1468-7

A catalogue record for this book is available from the British Library

Verlagskontakt:
Fresnostr. 2 D-48159 Münster
Tel. +49 (0) 2 51/620 32 - 22 Fax +49 (0) 2 51/922 60 99
e-Mail: lit@lit-verlag.de http://www.lit-verlag.de

Auslieferung:

Deutschland/Schweiz: LIT Verlag Fresnostr. 2, D-48159 Münster
Tel. +49 (0) 2 51/620 32 - 22, Fax +49 (0) 2 51/922 60 99, e-Mail: vertrieb@lit-verlag.de

Österreich: Medienlogistik Pichler-ÖBZ GmbH & Co KG
IZ-NÖ, Süd, Straße 1, Objekt 34, A-2355 Wiener Neudorf
Tel. +43 (0) 2236/63 535-290, +43 (0) 2236/63 535 - 243, mlo@medien-logistik.at

Distributed in the UK by: Global Book Marketing, 99B Wallis Rd, London, E9 5LN
Phone: +44 (0) 20 8533 5800 – Fax: +44 (0) 1600 775 663
http://www.centralbooks.co.uk/html

Distributed in North America by:

Transaction Publishers
New Brunswick (U.S.A.) and London (U.K.)

Transaction Publishers
Rutgers University
35 Berrue Circle
Piscataway, NJ 08854

Phone: +1 (732) 445 - 2280
Fax: + 1 (732) 445 - 3138
for orders (U. S. only):
toll free (888) 999 - 6778
e-mail:
orders@transactionspub.com

Contents

Introduction

Michael Fremerey

This volume contains a collection of articles based on six years (2000-2006) of empirical social science research in forest margin communities around the Lore Lindu National Park in Central Sulawesi, Indonesia. The research activities were part of a larger German-Indonesian multidisciplinary research programme on the "Stability of Rainforest Margins" (STORMA), jointly conducted by scholars from the universities of Göttingen, Kassel, Bogor (IPB) and Palu (UNTAD). As a "Collaborative Research Center" (SFB) it is funded by the German Research Association (DFG). Sociological research was lead-managed by the Institute for Socio-cultural Studies (ISOS), University of Kassel, however, in close partnership with the Center for Agrarian Studies, Agricultural University of Bogor (IPB). It was conducted under the title of "Social organization and processes of ecological stabilization and destabilization" (A2).[1] Throughout the six-year term, an unrestricted co-operative approach among the German and Indonesian participants, including joint responsibilities in drafting and executing research, was applied. Synergies emanating from this close collaboration have been clearly felt in the process of joint investigation (Fremerey and Panggabean 2004: 530), and they hopefully are perceptible from the contributions in this volume.

The book refers to a worldwide and particularly topical issue, i.e. the declining forest resources and man's role in the observed processes of nature degradation. However, it refrains from rather simplistic protectionist approaches which, openly or implicitly, boil down to a separation between man and nature in order to avoid the depletion of natural resources. The title of this volume, "A Matter of Mutual Survival", indicates that we emphasize the reciprocity in this relationship.

---

[1] Two scholars, who have contributed to this volume, M. Shohibuddin and K. Glenk, have participated in other sub-projects of the research scheme (A1 and A5 respectively). We are grateful that they have provided their papers to complement the perspectives in this volume.

The title may sound somewhat dramatic but, as the contributions will show, it is nothing less than that: social, economic and even physical survival of people living at the margin of the Lore Lindu National Park depend on the existence and use of forest resources. On the other hand, these resources will soon be depleted if people do not succeed in utilizing them in more sustainable ways.

The village communities under investigation are bordering a forest zone, which was formally ascribed the status of a National Park in 1993. This includes severe restrictions in access to and use of forest resources beyond an established borderline. In many places this borderline has been drawn without proper consultation of the villagers, neglecting their dependence on the resources of accessible forest areas.

The use of forest resources as, for instance, wood (for fuel and construction), rattan, resin, game or herbs is a long-standing tradition of communities bordering the forest. Besides that, the forested area became increasingly important as acreage for perennial crops, like coffee and cacao, long before the government declared the forest as a national park. In recent times, the use of forest resources and the transformation of forest reserves into cultivated areas have been largely driven by the growth of the population, the exhaustion of non-forested arable land, and, last but not least, the lack of opportunities for non-farm income. To a great extent, the population growth is to be attributed to in-migration. Many of these migrants, particularly the *Bugis* from South Sulawesi, are moving to the Lore Lindu region with the explicit intention (and skills) to use forested areas for growing perennials, in particular cacao.

The overt needs of the villagers and the resulting pressures and demands, the massive intervention and local involvement of some NGOs and, finally, the factual incapability of central or regional authorities to effectively control the activities of local people in the forest, have caused some reconsideration on the side of relevant official agencies, in particular the National Park Authority. This led to a revised policy which intends to give the local communities more freedom and rights to autonomously manage their forest resources. A set of new national laws and regulations strongly backs this policy, which prepares the ground toward a substantial political decentralization (Law 22/1999) and increased local autonomy in utilizing natural resources (Law 41/1999 and Min. of Agriculture Decree 5/1999).

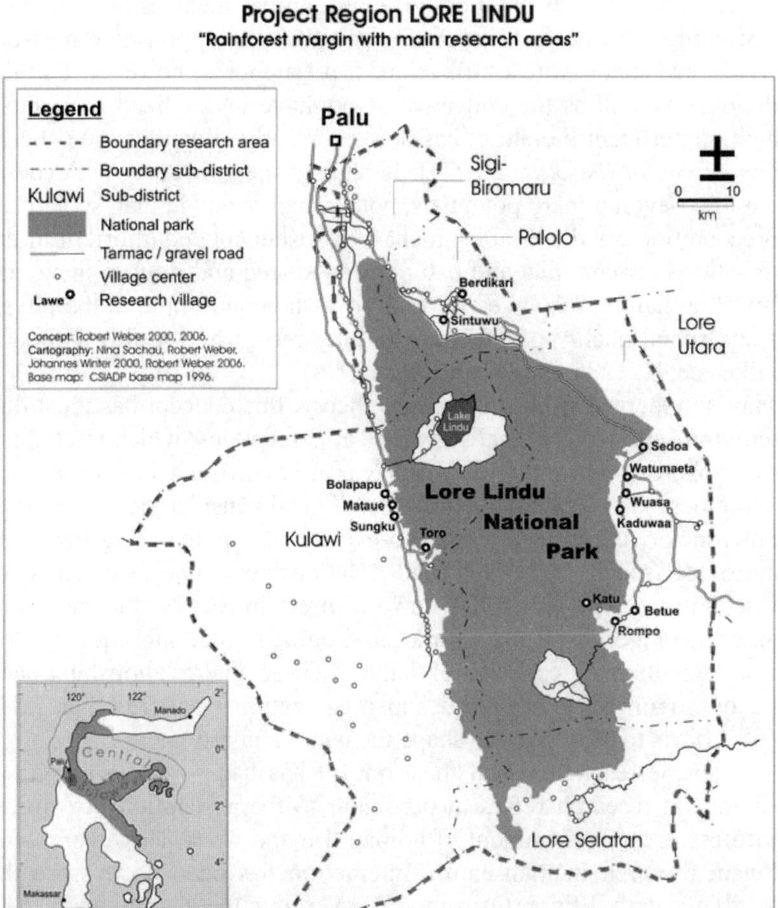

Figure 1. *The research region*

The outlined scenario elucidates that the forest margin around the Lore Lindu National Park is all but stable. Yet, in the light of the conditions and experiences mentioned above, the "accreditation" of local communities as largely autonomous management agencies of their natural resources, at first sight, seems to constitute an appropriate measure towards sustainable forest utilization. It presupposes, however, that the individual as well as the collective actors have internalized a principle which, in pertinent literature, has become widely known as the concept of *co-evolution* (Noorgard 1994). It regards the existence or development of co-evolutionary potentials, both in nature and human society, as a precondition for the establishment of a sustainable equilibrium in the interaction between man and nature. Man's need and right to make use of existing natural resources in order to sustain and improve the physical and social quality of life are explicitly recognized in this approach (Wilkens et al. 1985; Beck 1996; Bell 1998).

Having emerged from the systems theory, this concept has most distinctly transgressed classical sociological perceptions, which treated the social sphere as an entity *sui generis* (Durkheim) and left the natural environment largely untouched by sociological considerations. As far as human society is concerned, such co-evolutionary potentials grow in the context of social structures and relationships. These patterns of "structuration" (Giddens 1984) have a direct impact on the modes of man's interaction with his natural environment and, subsequently, on the co-evolutionary capacity of nature. These interrelationships shed light on cognitive, behavioural and organizational patterns that guide man's efforts to preserve *and* shape the natural environment.

The sparse resonance that this concept has had so far in empirical sociological research refers in particular to the perception of tropical rainforest as an environment of human life and work. However, some relevant research in man-nature interaction has been done since the late 80ies, with Elinor Ostrom's "Governing the Commons – The Evolution of Institutions for Collective Action" (1990), as a major reference book for an increasing number of case studies. Though still in a rather marginal position, the issue of man-forest relationship has not been untouched by this development, supported by a growing public awareness of the dramatic ecological destabilization of forests and forest margins (see in particular: Li 1999; Gibson et al. 2000; Ostrom et al. 2002; Colfer 2007).

The main focus in the series of recent research concerns *local* institutions and organizational patterns as relevant determinants of collective and individual modes of forest utilization. Hereby, a new paradigm is gradually established, which puts into perspective hitherto dominant explanatory patterns for human interventions in forest areas as, for instance, state policies, technology, market incentives or demography. This approach is closely linked to the idea of "social actor" (Long & v.d. Ploeg 1994). It focuses on "networks of actors", which frame the conditions of interpersonal exchange in a defined community. The individual actor is thus transformed into a "socially constructed" actor, whose attitudes and behaviour are to a considerable extent determined by the local culture in which he is embedded.

The contributions in this volume tie up to the outlined theoretical considerations. They deal with the paradigm of co-evolution from different perspectives, with a major emphasis on socio-economic living conditions, communal organization and social norms and values as determinant factors in man-nature interaction. A major theme throughout the book is to show how these factors, far from being autonomous forces, are embedded in and shaped by larger influences. The margin area of the Lore Lindu National Park provides an ideal empirical setting for this purpose. It exemplifies both the deterioration of forest resources due to human encroachment, and the dependence of local communities on the use of forest resources in order to survive.

In addition, the contributions will provide further empirical insights into modes and conditions of forest management as designed and executed by local "social actors". This does, of course, not exclude the consideration of these modes and conditions as being embedded in and shaped by larger social, political and economic influences. This view becomes most distinct in those parts of the book dealing with the *political ecology* of man-forest relations. In large part, however, the contributions refer to a huge variety of perspectives characterizing the local context in which social action grows. They all relate to a common denominator which is best denoted as *socio-economic security*. It encompasses aspects of access to resources, land tenure and relations of production, as well as the functioning of local organization, leadership, and the reliability of rules and sanctions. Moreover, as will be demonstrated, even issues of local knowledge and gender may have an impact on socio-economic security.

This book is not meant to provide ready-made tools to improve co-evolutionary potentials. However, by highlighting the actual dilemma and analyzing some of its basic roots, it tries to reveal such issues on which policy-making and practical action need to zoom in. It is neither our intention to objectively measure the degree of (in)security, and it may be questioned whether this is at all possible or meaningful. Instead, it is assumed that the perception of (in)security is the essential catalyst of individual or collective behaviour. And there is ample evidence that there is a positive correlation between feeling of insecurity and destructive or even fatal use of forest resources. By extending investigation into a large variety of security factors, the contributions to this volume identify multiple approaches towards corrective action by the local communities, state agencies or non-governmental organizations. It may be forestalled, however, that socio-economic insecurity nurtured by various factors has accumulated to a critical state in many communities around the Lore Lindu National Park. External interventions by state agencies and NGOs have contributed to this fact rather than attenuating it. Thus, the hope for a mutual survival of man and forest must be assessed as rather faint.

References

Beck, U. (1996): Risikogesellschaft. Auf dem Weg in eine andere Moderne. Frankfurt.

Bell, M. M. (1998): An Invitation to Environmental Sociology. Thousand Oaks.

Colfer, P. (2007): The Complex Forest. Bogor.

Fremerey, M. & Panggabean, H. (2004): Between Difference and Synergy: Cultural Issues in an International Research Scheme. In: Gerold, G., Fremerey, M.& Guhardja, E. (Eds.): Land Use, Nature Conservation and the Stability of Rainforest Margins in Southeast Asia: 523-533. Berlin, New York.

Gibson, C., McKean, M. & Ostrom, E. (Eds.) (2000): People and Forests. Communities, Institutions and Governance. London.

Giddens, A. (1984): The Constitution of Society: Outline of the Theory of Structuration. Cambridge.

Li, T. M. (Ed.) (1999): Transforming the Indonesian Uplands. Marginality, Power and Production. Halifax.

Long, N. & van der Ploeg, J.D. (1994): Heterogeneity, Actor and Structure: Towards a Reconstitution of the Concept of Structure. In: Booth, D. (Ed.): Rethinking Structural Development. Theory and Practice: 62-89. Essex.

Noorgard, R.(1994): Progress Betrayed: The Demise of Development and a Coevolutionary Revisioning of the Future. London.

Ostrom, E. (1990): Governing the Commons: The Evolution of Institutions for Collective Action. Cambridge.

Ostrom, E., Dietz, Th. & Dolšak, N. (Eds.) (2002): The Drama of the Commons. Washington.

Wilkens, H., Petersen, H. J., Schulz, S., Schumacher, D. (1985): Wirtschaftliche, soziale und politische Bedingungen von Entwicklung. Ein Beitrag zur Erklärung von Entwicklungserfolgen in Ländern der Dritten Welt. Forschungsberichte des Bundesministeriums für wirtschaftliche Zusammenarbeit. München, Köln.

Chapter I

"Stability" or "Sustainability"? Changing Conditions of Socio-economic Security and Processes of Deforestation in a Rain Forest Margin

Günter Burkard

Abstract

This article analyses the role of socio-economic security in the observed processes of ecological stabilization and destabilization in the rain forest margin. In Central Sulawesi, socio-economic security takes the form of three interrelated dimensions: security of land tenure, security in terms of the relationship between society and the natural environment and security in regard to relations of production and social organization, including issues of intergenerational access, local-migrant relationships and aspects of leadership. A major constraint in mitigating people's security concerns lies in the fact that different groups of the rural society put their priority on quite different conditions of security.

Socio-economic Security: Contexts and Definitions

There is no overall definition of "socio-economic security" due to its various dimensions as they appear in concrete empirical settings. First, having different preferences, experiences and interests, individuals and groups may put their priority on different conditions of security. Second, such conditions are not static, but in a constant process of transformation due to changing circumstances. Third, conditions of "security" cannot be treated in isolation from their respective "providers" and their reliability. In developing countries – and especially in rural contexts – socio-economic security is often obtained from various providers at the same time such as the state, the family network, reciprocal arrangements based on local traditions, member based organizations and the market (Jüttich 1999).

Security, in our conception, has three interrelated dimensions. First, security includes both, objective as well as subjective conditions. In objective terms, security refers to the absence of threats to acquired values, whereas in subjective terms it refers to the absence of **fear** that such values will be endangered by external arbitrariness (see Ibeanu 1999: 10). In peasant societies threats to acquired values are predominantly related to issues of "land tenure" which usually encompasses both, aspects of ownership and access to resources (Banana and Gombaya 1995: 93)

The second dimension refers to the relationship between human society and the natural environment. In this sense, security, defined by the capacity of nature to provide and sustain the physical needs of society, is linked to issues of productivity, stability and sustainability. "Productivity" refers to the yields of goods from agriculture and must be distinguished from "stability" which is related to the *reliability* of yields and incomes. "Sustainability" on the other hand refers to the viability of resources and the ecosystem to sustain the needs of the present without impairing those of the future. Thus, "stability" aims at "mid-term" survival, whereas "sustainability" is aimed at "long-term" survival.[1]

The third dimension refers to the relationship between the members of society. It involves (1) relations of production as they manifest themselves in agricultural arrangements and (2) issues of social organization and leadership in regard to their capacity to protect people from threats that endanger their security and their "responsiveness" towards people's security concerns.

After a short introduction to the research villages in section 2, the various security dimensions pointed out above will be discussed one

---

[1] This definition borrows from Eijkemans (1995) with a slight, but meaningful deviation. Eijkemans (ibid: 54) relates "stability" to "short-term survival". In our conception, "short-term" survival is a zero-sum-game that takes no investments for the future into account at all. Thus "short-term survival" is rather characterized by a lack of stability than by stability. Therefore we prefer to link "stability" to "mid-term survival" rather than to "short-term survival".

by one. Section 3 deals with tenure security in relation to access and land use systems. The focus of section 4 is on security as it is related to man-nature-relations, emphasizing the socio-cultural and institutional dimensions of security as they manifest themselves in patterns of agricultural intensification, market exposure and subsistence production. In the first place man-nature interactions will be discussed in regard to indigenous and migrant groups. Section 5 analyses the emergence of agrarian institutions in regard to their reliability and investigates the role of social organization and leadership as local custodians of security. Security providers represent a crosscutting dimension and will be discussed within the single sections before the various security dimensions will be linked to the overall problem of deforestation and resource degradation in section 6.

The Research Villages

The results presented below are based on fieldwork that was conducted in the village communities Sintuwu, Watumaeta and Rompo in the vicinity of Lore Lindu National Park, Central Sulawesi, Indonesia. The villages, located in three successive valleys on the eastern side of the Park – the Palolo, Napu and Besoa valleys respectively – are part of the overall STORMA-research sample.

Sintuwu (328 households) is located about 60 km from the provincial capital of Palu in the Palolo valley, which is the most northern of the three valleys. Founded in 1961 by resettled Kaili groups, Sintuwu became a migration target since the end of the sixties, when a steady flow of migrants from the Kulawi valley (located on the western side of the Park) and from other parts of Sulawesi moved into Sintuwu, of which the Bugis of Southern Sulawesi have become the most dominant group. Today, Sintuwu is an ethnically mixed community, with about 65% of its population consisting of indigenous groups, whereas about 35% are migrants.

Watumaeta (417 households) is located in the Napu valley South of Palolo, 102 km from Palu and was founded in 1926 by local Napu people who moved either from the neighboring village of Wuasa or from the dispersed long-house settlements located around Watumaeta.

From the thirties, migration from outside was mainly confined to some Toraja families. The first Bugis entered Watumaeta in the early sixties and their number had slightly risen again during the eighties, but it was not before the mid-nineties, that in-migration became a "political" factor in village life when a great number of Buginese and Makasarese migrants (and to a lesser extend Sundanese) entered Watumaeta, either directly from their homeland or from other places in Central Sulawesi. This migration stream can partly be attributed to the completion of the asphalt road from Palu in the early nineties, but the most significant pull-factor was the availability of large, flat land areas that were administered by a responsive village head who, in his plan to increase the number of inhabitants for political reasons, imposed almost no restrictions on in-migration and offered secondary forests to newcomers almost at will.[2] Watumaeta opened another frontier area for migrants in 1998, when a new wave of migrants came in search for land, partly influenced by the economic crisis and the high cacao prices at that time. In July 2001, the migrants had already outnumbered the local population and made up 63% of Watumaeta villagers.

Rompo (102 households), located in the Besoa valley and founded in 1923 by local families from nearby Parigi village, offers yet another situation. Notwithstanding the fact that some Buginese families had temporarily settled in Rompo during the seventies and eighties, until some years ago, ca. 95% of Rompo inhabitants were local Besoa people. This changed in 1999, when about 90 Toraja families moved to Rompo in search for agricultural land. However, their expectations of agricultural perspectives were unfulfilled, and about 60 of them had already left Rompo at the time of our research.

There is a clear continuum to be observed between Rompo, Watumaeta and Sintuwu in regard to land use patterns. Six major types of land use can be distinguished within the research area:

---

[2] The position of the village head was threatened by plans of the provincial government to form one village out of three due to low population rates. Increasing the population was one way to keep the administrative status of a village (*desa*). Further, the allocation of state subsidies to the villages is often based on the number of inhabitants.

1) Wet-rice fields (*sawah*), either semi-technically irrigated or rain-fed,

2) Dry fields (*ladang*) used for seasonal and annual food crops only (no perennial trees),

3) Intensive vegetable plots with seasonal crop rotation (*bedengan*),

4) Annual - perennial "mixed stands" in which seasonal plants and annuals (maize, beans, cassava) are grown as cover crops within perennial coffee or cacao trees,

5) Pure perennial stands (*kebun*) with coffee, cacao or other tree crops only,

6) Semi-cultivated secondary forests, including deliberate elements of forest gardens.

Types of Land Use (in %)						
Village	wet rice fields (sawah)	vegetable gardens (bedengan)	perennial stands (kebun I)	mixed stands (kebun II)	annual dry fields (ladang)	enriched forests (belukar)
Rompo	23,8	-	6,6	40,4	13,0	16,2
Watumaeta	13,2	7,3	14,5	50,2	3,5	7,3
Sintuwu	9,1	-	66,3	22,2	0,6	1,8

***Table 1**. Distribution of land use systems in the research villages*

Whereas *ladang* dry fields and enriched secondary forests declined in a linear manner, annual-perennial mixed stands make up half of the land worked in Watumaeta, but occur with a lower frequency in Rompo and Sintuwu. This reflects the fact that mixed stands occur only as an intermediary stage of agricultural development, when annual dry field cultivation is substituted by perennial crop farming. Perennial estates form over half of the land under cultivation in Sintuwu, but are still below 15% of all the land worked in Watumaeta and Rompo. Different from coffee *robusta* and *arabica*, which have been planted in the region for a long time, cacao is still a rather new crop in

Watumaeta and Rompo (since ‘95 and ‘97 respectively). This contrasts significantly with the situation in Sintuwu, where cacao has been planted for at least 25 years, and where the perennial estate forms an established land use system. While significant areas of wet rice fields have already been converted into perennial stands in Sintuwu, this has not yet happened in Watumaeta and Rompo (see Sitorus 2002).

Besides plots under cultivation, a second criterion for differentiation is the ratio of land under fallow to cultivated land in general. There is a certain problem in defining what constitutes “shifting cultivation” and what is already an “upland fallow system”.[3] As a “rule of thumb” Ruthenberg (1980) argues that “shifting cultivation” is replaced by a “fallow system” where the ratio of land under cultivation to land under fallow is higher than 33%; where it exceeds 70% fallow systems are transformed into “permanent upland systems”. The proportion of fallow land to land under cultivation is summarized in table 2.

Ratio of Land under Cultivation to Fallow Land (in %)					
Village	permanent cultivation	integral fallow (rotation)	partial fallow I (bush, forest)	partial fallow II (imperata)	fallow total
Rompo	34,3	15,8	19,6	31,3	65,7
Watumaeta	72,9	4,3	22,8	-	27,1
Sintuwu	81,5	-	18,5	-	18,5

Table 2. *Ratio of land under cultivation to fallow land*

[3] In general, shifting cultivation refers to any temporal and spatial cyclical agricultural system that involves repeated clearing of land with the assistance of fire (Brown and Schreckenberg 1998: 2). It is further distinguished from other systems by rather long fallow periods between relatively short periods of cultivation (Eijkemans 1995: 27). In economic terms, shifting cultivation is characterized by (1) maximization of product per labor force rather than by (2) maximization of product per unit of cultivated land. In both cases we deal with types of economic intensification, but in general it is only systems of type (2) that are recognized as “intensive systems”, whereas systems of type (1) are referred to as “extensive systems” (Straube 1967, Mischung 1984).

If we compare land under permanent cultivation with total land under fallow, Rompo is at a stage where shifting cultivation passed recently into a fallow system; Watumaeta passes from a fallow system to a permanent system, whereas Sintuwu is dominated by permanent land use. Regardless of the fact that it did not gain much scientific recognition, on first glance Ruthenberg's suggestion seems convincing. But despite that, it proved quite useful for taxonomic purposes on the community level, the "rule of thumb" turned out to be rather simplistic when applied to the level of households. For the individual household the practice of fallow periods is less a single stage in the evolution of agricultural production, but is better understood as a variable element within a mix of various farming systems including stable rotational systems, semi-cultivated secondary forests as well as permanent annual and perennial farming (see Brown and Schreckenberg 1998: 3). Presently, the local population named two major types of fallow land. One is called *lahan bero* and refers to rotational plots that lie fallow for one period after two or three periods of cultivation (mainly for purposes of soil regeneration). These plots form a stable element in farmers land use systems and can therefore best be described as "integral fallows". Further, they are clearly distinguished by the local population from fallow plots that lie dormant for longer periods (often many years) and that are addressed as "sleeping fields" (*lahan tidur*). Such fallows are perceived in terms of land reserves for the future rather than in terms of present household production. They are, therefore, better identified as "partial fallows". Whereas the role of the "integrated fallow" is declining in a linear continuum from Rompo to Watumaeta and Sintuwu, the portion of "partial fallows" has been remarkably stable in all villages.

Security in the Context of Land Tenure

There is a widespread belief that material poverty and insecure land tenure lead to a situation of short term planning horizons that do not take investments for a more sustainable land use into account. The argument is that only when farmers have secure tenure over their lands will the motivation to invest in long-term undertakings of sustainable resource use take place. Further, increased incentives for sustainable resource use are believed to develop most likely within a framework of private land ownership (see Vivian 1991, Banana and Gombaya-Ssembajjwe 2000). In general "tenure" refers to a set of rights a person or entity holds to land. Primarily, land access, formal recognition and the type of land use determine tenure security in the research region.

Tenure Security and Access to Land

Local language makes a clear distinction between the "vegetation type" and the "access status" of a resource. Residents distinguished no more than two types of plots, namely dry fields marked by earthen mounds (*bonde*) and wet-rice fields (*bonde keuwai*). The situation differs in the case of land under fallow and forests where people differentiated at least eight resource types:

- *holua* = fallow plots with secondary growth dominated by grass and bush vegetation,
- *bonde* = plot planted with annuals marked by trees. The *bonde*-plots are cultivated more systematically than "holua" and are defined by their edges marked,
- *lopo lehe* = "young" secondary forests lying fallow from two up to seven years,
- *lopo matua* = "old" secondary forests lying fallow at least seven up to 30 years,
- *pandulu* = primary forest on flat land that has never been cleared for agricultural cultivation, but was previously used to take out forest products and fire wood with trees selectively extracted. *pandulu* thus refers to "disturbed primary forests" in plains,
- *wanakiki* = undisturbed primary forests inside the forest, not "visible" from the village.
- *wanakiri* = primary forest patches with no tall tree growth, natural vegetation dominated by small trees and bamboo. Old and young secondary forests are identified and distinguished from *wankiri* and *pandulu* by the growth of a certain tree called "mayopo" that serves as an indicator that the forest has once been opened in the past,

- *udu* = secondary forests around Rompo located in small alluvial areas close to watersheds.[4]

Besides these classifications according to the age of the vegetation, and location, local terms divide plots into three major types according to "access":

- *mapandulu* relates to access by opening primary forest on ones own. Because of the mystical aspects associated with *pandulu*-forests – they contain the holy "beringin"-tree – and because of the hard work of opening them it was only the spiritually robust, strong and brave men (*karunya*) who opened *pandulu* that they often granted after clearing to others on a long-term borrowing basis.
- *rape bolo* refers to access on the basis of borrowing. Initially this was one of the most common ways of access when people obtained the right to use a plot opened by others.
- *sosora* is land that is obtained by an inheritance procedure. Traditionally the father gave the "mandate" to his eldest son to divide the land among his siblings. Only wet rice fields and plots planted with perennials could be inherited. Dry land *ladang*-plots not planted with at least some perennials could neither be inherited nor be claimed as *sosora*.

This is not to say that "all the younger generation worked on borrowed land" as Li (1995: 512) has observed among the marginal Hill Lauje of Central Sulawesi; but more important is the distinction between "owning" a plot and "borrowing" that was not infrequently elaborated under shifting cultivation conditions. The notion of "inheritance" was not an elaborated one either. Secondary forests (*lopo matua*) could revert to *pandulu* when left untouched for more than 30 years. Per definition *pandulu* never had an owner, whereas *lopo* always had one. Thus, with the course of time, secondary forests reverted to an open access resource. Whereas formerly the different kinds of forest corre-

---

4 In the past, old men retired for contemplation in the "udu" living in bamboo huts (bambaru). "Udu" are always marked with perennials and ownership is confined only to certain long-established families who can trace their descent from the village founders.

lated with large tracts of land that were located more or less in concentric circles around the settlement, the *holua, lopo* and *pandulu* types are meanwhile mixed up due to a shortening of the cultivation cycle and the fact that people have cut most *pandulu*-forests as well. The common pattern of forestland distribution is patches of secondary forest scattered throughout cultivated areas.

The most current forms of access are purchase, access by land grants from the village head and the opening of secondary forests that are claimed on the argument that family members once used them. Despite the fact that its availability has diminished, the opening of primary forest is still rather safe in terms of "exclusivity" because people still know who opened it and no claims can be established by others. Opening old secondary forest is rather safe as well. This is rooted in the fact that these forests have often reverted to a state almost similar to virgin stands and have laid fallow for periods long enough that the original clearer is not anymore known by the present population.

Besides opening primary forest, the inheritance of land is the safest way of access. Thus, it is not surprising that people will always sell land they opened before they will sell inherited plots. Selling of inherited land needs the permission of and consent among all family members. Anticipating this procedure as a difficult and time-consuming one, people seem often reluctant to start it. Whereas people agree that inherited land is still safe at the moment, changes in inheritance practices can undermine this traditional security mechanism in the near future. The practice of land division through the eldest son offered a formal framework that ensured ownership, with all heirs "knowing" their share. The substitution of this practice by a gradual "informal inheritance" process with children working land still owned by parents will probably lead to increasing conflicts. Thus, younger respondents working land still owned by their parents categorized these plots as "insecure" because they work and invest in those plots, but they cannot be sure if they will later receive them when the parents have died or when the land will be divided.

Buying land seems less secure with six conflicts occurring out of 18 land transactions. But five out of six conflicts occurred in Watumaeta when land locals sold land to migrants. It is typical for Watumaeta, to have a relatively high frequency of problems when land is delivered or sold by the village head. Because of the lack of village organizations with clear responsibilities due to weak monitoring and law enforcement, the village head exerts almost unlimited power in the distribution of resources. Four out of five farmers that had problems after opening primary forest obtained their land from the village head. The problems encountered (borders not fix, transaction unclear, insecure location in regard to the National Park) seem to be linked to this "source of acquisition" than to the "age" of the forest in question. Among those who opened primary forest without interference of the village head (locals) no one had to defend his claim against others or had to negotiate about the boundary of the newly opened plot.

Most conflicts arise when people open secondary forests on their own. In the past plots have often been abandoned without the establishment of permanent use rights. Frequently, different families cultivate the same plot at different times. Claims of secondary forests are intrinsically linked to a land status concept that seems to gain increasing importance. Secondary forests are often referred to as *budel*-land, a term which is generally used as a synonym for the Indonesian *lahan peninggalan* ("land left behind"). On a higher level of abstraction, it refers to plots that have been used by family members in the past without being inherited via the eldest brother. This is the "orthodox" explanation given by local *adat* leaders[5] in Watumaeta and Rompo who insist on the distinction between *budel* (the "family pool") and *sosora* (an individually inherited plot) as crucial in drawing the lines between both conceptions. On the other hand, "commoners" often explained *budel* in rather "heterodox" and diffuse ways, either as any secondary forest that was not opened the first time by the present user or (less often) as a synonym for the Indonesian *warisan* (land inherited). Whereas the term is definitively derived from Dutch language, most respondents referred to *budel* as an Indonesian term. Such varying representations of access should not be judged in the first place by the criterion of "ethnographic accuracy". As Li (1995: 501, 504) points out with special reference to Central Sulawesi, struggles over

[5] Indigenous people use the term *adat* to denote their local and regional practices and beliefs. In the Malay, world ethnic categories are mainly constructed around *adat* that often fulfills the role of an ethnic marker. In social science, *adat* is usually translated as "custom", which actually does not cover all of its meanings.

resources are often accompanied by struggles over "meaning" that are invested in means of production. Given the fact that land was freely available under shifting cultivation conditions, it was unusual to adhere to a "legitimacy concept" like *budel* to support existing land claims. Written guarantees are the safest form of land acquisition, particularly in the form of land certificates issued by the "Agrarian Board" (*Badan Pertanahan*) located in the regency capital. Certificates are important not just to secure claims to land, they are even more important as collateral to borrow money from the bank. Because rural credit institutions are either missing or do not function well, people often rely on loans from agricultural traders and rattan middlemen that are characterized by high interest rates and short time spans. Thus, bank loans are an important source if one wants to borrow money for long-term investments. Only certified land can function as collateral. Not accepted as collateral at the bank but safe in terms of tenure rights are the written "records on land transactions" (*akte jual-beli*) that are witnessed and stamped by the district head (*camat*). Next in rank are receipts on tax payment (STTP) that prove that the land is registered at the village office (*spop*) and that land taxes have been paid. Security is lower because there is no proof from the district office (*kecamatan*) and the formal right can (at least in theory) be impaired if one does not pay taxes continuously. Formal recognition in form of "ownership letters" issued by the village heads are perceived as most insecure, because much manipulation has happened in the past and village heads are perceived as being not very trustworthy.

It is not surprising that discourses over the "access status" of secondary forests become meaningful in a situation when access to resources increasingly pose a struggle.

Tenure Security and Formal Recognition

In general, indigenous people own fewer certificates compared to migrants. There are two major reasons for a relatively high concern with certification among the migrant population. First, they suffer most from the fact that their conflicts have not been solved or brought to an end. Because the local population often perceives migrants as second-rate villagers or intruders when land issues are at stake, they expect no help from the "customary village council" (*lembaga adat*). Further, being outsiders, their "bargaining power" must be rendered as rather low. They may not open a new plot unless through the acquired land, he will be uncertain that he can enjoy the fruits of his investments. Migrants now have clear preferences if they want to expand their cultivation area: they prefer to buy land already cultivated rather than secondary forests and they prefer to buy it from other migrants instead of locals. Further, because migrants devote most of their plots to perennial cultivation involving rather high investment costs, these are best assured by certificates.

Tenure Security and Types of Land Use

Irrespective of formal recognition, one way to secure land claims is permanent cultivation. Planting of perennials is perceived as being a stronger "proof" of ownership than planting annuals. Permanently planted plots located within large patches of cultivated dry land are therefore more safe against claims by others as isolated plots surrounded by uncultivated areas where ownership of neighboring uncultivated plots is not always clear and where boundaries can easily move due to missing "day-to-day" control. Unused land is "marked" by planting of trees inside partly enriched secondary forests.

Because fallow plots and secondary forests are rather insecure compared to permanently planted plots, and because claims on fallow land can only be upheld by the planting of perennials, aspects of tenure insecurity can even have a positive effect on sustainable resource use if the planting of perennials fosters traditional land rights under local conditions. The government overlooks this fact, because it is agroforests planted in rows (best monocultures) that the state recognizes (see Löffler 1996: 65). Not a single case of the national certification-program (*prona*) formalized the ownership of enriched secondary forests, despite the fact that this kind of land is one of the most sustainable because it leaves part of the forest cover untouched and imitates the natural surrounding by inter-planting of different types of shrubs and trees. Such enriched forests play an important role in terms of future land security. Instead, the government first certified home lots, wet rice fields and perennial stands; plots that enjoy the highest security of tenure anyway, irrespective of certification. Until now the incorporation of fallow land into legislative reforms does not seem to be the intention of the central government, given the probable loss of timber resources it would imply (Brown and Schreckenberger 1998: 7).

Most secure in terms of land use are wet rice fields (*sawah*). *Sawah* are located in "irrigation complexes" where *sawah* of many people are located. In order to control pests and destruction of harvests by birds and rats, individual plots are only rarely located in isolation from other *sawah*. This cultivation also requires cooperation among households. Land has to be leveled, ditches made and channels for irrigation must be maintained. In contrast to dry field cultivation, which does not follow a systematic seasonal cycle, and which is dominated by "staged planting" (crops are at a different stage of maturity at a certain time), wet rice cultivation is characterized by "simultaneous planting". When rice plants begin to ripen, people often stay in temporary huts (*bambaru*) close to their *sawah* at night to safeguard them against birds and rats. If conflicts arise, they can be discussed and solved in time before they require the intervention of some village institution.

It has already been pointed out that unmarked fallow plots are most insecure. Most of the conflicts reported involved grass-land, bush vegetation and young secondary forests. On an aggregate level, the frequency of conflicts over land must be rated as high. Within the time period from 1998 to 2001, about one-third of our respondents experienced conflicts. The frequency of conflicts however, differs significantly between the villages.

Frequency	**Type of conflict**			
	Conflicts about plot borders	Others claim ones plot (or part of it)	Plot claimed by third after purchase	Problems of location and legal status
Rompo	4	1	-	-
Watumaeta	4	5	5	2
Sintuwu	1	1	1	-

***Table 3**. Frequency of land conflicts according to research villages*

These figures could lead to the assumption that land conflicts constitute a major factor for the decreasing forest cover in the region when people feel themselves, and are forced to substitute "conflict land" with new plots in order to avoid troubles, time-consuming discussions and difficult procedures. Such a conclusion cannot be drawn from our data. In only seven cases did conflicts over land result in temporary land losses or in the loss of a substantial part of the plot involved. Those losses exert no relevant impact on whether people have opened new land afterwards or not and are independent of the size of opened plots. Even the existence of conflicts (irrespective of land losses) is not associated in a systematic manner with forest clearance. The higher rate of conflicts in Watumaeta is linked to the fact that more land in opened in Watumaeta as compared to Rompo and Sintuwu. As far as Rompo is concerned, troubles are mainly confined to issues of plot borders whereas claims of whole plots by others seem to rarely happen in this community. In Sintuwu on the other hand, where population increased only slightly within the last years and where permanent land use systems are the rule, not much conflicts over land seem to happen. Established plots, as long as they are permanently cultivated or sufficiently demarcated, are rather safe in all villages.

In general the number of plots demarcated has substantially increased within the last years, especially in the Napu valley. This tendency is a product of incidence rather than conscious planning. Introduced first by Buginese migrants before CARE and the regional extension services propagated it officially *gliricida sepium* (*gamal*) became the preferred shadow tree in cacao cultivation. In comparison to the *erythrina* species (*dadap*) that have sporadically been used by locals as shadow trees for coffee, *gliricida* offers five major advantages: (1) it does not throw off substantial amounts of leaves in the dry periods, (2) it is fast growing, (3) it is easy to remove when shadow trees must be thinned out, (4) it is an excellent fire wood and (5) it is cheap and easy to reproduce. Being an inexpensive substitute for perennials, *gamal* has become the most popular plot marker. During the field survey, an increasing use of *gamal*-trees in the demarcation of annual fields (*ladang*) and vegetable gardens (*bedengan*) in Watumaeta could be observed. The substitution of perennials (usually coffee *robusta*) by *gamal* as "plot markers", is more than the substitution of one plant with another. Whereas perennials have often been planted within the plots, *gamal* is planted along the plot edges. Thus, "ownership marking" as such, is replaced by "border demarcation". The primary advantage of border marking is its social function that allows for public and impartial notification of boundaries (McKean 2000: 44).

We conclude that "tenure security" is not a sufficient explanation for the decreasing forest cover in the region at the moment. Security of tenure in itself may not even be a sufficient condition to safeguard sustainable resource use, as there is a bundle of other factors such as risk and market incentives that seem more important in the development of current land management practices.

Security in the Context of Man-Nature Interactions

Our conception distinguishes three major context variables in regard to human-nature relations. Variable (1) is linked to the relation between societies and the environment as it manifests itself in culturally defined concepts of property, access and security. Variable (2) is related to the preferred forms of agricultural production in regard to subsistence and market production. Whereas variables (1) and (2) can be addressed as "socio-cultural factors" in the sense that they refer to decisions within the scope of individual peasants (or households), context variable (3) refers to the "institutional environment" and its impact on peasant interaction with natural resources in general.

In his analysis of social security among the Ogoni in Nigeria, Ibeanu defines security as "the extent to which members of society understand the laws of nature (science) and use this understanding to create tools (technology), thereby enhancing their capacity to derive their physical needs from nature; and their capacity to exploit nature efficiently" (Ibeanu 1999: 11). He further argues that efficiency should be achieved by "sustainable exploitation" that aims at a balance between immediate needs on the one hand and future needs on the other (ibid: 12). This definition alleges implicitly that the "quality" of man-nature interactions is predominantly determined by the stage of technological development. Although such a connection has been found in various African settings for a long time (e. g. Straube 1967), it cannot be taken as a relationship of universal validity. With special reference to mainland South East Asia, Mischung (1980, 1984) has pointed out that man-nature relationships are less related to issues of technology, but are intrinsically linked to culturally defined tendencies of behavior, in which the cultural perceptions of the environment play a crucial role. Our concern with socio-cultural factors in man-nature relationships should not be equated with "possibilism" or "determinism". It is merely to recall that people do not behave randomly or in isolation from deeply-rooted values and practices, but base their actions on former experiences and points of reference.

Eijkemans (1995: 13) made a significant distinction by pointing out that "maximum production" and "optimal production" are not equivalent in the eyes of the peasant in dealing with risks and that peasant decisions are often guided by "risk-rational" considerations that deter from a viewpoint that is aimed at maximum production by opting for crop mixes that are less optimal in terms of product value. Whereas this finding holds true for part of the society under investigation, it is not a uniform phenomenon. The point is, that in dealing with risks, local groups (Kaili, Napu, Besoa) follow rather a strategy of "mixed cropping" whereas most of the Buginese and Sundanese migrants often rely on a single or a few agricultural products of high economic value and cope with uncertainty by intensification and by enlarging their cultivation area to a point where possible risks through price decline or part-harvest failures are overcome by the bulk of the product produced.

Man-Nature Interactions in the Context of Environmental Perceptions

In the past, the relation of local people and the forest environment was governed by the principle of "open access". The "quasi natural right" (Fremerey 2002: 4) to convert forest into agricultural land was limited by personal industriousness rather than by local regulations. One had only to ask formal permission from the village leader (or the *long-house-chief* in Watumaeta) if one wanted to cut primary forest and the wish to open new land was rejected only when certain "holy" places within the forest were affected. These places were scattered inside the forest within rather narrow boundaries and they did not constitute a preservation shelter over large forest areas. The opening of land often marked individual life cycles. Primary forest was usually opened in case a son was born or after a marriage agreement in order to equip the newlyweds with the necessary resources. In such a conception, "land" is not primarily perceived as a production factor, but as an overall *security provider*. This conception is contrary to the conception of migrants that view their plots mainly as a production factor that must first of all be secured by legal titles.[6]

Whereas migrants derive their security predominantly from the market, locals rely first of all on land as a "security commodity". The wish to secure land for children, or as a capital in order to deal with negative events, occurs more frequently as a motivation to clear new land among the locals than among the migrants.

The difference between locals and migrants regarding their relation to the environment is most clearly reflected in the perception of fallow land. Whereas most migrants have planted recently opened plots with export crops (cacao) directly after ownership rights were established, only 37,5% of the locals had already planted their new plots completely. Given the fact that the reciprocal principles of the "moral economy" (Scott 1976) are only weakly developed in the region, land remains as the sole security factor in cases of misfortune if one does not want to rely on the market. Some 60% of all land sales in the sample were used to cover the costs of medical care and ¾ of all plots sold have been fallow plots of different ages. This indigenous security conception seems not to be sufficiently recognized by village officials and the migrant community. Rompo made a village rule that stated that all fallow plots can be withdrawn by the village administration if they are not actively cultivated. The rule had to be dropped after the officials realized that it posed a real "threat" to the farmers. From the viewpoint of the Indonesian government it has similarly been argued that secondary forests are "empty land" (*lahan kosong*) that should be converted into agricultural land. This view seems not only to be held by village heads who in the past often granted land rights over secondary forests to outsiders, but by the migrant community as well. Bugis in Watumaeta point out that the secondary forests (*belukar*) granted by

---

6 Migrants, unlike the local population, only rarely enter the National Park for purposes of converting forest. Being outsiders with only a limited "bargaining power" at hand, compliance with the law in order to avoid legal and social troubles is a main condition of migrant security.

the village head that have been cleared since 1995 have had no owner. This conflicts with the local perception that forests of the *belukar*-type (dominated by small trees) are always secondary stands, younger than at least 20 years and therefore always have an individual owner.

In its extreme, but nevertheless typical from the Buginese and Sundanese economy is dominated by the export sector, whereas the subsistence sector is of little importance. In their market orientation migrants want other owners to use their land compatible with their own, in order to minimize crop failures and to optimize productivity. It was the concern of the Bugis in Watumaeta that fallow plots surrounded by cultivated land should be immediately cultivated. Due to harvest failures and crop devastation caused by fires from neighboring plots being cleared, the awareness for mutual regulation of land use among the migrant community has risen. On the other hand, their suggestion to prescribe the active cultivation of fallow plots surrounded by cultivated land in the village rules and to sanction non-compliance with a temporary withdrawal of ownership rights was not approved by local farmers who use these fallow plots often as a reservoir for fire wood and fodder.

Man-Nature Interactions in the Context of Agricultural Production

It has already been stated, that from the official point of view, "maximum production" and "optimal production" are not always compatible and that in dealing with risks "stability", and "sustainability" are not the same things, but often conflict with each other. There is no doubt that farmers in the research area are primarily concerned with "stability" in the sense that they want to raise present incomes, improve their living conditions and secure primary education for their children and that they are less interested in matters of ecological sustainability. In this context, the non-agricultural sector (formal as well as informal) is not capable of absorbing a substantial number of people. Sixty-one percent of the farmers depend exclusively on agriculture for their livelihood and among those who have access to other sources, with the exception of public servants, the relative share of non-agricultural income is small.

People give various reasons for cutting forest, but the reasons given above cannot be treated as sufficient explanations without taking social, cultural and economic factors into account that are directly linked to cropping systems and types of land use. Irrespective of the various decision-making theories that have been developed to explain "peasants' rationality" it is an established fact that peasants in developing countries have to deal with risks and uncertainties of various kinds, like market fluctuations, variations of climate, natural hazards and plant diseases resulting in harvest failures, limited availability of resources (that are not confined to "material resources", but include "social resources" and "information resources"), and that peasants have developed viable strategies to cope with such uncertainties (Ellis 1988, Eijkemans 1995).

Perennial Expansion and the Mixed Cropping Balance: Strategies of Migrants and Locals

Within the time period of three years 68, 2% of all sample households, have cut secondary forest older than five years or opened primary forest. The decreasing forest cover in the area is certainly linked to the introduction of perennial crops (cacao boom since 1997). But the spread of perennials does not follow a uniform pattern. It is true that at least in Napu a significant amount of secondary forest has been cut since 1995, when a vast wave of immigrants entered the area, lured by relatively low land prices and the promising prospects of cacao. The strategy of migrants is such that newly acquired plots are planted with cacao shortly after ownership rights have been established. In the case of locals, it seems that only a part of them followed the migrants strategy of opening new areas (*lahan bukaan baru*) to plant cacao. Instead, cacao and coffee varieties were often planted first in home lots, on the edges of dry fields still used for seasonal crops, within existing maize fields or within secondary forests. In general, perennial expansion in the area occurs within an evolutionary process, in which the area for perennial crops is first extended with annual crops being inter-planted. Cover crops are mainly used for subsistence purposes. In the course of time, perennials increasingly dominated the mixed cropping stands and inter-cropping decreased. There is no doubt that with resource degradation, annual farming is less favorable than mixed cropping and perennial cultivation. Soil fertility is maintained by the planting of perennials. Furthermore, the soil is shaded and the root systems of shrubs and trees prevent erosion. If this doesn't prevent progressive degradation of the soils, it is at least delayed (Eijkemans 1995: 41). The ratio of annual plants to tree crops within mixed stands varies significantly between households and villages. In general, locals plant fewer perennials in mixed plots than migrants, keeping a slightly higher planting distance between the trees.

Whether a household opts for "perennial specialization" or "mixed cropping", depends first of all on the household size. Migrant households are almost exclusively of the nuclear and regular extended types, whereas irregular extended compositions dominate local households, resulting in a often higher household size compared to migrants. In general, households can be divided into four major types:

- *Nuclear households* are residence units consisting of the married couple and their children.
- *Regular extended households* are households consisting of two linear related generations
- *Irregular extended households* comprise other family members (i. e. cousins etc.) and non-family members as well. This category also includes cases in which children of kinsmen have been fostered (*anak tinggal*).
- *Multiple family households* comprise at least two nuclear families that are not linear related. As has been reported from other areas in Central Sulawesi (Shrauwers 1999) household compositions are very flexible. It seems that "bi-local" persons (shifting their household of residence according to circumstances) are quite often to be found in the area. In other words: the rather blurred conception of household composition gives locals the flexibility to redistribute kin (and kin labor) according to economic circumstances.

There is a clear relationship between household size and the area of a holding that is devoted to annual-perennial mixed stands ($r = 0.40$, statistically significant with 0.01). The more people per household that have to be fed, the more are risks controlled by diversification of products. Especially local families have been hit by various harvest failures in non-perennial agriculture within the last years, foremost in wet rice, maize and beans. Being less successful in perennial cultivation, they also suffered more failures in perennial cultivation than migrants. Accordingly, many local families in Watumaeta and Rompo will try to minimize risks by diversified cropping systems that guarantee at least a successful harvest of some of their products. Given limited knowledge and limited resources among part of the farmers, starting perennial (cacao) cultivation for the first time is often a risk-loaded exercise (see Eijkemans 1995: 164). First, there is the life cycle of perennial crops because during the first years, as long as trees do not bear fruits, investment costs will probably be higher than the benefits of the first harvests. Besides, after some years, yields will start to decrease and the process of rejuvenation starts. This is especially the case with new coffee varieties (*kopi kate*) that must be replaced after five years. Therefore, many farmers continue with the established long-living *robusta* and *arabica* varieties though they are often less productive and more difficult to harvest. Migrant farmers in Sintuwu and Watumaeta sometimes do not use cover crops at all because they may influence the development of the tree crops. But for local farmers (often inexperienced with cacao cultivation) and migrant newcomers (with no perennial income the first 2-3 years), fast-yielding annuals are vital to overcome the first unproductive years of perennials. Given the higher productivity of their perennial plots and the smaller number of mouths to be fed, migrants owning exclusively perennial cacao stands (no inter-cropping) that are well cared for are still well-endowed, if only a part of the product can be harvested or sold. Cacao has suffered several pod-borer attacks (*lepitoptera*) with harvest damages reaching up to a maximum of 30%. With an average holding of 2-3 ha (equivalent to 2000-3000 trees) perennial farmers could cope with this misfortune easily because the selling of 70% of their harvest was more than sufficient to keep their income stable.

The key question now is how these findings relate to the forest cover. Of course, migrants want to extend the perennial cacao area. The finding of Anzdec (1997: 17) however, that encroachment occurs mainly because forest is converted for the planting of perennials is only partly proved by our data. Planting cacao is the plan of many locals as well as long as annual cultivation is safeguarded by other available land resources. But almost one third of the local respondents cleared forest to plant *palawija*-crops (maize, peanuts, cassava) on those plots, with the planting of perennials postponed or not envisaged at all (see table 4). This occurs when perennials in mixed stands have reached the point where their increasing shade does no longer allow for annual cover crops. To keep the "mixed-cropping-stability", new plots must be established to plant maize and other annuals. Thus, the socio-cultural reason to clear forest, because current disposable land is perceived as insufficient, is linked to the motivation to secure "stability" by keeping the mixed cropping balance. Whereas both strategies, securing mixed stability and expanding perennial production, may often result in the same land use type (temporary mixed stands), the rationale behind it can be a quite different one.

Primary crop type	**Locals**	**Migrants**
Annuals crops (*palawija*)	30,4	6,3
Cacao	36,4	61,8
Coffee	18,7	10,3
Coffee and Cacao	12,1	2,4
High Value Vegetables	2,4	22,2

Table 4. *Primary crop type planted in recently opened plots by locals and migrants (in % of respondents)*

Man-Nature Interactions and the Institutional Environment

As far as Watumaeta and Sintuwu are concerned, most forest reserves are now located within the "prohibited area" of the National Park. Conflicts of interest between the state and the villagers involve first of all issues on Park borders and "resource sovereignty". Notwithstanding the fact that state control is rather fragmented, a "fundamental asymmetry" (Dyck 1992, cited in Lane 2001: 1) between the parties involved persists. Whereas indigenous land claims are perceived by the state not only in terms of "political control" but first of all in terms of "public interest", local populations perceive their claims as crucial to "community security". Justified by collective concerns like resource preservation or climatic stability this "public interest" is portrayed as having higher legitimacy than local land claims whose protagonists are often conceived as merely "appropriators".

Following Nygren (2000), deforestation is first of all a process of change in which productive resources, property rights, and authority are struggled over. In this struggle, local people "alter their production strategies, as well as their perceptions of the environment, within a context which is structured, but not determined" (Nygren 2000:14). Land inside the National Park is only reclaimed in Sintuwu and Watumaeta. Because communal forest resources are still accessible in Rompo and no encroachment into the protected area occurs, this village formulates no land claims. Whereas in Sintuwu, land has been reclaimed by the village head in order to enhance his political power (Sitorus 2002), in Watumaeta it is first the "customary village council" (*Lembaga Adat*) that acts as the "voice" of local people and articulates community interests. The council legitimates its claims with the alleged existence of "ancestral land" (*tanah adat*) located inside the Park. Various proof of former settlements (planted bamboo and coffee stands, stone mortars, relicts of housings, leguminous plants that are used as home lot markers) have been "found" inside the protected area. Thus, the actual conflict arises primarily between the peoples' security claims to a "domesticated forest" and the state's objective of forest and bio-diversity conservation.

A second aspect of the dilemma is linked to insufficient monitoring and a weak enforcement of sanctions. As long as access to the forest resource is not effectively restricted, and the future legal situation is unpredictable (i.e. will there be a stronger enforcement of sanctions or not, will part of the protected area be returned or not?), people will "lose their incentive to protect it or to report on others that invade it, but are rather in a "race" to compete with others to extract as much as possible from it" (McKean 2000: 35). This is especially to be expected in a situation where land claims are to a high degree linked to issues of "intergenerational access". Furthermore, it seems that "legal unpredictability" precedes "ecological unpredictability". The second is closely linked to social issues like "trust" and "reciprocity". If a high number of "free riders" is to be expected under the given circumstances, the predictability of an "undisturbed" forest that is worth to protect is rather low and people will not feel much incentive to support its preservation (Gibson et al. 2000: 213).

Security in the Context of Relations of Production and Social Organization

Relations of Production: Security and the Evolution of Agrarian Arrangements

Above all other disciplines, it was anthropology that recognized the intrinsically social nature of property that is less a relationship between people and things, but primarily a relationship among people (Li 1995: 502). "Ownership" as such, without taking relations of people into account, is a one-sided means to measure access to land, because through various arrangements, people often earn income from land over which they do not exert ownership rights (Hart 1986). There are five general types of agrarian arrangements to be found within the research area: (1) lending out/borrowing in (*pinjam garap*), (2) share tenancy (*bagi hasil*), (3) fixed-rent tenancy (*bapajak*), (4) landsharing tenancy (*bagi tanah*) and (5) wet-rice fields (*sawah*) which are worked by tenants based on the promise of future harvest shares (*bapetak*). For the parties involved, individual agreements differ widely in terms of responsibilities, conditions and reliability according to the type of arrangement, the relationship between the actors involved and the general normative principles adopted by the communities in which they take place.

As stated above, the differentiation between "land owned" and "land borrowed" dry land was a rather blurred one under the conditions of shifting cultivation. Given the fact that least in theory, one could work a borrowed plot as long as one wished, "security" of borrowed plots was not an issue for those who worked them. Given the principle of "open access" it was not a security matter for those who lent out their plots either. The situation was different in the case of wet rice fields. Individual ownership rights were not only much more emphasized; *sawah* were also lent out for rather limited time spans – usually for one or two seasons – since they had been established in the forties and fifties. The fact that more pronounced ownership conceptions emerge as a response to the introduction of more permanent land use systems is not confined to Central Sulawesi; it has been shown at least for West Africa, the Philippines and (with special reference to Indonesia) for the Toraja of South Sulawesi and the Batak people of

Northern Sumatra (Seitz 1990). In the investigated villages, the seasonally confined lending out of dry plots did not take place before dry land cultivation moved from the "shifting type" to the "upland fallow type".

Being the "prototype" of agrarian arrangements, until recently the lending out of plots served primarily as a kind of rural welfare mechanism by which plots were temporarily redistributed to the poorer stratum of the villagers. The arrangement implies that the tenant is only allowed to plant annual crops (or rice in *sawah*). Inputs in form of fertilizer, herbicides, pesticides, costs for work force and technical equipment are fully taken over by the tenant. In principle the landowner should obtain no compensation for his generosity. In the past, as a rule, *pinjam garap*-agreements were open to *all* community members irrespective of personal industriousness. Kin relations did not play a crucial role in such agreements. This "moral" dimension is still fulfilled in Rompo where access to arrangements is open for all (*tidak memandang orang*) and where borrowed plots are still perceived as safe, at least for the time period they are borrowed. However, different from Watumaeta and Sintuwu, *pinjam garap* is still the exclusive kind of arrangement in Rompo.

In general, such rather generous practices can only be maintained in situations where the supply of land exceeds the demand. In Watumaeta and Sintuwu for instance, arrangements between locals seem more and more to be confined to close kin, whereas arrangements between migrants and locals as well as between migrants themselves often contain "disguised" advantages for the landowner. In *pinjam garap*-arrangements based on family solidarity (mostly small plots of land) as a rule the tenant receives a plot that is already cleared and ready for cultivation. The situation is different if one of the parties involved is a migrant. In about half of all cases migrant tenants borrowed land from locals or from their fellows, the land involved consisted of secondary forest, which the tenant had to clear on his own costs before. After the arrangement is finished, the owner receives his land in cleared condition. In cases where the land is already cleared by the owner, but time allocation and lack of money do not allow for cultivation of all fields for the moment, land owners often lend out part of their plots for one or two seasons in order to keep them clean (*bersih*) from secondary bush and forest growth. In such cases it was mostly the land-owner who searched for a tenant to borrow his land and production inputs have, in some cases, been divided between the two parties. Given that today permanent use of fields is crucial to secure land claims, the lending out of uncultivated plots has become a proper means to enhance the security of owner's tenure. Besides purchase, *pinjam garap* forms one of the most important ways of land access for migrants after arrival. In our sample, 28% of the migrants worked the first land in their new location on the basis of *pinjam garap*. Thus, borrowing land is often the first step of land transfers.

Whereas in Rompo *pinjam garap* is still enshrined in the normative vision of the community, it is perceived as an increasingly insecure arrangement in Sintuwu and especially in Watumaeta. Tenants point out that the arrangement can be dissolved at any time after harvest and that land-owners will never give written guarantees. Thus, the borrower will always be afraid that he will have to return the plot after the next harvest, because fixed time spans are seldom arranged in advance. In several cases, the tenant claimed borrowed plots as "ownership" after he started to work them. The security of borrowing-agreements is not linked to written guarantees in the first place, as the Rompo example shows. It seems rather that written guarantees become a crucial condition for tenants, and (to a lesser extent) owners' security when the fear of arbitrariness becomes paramount.

The widespread belief that borrowers are rather reluctant to apply methods to increase productivity and soil conservation measures because of uncertainty, and thus contribute to exhaustion of the soil and ecological degradation is not *a priori* given, but depends on the circumstances under which the arrangement takes place. The vegetable gardens (*bedengan*) in Watumaeta, for instance, are worked intensively, despite the fact that most of them are established on borrowed land. But this is a special situation where land is not borrowed from village fellows, but from the "public seed center" (*Balai Benih Induk*), so that borrowers can count on longer periods of land use than in other cases. In private arrangements, the fear of investing labor and money in vain poses a threat that becomes increasingly tense among tenants and owners.

Limited access to land and agrarian arrangements are not only reflected as the conditions for borrowing land become successively narrowed down, but also in the emergence and invention of other, more restricted types of arrangements. Share tenancy (*bagi hasil*) and fixed rent tenancy (*bapajak*) have been the dominant tenancy forms in Java since long, but are less often to be found in the outer islands of eastern Indonesia (Löffler 1996: 21). Whereas in Sintuwu, share tenancy is established in the case of *sawah* cultivation for at least two decades, it has only recently been adopted in the cultivation of dry fields. In Watumaeta it is still rare, and is mainly confined to wet rice cultivation, whereas it is not yet practiced in Rompo. The same holds true for fixed-rent tenancy, which is meanwhile quite common in Sintuwu, but appears only sporadically in Watumaeta in combination with "absentee landlordism". Both arrangements, share tenancy and fixed rent tenancy, were not developed within the local communities, but were first introduced by Buginese migrants. As the Indonesian term *bagi hasil* (distribution of the harvest) points out, in share tenancy the tenant cultivates land rented out by the landowner for an agreed share of the yield. Two types are distinguished according to the distribution ratio: (1) in the *maro*-system – practiced in wet-rice cultivation – harvests are divided 50:50 between the two parties with input costs being equally shared as well, whereas in (2) the *mertelu*-system – practiced in dry land cultivation – the tenant receives 2/3 of the yield with providing all inputs on his own. As far as the minimum security conditions for tenants are concerned, share tenancy arrangements do not fulfill the requirements of the law; neither the regulations of the Basic Agrarian Law (UU 5/1960), nor the regulations of the "Presidential Instruction" from 1980 (No. 13). The first states clearly that the sharecropping contract must be drawn up in the presence of the village head. No case of written guarantees was found in our sample. Furthermore, the land law prescribes that the length of the contract must be at least three years for *sawah* and five years for dry land. In opposition to the law fixed time spans were never arranged in advance and no guarantee in terms of cropping time were given.

In fixed-rent tenancy, land is rented from the land-owner for a defined, agreed period. This is usually one year with a fixed amount of rent in the form of cash before the land is cultivated. The advantage of renting out land against cash seems to have not yet entered the minds of the local community members in Rompo and Watumaeta. Villagers in Rompo view this kind of arrangement as even incompatible with local *adat*. Whereas fixed-rent tenancy is less favorable for the tenant, it implies far-reaching security conditions for the local population in general. Indigenous people have sold large areas of land to migrants in Watumaeta in order to cover costs for medical treatment, school fees and marriages. The renting out of plots against cash for limited time spans may well be an alternative to the selling of land once and for all.

Invented by Buginese migrants in Watumaeta, land-sharing tenancy (*bagi tanah*) is the most secure arrangement. In difference to borrowing and share tenancy, *bagi tanah* fits ideally the requirements of perennial cultivation. At the same time it recognizes migrants' concerns with plot identification and clear tenure rights. Whereas the planting of perennials is not allowed in *pinjam garap,* the tenant prescribes that perennials are to be planted by in *bagi tanah,* with the plot being divided afterwards according to the individual arrangement. Tenants prefer this kind of arrangement more than other plans because it implies the written guarantee that they will obtain part of the land. The legal requirements listed above are explicitly fulfilled in the contracts with a "model contract" being kept in the house of the head of the Buginese hamlet (*kepala dusun*). The contract includes a clear demarcation of the plot borders, a detailed listing of input sharing, the number and kind of perennials and shadow trees to be planted an a sketch of the envisaged planting pattern. The contract is signed by the village and hamlet heads as well as by two neutral witnesses. So far, the involvement of locals in land-share tenancies is confined to the ruling family of Watumaeta who made several *bagi tanah*-arrangements with Buginese families who are preferred over locals because they are more experienced in establishing perennial gardens and have the image of being more industrious than locals. Whereas the new arrangement fosters security and reliability for migrant tenants, on the part of the locals it seems to support the land accumulation of the ruling family. Common local people on the other hand, due to their shortcomings in perennial experience, are excluded from the possibility of obtaining land via land-share tenancy-arrangements.

Whereas it was not found in the Watumaeta sample, the practice of planting and working wet rice fields on the basis of future harvest participation (*bapetak*) is occurring as a constant variable within the Sintuwu sample. The *bapetak* is similar to the *ceblokan* system found in Java (Hart 1986, Harjono 1997) and is based on the following principle: a tenant plants the *sawah* of a land-owner under the promise of later receiving a harvest share (see Burkard, this volume). Only landless (local) Kaili and Kulawi people work as tenants on a *bapetak*-basis. In general, a system like the *bapetak* only occurs when access to land is extremely limited because the tenant receives no compensation for the period he works the land, and harvest shares are much lower than those obtained from share tenancy arrangements. The tenant's right to harvest the plot is exclusive. *He* is the privileged worker who enjoys the sole right to obtain income from the plot, whereas harvest participation of others is strictly excluded by the arrangement. Due to the large-scale conversion of *sawah* into cacao plots, the wet rice reserves of local families have been severely reduced with the effect that inheritance is significantly delayed. Young people now work the plots of their parents whom they have to sustain as well. As a consequence of the severely diminished wet-rice area, the arrangements have become almost an exclusive internal family affair. Thus, people have not been "pulled" out of agriculture, as in rural Java, by new opportunities outside the agricultural sector, but have rather been "pushed" out of agriculture by the processes of change induced by cacao penetration and large-scale conversion.

Security in Relation to Issues of Social Organization and Leadership

Deforestation and resource degradation have been linked to various conditions, i.e. population pressures, market incentives, technological factors and state policies. Given the fact that the effects of such "context variables" is not a linear one, but can be quite different from one community to another, environmental studies have increasingly turned their attention to local institutions that "mediate" between external factors and the environment (Agrawal and Yadama 1997, Gibson et al. 2000). Given the fact that no indigenous grass root organizations in regard to resource management exist in the area, there is less concern with "institutions" in the sense of persistent structural entities. Instead, we shift our focus on organizational "profiles" that are better understood in terms of peoples interactions than in terms of tangible structural factors. Notwithstanding the fact that human relations can be highly complex and multi-layered even within small and "homogenous" communities, socio-economic security as it relates to social organization will be discussed in regard to four major relationships and lines of conflict: (1) the relationship between generations which is implicitly linked to questions of intergenerational access, (2) the relationship between the common peasant and the village authorities which is related to issues of legitimacy of leadership and power, (3) relations between the members of the local population with special attention to matters of social cohesion and (4) the relation between the local and migrant populations as they manifest themselves in terms of trust and reciprocity.

Security and Intergenerational Access

By definition, sustainable resource management must include an intergenerational element. In two aspects, i.e. the clearance of forest and the selling of land, intergenerational access plays a crucial role. Encroachment and forest clearing is primarily perceived as a result of increasing land scarcity. Whereas there is a high degree of consensus in regard to this "intervening" factor, land scarcity itself is often linked to different reasons. Whereas older people blamed the arrival of migrants and the establishment of the protection forest to be responsible for current land scarcity, the younger generation often accused their parents of having sold land too fast and at prices that were too low to outsiders. Young people complain that they have often not been involved in the decision-making process, whereas it is their concern when family land is sold.

Within the representative sample, 40% of all local households sold land within the last five years. In most cases land was indeed sold by the parents or parents-in-law of the present household head. There is no doubt that the high frequency of land sales is linked to the general situation of resource unpredictability. Either in the conviction that parts of the National Park will be returned to the village communities or in the belief that sanctions will not be enforced in case of encroachment into the protected area, people often sold land in case of marriage or sickness under the assumption that new plots can still safely be opened in the future. Vaguely defined park borders that were insufficiently communicated to the villagers when the Park was established exemplified the prevalence of this attitude. Key informants point out that this perception prevailed until around 1998; after that the communication of issues related to the Park seems to have somehow improved with the frequency of land sales declining. Whereas about half of our respondents perceived the size of their holdings as sufficient for the present, more than 80% perceived them as insufficient to equip their children with necessary resources.

Security in the Context of Local-Migrant-Relationships

Tensions between locals and migrants can be observed in all villages, but their articulation is more pronounced in Watumaeta than in Sintuwu; whereas there seem to be little tension in Rompo. These differences are first linked to the outer circumstances under which migrants entered the villages. Sintuwu is inhabited by a significant number of migrants for at least two decades with people entering the village in several phases. Thus, the number of newcomers at a given point of time was easy to survey for the local population, which implies that locals and migrants had sufficient time to recognize their presence and to adapt to each other. This is different from the situation in Watumaeta where a significantly higher number of migrants entered the village within a very short time period, probably too short for adaptation. Because the population data record is kept under lock by the village head, other village officials have no reliable information on the actual number of migrants living in their community. Neither village officials, nor the local population seem to realize how far the Bugis and Sundanese have already established themselves as a large and economically powerful group in Watumaeta. The village secretary was really shocked after he obtained the complete population data record that was made available to him via our co-researchers from subproject A1 (Melanie Abdulkadir, personal communication). He talked to about five or ten Buginese families living in the new frontier area (RT 9) that was opened for settlement two years before, without realizing that they had already grown to more than 100 people. Most migrants in Watumaeta live in separate village clusters. "Iteration" between locals and migrants is almost missing and – different from Sintuwu – no intermarriage occurs. The recognition of the need to balance the number of people with the availability of natural resources is a basic precondition for the development of sustainable resource use (see Colfer et al. 1998). This is difficult to achieve when the number of people living in a community is unknown by the village officials and the indigenous population.

In Rompo, on the other hand, migrants (exclusively of Toraja ethnic origin) are well integrated into the local community. In general, ethnic boundaries are not emphasized between the Toraja and the indigenous groups of Central Sulawesi; both view each other rather in terms of different subcultures of the Sulawesi "highland people" than in terms of distinct ethnic groups. Trade connections between the Besoa valley and the Toraja highlands have been established for generations. Further, both adhere to Christianity. Given the religious conflicts that shattered the regency's capital Poso within the last years, migrants and locals find themselves in a similar "structural position" in relation to the Muslim groups, especially the Bugis who have become one of the most dominant groups in the Lore Lindu region. After their arrival many Toraja lived in the houses of local families with formerly nuclear and extended households being transformed into multiple-family households. "Iteration" between locals and migrants is sometimes higher than among the locals themselves. Given the continued existence of village forest resources and the high degree of acculturation and integration on behalf of the migrants, it is not surprising that the subjective "security conditions" of the local population are less affected by in-migration in this community. Besides, migrants participate on a regular basis in the weekly "church meetings" that function as an informal forum, where minor land conflicts can be discussed in a familiar atmosphere without the interference of the village apparatus.

Apart from the degree, to which migrants are integrated into the local community, the procedures and conditions under which migrants obtain access to land is crucial. Local people in Watumaeta are not only ignorant about the actual number of migrants living in their village, but they are also ignorant about the real extent and the location of secondary forests which was delivered to migrants by the village head. Because no transparency in this matter exists, locals complain that too much secondary forest has been sold or granted to outsiders whereas local people are not obtaining much. If access to forest resources is perceived locally to be unfair, people will take what they need and legitimize individual encroachment. Economic conflicts within the village about the struggle for security are unavoidable. In Rompo on the other hand, the boundaries of the resource distributed to migrants was rather clear and agreed upon among the local farmers. The amount of land delivered to each migrant household head was confined to two hectares and no full ownership rights have been granted in the first place. Instead, land was granted on a usufruct basis with the condition that full ownership rights are only given to migrants in case they decide to become long-term residents in Rompo. In other words: the degree of land rights migrants enjoy was dependent on their willingness to identify themselves with the village community. Such arrangements are important to control outsiders' access to land and to reduce the fears of local people related to in-migration.

In a lucid article, Sitorus (2002) portrays the pattern of "economic succession" in Sintuwu where Buginese migrants first worked on land borrowed from locals. Before that, they accumulated land by purchase and, finally, become the richest peasants and biggest landowners in this village. However, the local groups have often descended from the status of owners to the status of tenants and wage workers. This trend can already partly be observed in Watumaeta where an increasing number of local people work on the land of Buginese farmers. Locals point out that many of their fellows have lent money from migrants within the last years. It is difficult to estimate in how far the security mechanisms based on reciprocal relationships are being replaced by new ones based on contacts with migrants. Locals often claim that land is given as a collateral to migrants. Whereas this pattern is portrayed by the respondents to be a widespread phenomenon within the village, its relevance is not that obvious from the data of the representative household sample. But "statistical correctness" is not the criterion by which such representations of community threats should be judged. For the population concerned, the subjective fear that their well being is threatened by migrants' economic dominance is much more crucial for their security feeling than the statistical expression of frequencies.

Whereas local security concerns center first around issues of migrants economic dominance and access to land, security concerns of migrants are linked to issues of secure ownership rights that allow for long-term investments in their plots. The increasing concern of migrants with plot identification and certification is first of all linked to the negative experiences they have made in the past. In Watumaeta as well as in Sintuwu plots sold by the village head were often located within the protected area; whereas plots sold by local families to migrants in Watumaeta were either incorrectly defined in terms of boundaries or were not owned by the person who sold the plot. In some cases plots have even been sold several times. Such practices are hardly accepted if they happen within one's own social group, but they are obviously tolerated in the interaction with outsiders. The anthropologists call this "negative reciprocity". It is important to point out that at the time of their arrival migrants did not have necessarily an advantage over locals in terms of material welfare. Many of them arrived with literally no more than the cloths they wore and a bush knife in their hands. The advantage they had was in terms of agricultural knowledge, organizational capacity and in terms of experience in dealing with state authorities.[7] This advantage enabled them to react relatively effective in the anticipation of troubles. Reporting to the village head shortly after conflicts arise, organizing land certificates directly after plots are purchased and the use of third persons to act as witnesses of land transactions have become the major strategies to anticipate troubles and to enhance social security in the "diaspora".

General Lack of Social Cohesion among the Local Population

Irrespective of the struggles over resources that characterize local-migrant relationships, there is also a high degree of distrust and envy among local people to be observed. In general it seems that traditional security arrangements have weakened with people often pursuing their own business. The weakness of "social cohesion" is further reflected in the fact that theft of perennial plants and perennial harvests occurs often in all villages. Theft of young cacao plants is a common threat since locals have discovered the economic value of the crop. Meanwhile, some farmers have transplanted cacao from the fields to their home gardens to reduce the burden of continuous "watching". Theft of fish forces many farmers to spend the night in huts close to their pond

---

7 During the time I lived in the house of the village head of Watumaeta, I was astonished about the high number of migrants who visited him in his house to discuss some agricultural issues (conflicts, land registration, taxes, certification). No local inhabitant came during that time period, with the exception of relatives that came for reasons of family visits only.

in order to watch out for thieves. New fishponds are almost always established in home lots. Several maize crops have been damaged in Rompo, because people did not take care of their pigs and left them to roam freely without taking the consequences for fellow villagers into account. It had to be made explicit in the village regulations that live-stock must be kept in fences and that sanctions would be enforced if animals roamed freely.

Besides mutual distrust, people often accuse each other of being lazy, a phenomenon well known from other Indonesian settings (Sherman 1982, Eijkemans 1995). The village regulations in Rompo and Watumaeta state that everybody has to work his plots actively. Otherwise it can be withdrawn and lent out to others by the village authorities. This regulation is the one most difficult to enforce, village officials, *adat* leaders and BPD members defend this rule not preliminary in terms of "sustainable resource use" or as a means to reduce i.e. rattan gathering activities, but first of all as a means to eradicate drinking habits and laziness in the villages. In general the number of "free riders“ seems quite high. As Popkin (1980: 446) defines them, "free riders" are individuals who withdrew from their social obligations, because they believe that they will obtain the necessary assistance anyway when they need it. Local families who opened small kiosks (*warung*) in Watumaeta had to close them down after a while, because too many kinsmen borrowed money or asked for food and cigarettes on tick, which they never paid back. Consequently, most *warung* in Watumaeta are now owned by Buginese migrants.

"Free riders“ make it difficult for others to work together as well, especially when their behaviour endangers the persistence of groups and organizations. The "farmers contact group“ (*kelompok tani*) in Watumaeta – which is organized on behalf of the state to co-ordinate agricultural work and to disseminate agricultural knowledge among the farmers – functions as the major medium through which government subsidies for wet rice cultivation and annual farming are transferred to the peasants. Formally, at least in the village monographs, no less than six "farmers groups" exist in Watumaeta; all of them are headed by lineal and collateral relatives of the village head. There are no activities to be observed, and no clearly defined boundaries of membership exist. When government programs are at stake, groups emerge on an ad hoc basis, but they dissolve very quickly after the subsidies have entered the village. Such conditions engender resistance against the establishment of social organizations that manage their resources in cooperation and in a sustainable manner, when individualism grows at the cost of reciprocity, and social relationships are increasingly governed by mutual distrust.

Security in the Context of Village Leadership

In order to deal with risks and uncertainties, peasants "invest" in social arrangements that provide them with a minimal subsistence in times of misfortune. People establish good relationships ("security ties") with the well-to-do and more powerful strata in the village in order to lend money or to ask their support in dealing with conflicts. According to the representatives of the "moral economy"-approach (Scott 1976), it is an inherent principle of such arrangements that they are open for *all* villagers irrespective of social status or individual industriousness. The political economists (Keyes 1983, Popkin 1980) on the other hand deny this viewpoint as an idealistic representation of village life that overestimates the degree of peasant participation and point to the small size of decision-making groups in rural communities (see Eijkemans 1995: 68). Irrespective of the fact that power abuse is a widespread phenomenon all over Indonesia and that this is probably inherent to such systems, "power abuse" in Watumaeta and Sintuwu goes beyond what is tolerable for villagers, and the villages have an image as "worst cases".

Despite the fact that the inhabitants of Sintuwu perceive the "quality" of leadership in their village as rather bad, they never question the "legitimacy" of the village head. In Watumaeta on the other hand, until the election of 2002, poor quality leadership complemented a serious lack of "legitimacy". The former village head had not been elected by the population for two periods, but was installed by the district head (*camat*) despite the fact that he did not the win the elections. In Watumaeta, a sort of "family rule" emerged with members of the leading family controlling lots of official positions that include the hamlet heads (*kepala dusun*), the heads of the "operational activities" and the "administrative activities", the public security official (*hansip*) and the heads of the farmers group. The village head and his family are not only accused of shameless land accumulation, but also of selling large areas of secondary forests to outsiders that are claimed by local families (see Burkard, this volume). The most serious reproach is that the leader sold "ancestral land" (*tanah adat*) to outsiders against the security interests of the local population. Further, coffee stands within the National Park have been sold to migrants, in spite of the fact that local families obtained the permission from the Park management to use them. Several *bagi-tanah* arrangements with Buginese newcomers have been established on large tracts of fallow land, but only a few local families related to the village head have been given access to the newly opened frontier area. As pointed out above, though, people complain that locals are often indebted to migrants and must hand over their land because they cannot pay their debts. In his unwillingness to restrict the number of migrants, his practice of dividing land among outsiders and in accumulating new plots by arrangements with migrants, the overall accusation is that the leader does not protect the local population from land losses and outside threats.

Despite access to land that was permitted with almost no restriction, the experiences of migransts with village heads are not encouraging either. In Sintuwu and Watumaeta it was often the village leaders themselves who sold "conflict land" of unclear status with the effect that migrants are not only more frequently involved in land conflicts than locals, but also suffered more often from unsolved problems. Those conflicts prevailed either because people from different villages claimed the land, or because the village head was not interested in solving the problem. In cases where land was sold to newcomers by the village heads in Watumaeta and Sintuwu, they often profited by issuing an "ownership letter" from the village head for which an "administration fee" had to be paid. Such letters as a proof of land ownership are of low value, and they are almost useless to prove claims if the plot is located in the National Park. The former village head of Rompo has also sold land to outsiders living in Palu and is accused of several "corruption cases". But fortunately for Rompo, changes in leadership and the certification of most plots by the Prona-certification-program *before* they could be sold by village officials have prevented the emergence of situations like Watumaeta and Sintuwu.

Summary and Conclusion

The villages under study represent a clear continuum in a process where shifting cultivation is successively replaced by upland fallow systems and permanent agriculture. In the end of the process, "integrated" fallow periods disappear where the plots are cultivated permanently. However, parts of the land lay under fallow for longer periods. These fallow plots are rather irrelevant as a production factor but form a major security condition for local people in case of unexpected disasters. The transition from shifting cultivation to permanent land use is accompanied by the emergence of agrarian arrangements that become more restricted, the more perennial agriculture dominates the village economy. The common believe that the transition to more intensive systems occurs only under conditions of population pressure and a resulting "lack of choice" (Boserup 1965, Ruthenberg 1980, Brown and Schreckenberger 1998) does not hold true in the case of Central Sulawesi. Permanent agriculture is not linked to population pressure in the first place; rather, it is related to the registration of land (taxation liability) and the penetration of market incentives.

In the research region, the transition to permanent agriculture is further accompanied by more defined ownership rights. Security of land tenure is not an issue as far as permanently cultivated land is concerned. In this context, migrants have enhanced tenure security by cautious plot identification, land certification, and by the introduction of written arrangements. The existence of a sort of "legal pluralism" that leads to individualization of ownership conceptions cannot be upheld. More crucial is the fact that locals and migrants put their priority on different conditions of security. The security perception of migrants is linked primarily to legal security of their plots. Because migrants rank as outsiders or even intruders, keeping a low profile and the avoiding conflicts with the law are their major concern. Relying primarily on the market as their major security provider, maximization of product value is crucial. Thus, compatible use of adjacent plots is a central security condition for Buginese and Sundanese farmers. Security of locals, on the other hand, is intrinsically linked to the availability of sufficient land reserves (partial fallows) and mixed cropping strategies. In general, lack of security is not due to the absence or insecurity of private land ownership as such, but rather to the fact that existing social structures, unclear plot borders and missing indigenous or state control mechanisms allow those in power to transfer use and ownership rights to others randomly and to grant secondary forests formerly used by local families to newcomers at will. As the Rompo example shows, which kind of resource is given to migrants under which conditions is important for the subjective security feelings of the local population. The ability to restrict and condition outsiders' access is essential.

Furthermore, secure tenure in itself may not even be a sufficient condition for sustainable resource use as long as other factors like risk control and market incentives are more conductive in land management decisions. Even if they adhere to a principle of mixed cropping or if they prefer to secure their incomes by intensive specialization, peasants need to predict future developments. Conflicts between state-claims that are articulated in terms of "public interest", and local claims to a domesticated forest in terms of "community security" coupled with weak enforcement of laws and sanctions, result in a high degree of legal unpredictability. Such conditions sponsor low compliance with rules, resulting in continuously degrading forest resources. Thus, mid-term stability and present income security are given priority over long-term sustainability, even if farmers are aware that this will endanger the resource base in the long run.

The trend to perennial farming as the main custodian against deforestation implies two risk-loaded side effects: (1) peasants will increasingly rely on the market as their primary security provider. It does not seem that local people are willing to substitute their two major traditional security conditions, mixed cropping and sufficient availability of forest reserves, with market security as long as they are less experienced with cacao cultivation and marketing strategies. Given the fact that traditional security mechanisms based on reciprocity and family solidarity are fading away, lack of social cohesion poses a major barrier against the establishment of sustainable "resource management organizations". Traditional arrangements have already eroded by modernization processes before new support systems have come into being. One possible developmental effort could be the establishment of member-based social organizations with clearly defined rights and responsibilities for their members. (2) Perennial agriculture will not be sufficient to provide people with the necessary security, nor will it relieve the forest margin as long as monitoring and law enforcement is weak and a situation of *de facto* "open access" prevails. Rather, less working hours and higher income through perennial farming are invested into new land clearings.

Due to the absence of clear regulations and organizations related to natural resource use, village leaders in the past could exert almost absolute power in the distribution of resources. This is changing at the moment with the formation of new councils (BPD) that are given wide legislative autonomy and that can challenge the leader's power. Their willingness to learn from past experience or even to develop "learning organizations" will highly depend on their responsiveness towards the varying security concerns existing in the villages and to rise above the different conditions of security in acting as an impartial arbiter in case of conflict. The precondition for this is that the legitimacy of the new leaders is sufficient and that the villagers are willing to share risks and accept sanctions for the members of their own group in case of rule infractions.

References

Agrawal, A. and Yadama, N. (1997): How do Local Institutions Mediate Market and Population Pressures on Resources? Forest *panchayats* in Kumaon, India. *Development and Change*, Vol.28: 435-65.

ANZDEC (1997): Report on the Central Sulawesi Integrated Area Development and Conservation Project. Palu, Sulawesi.

Banana, A. Y. and Gombaya-Ssembajjwe (2000): Successful Forest Management: The Importance of Security of Tenure and Rule Enforcement in Ugandan Forests. In: Gibson, C., McKean, M. and E. Ostrom (Eds): People and Forests. Communities, Institutions, and Governance: 87-98. London.

Boserup, E. (1965): The Conditions of Agricultural Growth: The Economics of Agrarian Change under Population Pressure. London.

Brown, D. and Schreckenberger, K. (1998): Shifting Cultivators as Agents of Deforestation: Assessing the Evidence. *Natural Resource Perspectives*, no.29: 1-9.

Colfer, J. P. et al. (1998) : The Grab Bag: Additional Methods for Assessing Human Well-Being. Bogor.

Eijkemans, C. H. (1995): Profitability or Security. Decision-Making on Land Use among Toba Batak Peasants in North Sumatra, Indonesia. Nijmegen.

Ellis, F. (1988): Peasant Economics. Farm Households and Agrarian Development. Cambridge.

Fremerey, M. (2002): Local Communities as Learning Organizations. The Case of the Village of Toro, Central Sulawesi, Indonesia. *Storma discussion paper series*, no. 6. Bogor.

Gibson, C., Ostrom, E. and McKean, M. (2000): Forests, People, and Governance: Some Initial Lessons. In: Gibson, C., McKean, M. and E. Ostrom (Eds): People and Forests. Communities, Institutions, and Governance: 227-242. London.

Gibson, C., Ostrom, E. and McKean, M. (2000): Explaining Deforestation: The Role of Local Institutions. In: Gibson, C., McKean, M. and E. Ostrom (Eds): People and Forests. Communities, Institutions, and Governance: 1-26. London.

Harjono, J. (1990): Land, Labor and Livelihood in a West Java village. Yogyakarta.

Hart, G. (1986): Power, Labor and Livelihood. Processes of Change in Rural Java. Berkeley.

Ibeanu, O. (1999): Ogoni-Oil, Resource Flow and Conflict in Rural Nigeria. In: Granfeld, T. (Ed.): Managing the Globalized Environment. Local Strategies to Secure Livelihoods. Trowbridge

Jüttich, J. (1999): Strengthening Social Security Systems in Rural Areas of Developing Countries. *ZEF Discussion Papers on Development Policy*, no 9. Bonn.

Keyes, C. F. (1983): Peasant Strategies in Asian Societies: Moral and Rational Economic Approaches. *Journal of Asian Studies*, Vol. XLII, No. 4: 753-768.

Lane, M. (2001): Indigenous Land and Community Security: A (Radical) planning Agenda. *Working paper*, No. 45. Land Tenure Center, University Wisconsin. Madison

Löffler, U. (1996): Land Tenure Developments in Indonesia. (GTZ). Bonn.

McKean, M.(2000): Common Property: What is it, What is it Good For, What Makes it Work? In: Gibson, C., McKean, M. and E. Ostrom (2000): People and Forests. Communities, Institutions, and Governance: 27-56. London.

Mischung, R. (1980): Meo und Karen: Die Umwelt-Anpassung zweier hinterindischer Bergvölker. Ein Beitrag zur ethnologischen Ökologie-Diskussion. *Paideuma* 26: 141-56.

Mischung, R. (1984): Sesshaftigkeit und Intensivierung beim Brandrodunsgfeldbau. Eine Fallstudie aus Nordwest-Thailand. *Paideuma* 30: 242-255.

Li, M. T. (1995): Images of Community: Discourse and Strategy in Property Relations. *Development and Change*, Vol. 27: 501-527.

Nygren, A. (2000): Development Discourses and Peasant-Forest Relations: Natural Resource Utilization as Social Process. *Development and Change* Vol. 31: 11-34.

Popkin, S. L. (1980): The Rational Peasant: The Political Economy of Peasant Society. *Theory and Society* 9: 411-471.

Ruthenberg, H. (1980): Farming Systems in the Tropics. Oxford.

Scott, J. (1976): The Moral Economy of the Peasant. Rebellion and Subsistence in SE-Asia. Yale.

Seitz, S. (1990): Zur Frage der Herkunft und der sozioökonomischen Auswirkungen des Naßreisanbaus in traditionellen Gesellschaften. *Freiburger Geographische Hefte* 30: 119-132.

Sherman, D. G. (1982): Social Organization of Samosir Batak Livelihood in Relation to their System of Beliefs and Values. Cornell University.

Shrauwers, F. (1999): "Its not Economical": The Market Roots of a Moral Economy in Highland Sulawesi. In: Li, M. T. (Ed): Transforming the Indonesian Uplands: Marginality, Power and Production: 105-130. Halifax

Sitorus, F. (2002): Land, Ethnicity and the Competing Power. Agrarian Dynamics in Forest Margin Communities in Central Celebes, Indonesia. Storma Discussion Paper No. 5. Bogor.

Straube, H. (1967): Der agrarische Intensivierungskomplex in Nordost-Afrika. *Paideuma* 13: 198-222.

Vivian, J.(1991): Greening at the Grassroots: People's Participation in Sustainable Development. *UNRISD Discussion Paper No. 22.* Geneva.

Chapter II

What is Community in Community-Based Natural Resource Management?

Satyawan Sunito

Abstract

This paper asserts that theoretical assumptions behind the concept of community based sustainable development have misperceived the nature of village communities living near and inside the forest as homogenous, subsistence-based communities. Instead, processes of incorporation and social differentiation have fundamentally changed the communities in almost all their aspects. This misperception raises the question of whether Community Based Natural Resource Management programs based on a false understanding of the nature of the indigenous communities can succeed in reaching their objectives. Alternatively, this raises questions on better-conceived approaches to avoid misunderstandings of the nature of local communities. This implies the need to change established ideas on forest protection and conservation.

Introduction

The objective of this paper is to highlight the gap between the theoretical assumptions behind the concept of *Community-Based Sustainable Development* (CBFM) and the reality of the differentiation process of the communities that are the subject of the community programs. To illustrate the standpoint above, the author invites readers to look at the process of establishing community-based forest conservation in a village community which has experienced profound changes in almost all aspects of life, a fate commonly shared by many communities living at the forest margin or inside the forest. The community is part of the upland village communities that share administrative borders with the Lore Lindu National Park in Central Sulawesi, Indonesia. The data used in this chapter were collected during several field studies in 2002 and 2003. The first part looks at the basic assumptions of the concept of *Community-Based Natural Resource Management* (CBNRM), under which we consider also concepts such as *Community-Based Forest Management* (CBFM), *Community-Based Forest Conservation* and *Social Forestry*. This includes notions of customary law and local knowledge, which are perceived as "conserving" in character. The second part looks into the process of social differentiation of a local village community that dates back to the start of the colonial intervention in the beginning of the 20th century. The third part illustrates the problems linked to the establishment of a community-based forest conservation program, initiated by the National Park Authority. The chapter closes with some considerations on methodological aspects.

Basic Assumptions of CBFM

The idea to delegate a certain amount of responsibility and authority in natural resource management to local communities was inspired by several factors. The first was the failure of the "modernization" approach that started after the end of the Second World War, focusing on large-scale and centrally managed programs to alleviate rural poverty and reduce income disparities in developing countries. One effect of this development approach was the immediate degradation of the environment, due to large-scale depletion of natural resources. This occurred even in previously isolated hinterlands. The second factor was the development of a new paradigm of planned intervention, which emphasized a bottom up approach through participation of those people who were the targets of development. The third factor was the surge of human rights' and indigenous peoples' movements, which was partly a reaction to the modernization approach that in most instances was implemented without much consultation and did not benefit the great majority of the poor. Under the rhetoric of national development and economic growth, access to natural resources was granted to large companies for large-scale exploitation putting local communities even further away from access to natural resources. Under international as well as internal pressure a process started that brought a paradigm shift from a centralistic approach to a more integrated approach, which combined environmental sustainability, the acknowledgement of local community rights, and, as a consequence, the devolution[1] of natural resource management and socio-economic development (Western and Wright 1995, Hall 1988, Meinzen-Dick et all 1999, Li 2002). The development of new concepts and strategies became part of the language used by activists and government policy makers alike, such as *Community-Based Sustainable Development* (CBSD), *Community-Based Resource Management* (CBRM), *Community Based Forest Management* (CBFM), and *Community Based Conservation* (CBC).

The popularity of the devolution concept of natural resource management however, owns its fast growth to an oversimplification of the nature of rural communities and their interaction with natural resources. Leach, Mearns and Scoones (1997) launched a strong critique of the basic CBSD assumptions concerning the communities involved and their interaction with the environment. In this assumption the communities are viewed as closely interwoven, homogenous in character and governed by customary law. It is assumed that these communities originally lived in harmony with the surrounding nature, developed a rich knowledge and sustainable systems of interaction with their natural environment. These local communities were considered to be conservationist by nature. In this view, environmental degradation was considered as caused by external forces. Thus, the process of degradation could be halted and the original state could be restored by restoring the rights of the local communities over their environment and the re-establishment of the local institutions (and local knowledge) for the management of the environment. Alterna-

[1] The term *devolution* indicates the transfer of responsibility and authority over natural resources from the state to non-governmental bodies, particularly user groups. *Decentralization* refers to authority and management transfer to lower levels of government. *Deconcentration*, on the other hand, describes the reallocation of administrative duties – without transfer of power and authority – from ministry or department headquarters to branch offices of the central government (Meinzen-Dick, Knox, Di Gregorio, 1999: 3-4).

tively, new local institutions for regulating collective action in the management of the human – nature interaction could be developed.

An important aspect of the discussion on the nature of local communities is the phenomenon of 'indigenous knowledge' (IK). As described by Ellen and Harris (2000), a new interest and appreciation for IK grew in the 1960s as a reaction against the "increasing remoteness", and ".... perceived arrogance and negative technological outcomes" of modern science (ibid. 12). Part of this reaction, as described before, is the idealization of the so-called traditional, indigenous, primitive people or communities. In correspondence with the notion of communities that lived in harmony with nature, IK is described as intrinsically conservationist and promoting a fairer social and economic structure (Li, 2000). IK does not deal only with systems for adaptation; it also contains the people's perception of natural resources and environment. There has been much criticism of this view of the homogeneous, conservationist indigenous community, living in harmony with its natural environment. Three forms of critique will be highlighted here, namely: 1) the confusion between sustainable adaptation systems and conservation; 2) the theoretical roots of this assumption; 3) the critique on the assumption of the homogeneity of communities.

Sustainability can be a product of one or a combination of several factors such as low human population density, low demand for resources, or limited technology choice. Within this context, conservation of biodiversity can be an incidental by-product of sustainability. From the same point of view, a change in population density, increasing demand and competition for resources because of the availability of technology could endanger the sustainability of the system. Voicing the consensus among academicians, Smith and Wishnie (2000: 501) concluded that to qualify as conservation, any action and practice should satisfy two criteria: 1) it should prevent or mitigate resource depletion, species extirpation or habitat degradation; 2) it should be designed to do so. In view of these criteria, Little (1994:350, cited by Smith and Wishnie, 2000: 516), after reviewing a large number of community-based conservation programs, concluded that "cases in which local communities in low-income regions manage their resource bases with the prime objective of conservation, rather than improved social and economic welfare, are virtually nonexistent".

Instead, indigenous resource management is based on the enhancement of livelihood resources, to safeguard resources from exploitation by outsiders, and to allocate subsistence efforts to the most rewarding areas and resources available (Smith and Wishnie, 2000: 516). An extreme example is posed by Tim Flannery in his book *The Future Eaters* (1998) where the confrontation between human beings and the Australian ecosystem starting some 40,000 years ago resulted in the destruction of the latter, long before the arrival of white people. More specifically Li asserted that instead of engaging in subsistence economic activity, indigenous people have a long history of involvement in the market economy. In this context local forest use could lead to conversion of forest into agriculture land and intensification may lead to further deforestation. This historical approach to the understanding of the interaction between local community and environment, made the assertion of the intrinsically conservationist nature of indigenous people questionable (Li, 2002: 269-270).

The view of a socially and culturally homogenous community living in harmony with nature had its roots in the functionalism in anthropology that views society as functionally integrated with others within its natural environment through its adaptation system. As stated by Agrawal and Gibson (2001: 2-3), the characterization of the community as a *Gemeinschaft* that functions as an organic whole in contrast to society or *Gesellschaft* which is differentiated in character, continues to influence the conservationists view of rural or indigenous communities. Modernization theory, influential since the 1950s, shares the same evolutionary idea of societies. In this view, societies developed from traditional to modern, where industrialized societies typify the modern stage of development. This view of a closed, self sustained and inwardly looking village community has always attracted and served the interest of different, even opposing, parties. Previously, colonial powers constructed an idealized village community as the foundation of their colonial society, from time to time positioning it as the loyal subject of the empire against the radical nationalists. The same village community was hailed by the nationalists as the cradle of national culture and identity. After gaining political independence, the process of nation building was built up on populist ideas where perceived village community institutions and virtues such as mutual help, consensus/*musyawarah* were used to legitimize state leadership. In most cases however, the populist rhetoric was used to continue the exploitation of village communities and indigenous people for the benefit of the upcoming urban middle class (Breman, Kloos and Saith, 1997; Wertheim, 1964).

In this context, the critique expressed by Tania Murray Li (1999) of the concept of marginality, a characterization frequently applied to communities living in the uplands and near forests, seems appropriate because the same connotations are attached to the term "indigenous community". She states that differences in cultures between the marginalized communities and the urbanized centers of industrialized societies have been explained in terms of a time dimension. These communities are assumed to be relics from the past. In other words: indigenous, marginal communities are perceived as – to use Eric Wolf's expression – "societies without history" (cited by Li, 1999:5 and Schrauwers, 2000:64). Such a view totally ignores the historical forces since pre-colonial times that have not left any community or society untouched, and which connect communities with development at the larger scale. This interrelatedness between the village community and the larger economic and political changes, especially in the last decades, has dramatically changed village communities in all their aspects. Thanks to the increased organizing power of nation states, companies and to the development of technologies, relatively isolated hinterlands have been opened up to loggers, hydroelectric power projects, oil mining, large scale plantations and to land hungry farmers pushed out from areas with high population density. In the wake of these processes, rural communities have become progressively differentiated and local institutions eroded or ruined, changed or replaced by new ones (Hirch and Warren, 1998; Li, 1999; Western and Wright, 1995; Osseweijer, 2000). Focusing on the uplands, Li speaks of the reality of "agrarian transformation" of the uplands as counterpart of the Green Revolution in the low lands, with the critical difference that in the uplands the transformation has been largely the result of local smallholder initiatives rather than state-sponsored programs (Li, 2002: 269; Li, 2002b; Sitorus, 2004).

The Differentiation Process of an Upland Community

Ngamba and Kakao' People

The community that functions as an example of differentiation is commonly known as *Napu*. Napu people live in the highlands, on the eastern border of the Lore Lindu National Park. Only lately, the community has begun to shake off the term Napu that was forced on them by outsiders, and changed it with their original name, *Pekurehua*. The legend of the origin of the Pekurehua people tells about the relation between two societies with different adaptation systems: the *Ngamba* people from the valley and the *Kakao'* from the forested uplands. The Ngamba people lived in the upland valleys 1200 meters above sea level, an ecosystem dominated by grassland and bordered by forested hills. The Kakao' people on the other hand lived in the adjacent forested hills. The legend describes the continuous conflicts between the two populations, each representing an adaptation to different ecosystems.

The legend narrates that a princess from a neighboring village – *Winua* – offered her services as a mediator. She organized competitions for the strongest members of both groups to decide the winner and to settle the conflict once and for all. After several competitions, however, it appeared that the *Ngamba* and the *Kakao*' people were equally strong and smart. The princess interpreted this to mean that the two groups are actually brothers and destined to live together in peace. There is little historical knowledge on the adaptation system of both categories and the relationship between these adaptation systems. In the southern part of the valley, however, people still differentiate between the lowland and upland regions: *Behoa Ngamba* and *Behoa Kakao'*.

Colonial Expansion and External Induced Social Differentiation

The history of Wuasa, the village of our study, started around the end of 19th century with the decision of a tribal chief and his followers to split from the original village and to erect a new settlement called Wuasa, named after a mango species which grows abundantly in that location. In 1905 the Dutch colonial power decided to put the region more directly under its authority. The subjugation of the village federations of highland Central Sulawesi was completed between 1905 and 1907. As part of the Dutch strategy to control the highlands, the fortified villages located at secure and strategic places were resettled into combined and bigger villages in the valleys, which were better accessible for the Dutch. The political organization envisioned by the Dutch for the people in the highland after their subjugation was a form of federation of "self-governors" under a hierarchical administration system appointed by the Dutch. This political structure was established by the colonial authority without much physical presence of the Dutch themselves – a process that Schrauwers referred to as "bureaucratic amalgamation". The resettlement of communities in the highlands of lake Poso took place around 1906-1908 (Schrauwers, 2000:48). In the case of the resettlement process of the Pekurehua, inhabitants of two villages – Lamba and Lengaro – were forced to settle in a newly built village situated near the old village of Wuasa. This new settlement became the present Wuasa.

The intervention of the Dutch into the life of the Pekurehua did not stop with the resettlement of the original villages. As described masterfully by Schrauwers (2000), the character of the Dutch expansion into Central Sulawesi was influenced by the Christian mission that was involved closely in the bureaucratic expansion due to the shortage of manpower and financial means of the Dutch at that time, as well as by its imperialistic interests. The Dutch intervention into the life of the Pekurehua was so intense that commenting on the condition in the second decade of the twentieth century the ethnologist Kaudern (Vol. I: 29-30) wrote: "Later development proceeded so quickly that at present very little is left of the original culture. Villages and heathen temples have been leveled with the ground, and new villages have been built after modern principles. The old dresses, the weapons, the adornments, the heathen feasts and much more belong to bygone times".[2]

After independence, the state intervention into the local cultural expression continued. Through authoritarian governance and top-down planning, the government has forced concepts of settlement structures

---

[2] Kaudern succeeded to save thousands of artifacts from the highlands of Central Sulawesi – ranging from complete indigenous houses to daily utensils – when he left Central Sulawesi.

and dwelling upon the local population.[3] In this context, starting at the second half of the 1960s, houses on stilts were replaced by urban style single-family houses on ground level, each with a garden surrounded by a neat bamboo fence. Administratively, the formerly pluralistic form of local self-rule with its own local laws and regulations (*adat*) was replaced by a village government system, uniform throughout the republic. The local *adat* council was reduced to a body of prominent elders of the community with authority limited to social and cultural life. In 1985 a good road connection to the district capital Poso and to the province capital Palu was built. Permanent and temporal immigration of farmers and traders from South Sulawesi have been reported since the beginning of the 20th century. The in-migration from the south gained momentum, however, since the outbreak of the economic crises in 1997. Motivated by the declining value of the rupiah and the high prices of agricultural export products and using the newly built road, land hungry farmers from the south poured in looking for land to grow coffee and cacao. A survey conducted by Yayasan Kayu Riva in 2001 reported an in-migrant population from Buginese[4] origin of 18.5% in Wuasa. This immigration brought about an acceleration of agricultural development, particularly by opening new land for horticulture and perennials such as cacao. This process increased trade and transport between the Napu valley and Palu and Poso substantially.

---

[3] Village resettlement has been continued by the Indonesian government in the context of a policy aiming to bring the geographically-culturally-economically isolated tribal communities into the fold of the Indonesian culture. Village resettlement activities took on a massive form since 1970 when large scale natural resource extraction started, under two government programs: "*Proyek Resettlement Desa*"/ Village Resettlement Project of the Directorate General of Village Development of the Ministry of Interior; and the "*Proyek Penanggulangan Peladang Berpindah*"/Project to Overcome Shifting Cultivation of the Ministry of Social Affairs. The consequence is that there are almost no original tribal community villages left in the entire islands of Sumatra, Kalimantan and Sulawesi, a phenomenon that has not received adequate attention by academicians studying social and cultural change. In the second part of the 1980ies a new concept was introduced which emphasized the in-situ – without resettlement – development for the so-called isolated tribal communities.

[4] The term Buginese used in the survey has to be read as a generalized term for all in-migrants from south Sulawesi

By the time Wuasa village was established by the Dutch, the community was constituted by two sub-communities originating from the two villages, which were merged into the new village of Wuasa: Lengaro and Lamba. In the course of time, with the immigration of people from other parts of Sulawesi, the population of Wuasa diversified further. This process peaked out when Wuasa gained importance as a center of the sub-district administration. Most outsiders, however, will observe only the difference between the (Christian) locals and the (Muslim) migrants from the south. This is obviously too simplistic a picture of the society in Napu valley and the highland of Central Sulawesi in general. The pluralistic character of the village community is un-observable for the outsider because a large category of the population forming this pluralistic community came from ethnic groups that shared many cultural expressions, and most of these ethnic groups were also Christians. There is always a special bond between people of the same village of origin or of the same ethnic background. This special bond is expressed in mutual help in different fields of activities, economic as well as social and cultural. Mutual help in economic activities is practiced, among others, in *mapalus* groups, where members alternately have the right to use the labor of the group for different purposes, although mostly for land preparation and harvest. In 1986 however, following the many organizations introduced by the government, the solidarity group of people originated from the old village of Lamba turned into a more formal organization, calling itself *Kerukunan Keluarga Lamba* / The Association of Lamba Families. Soon, people originating from Lengaro and other solidarity groups followed this step. Besides the Lamba and Lengaro associations, Wuasa includes at least five other groupings. They represent people of Mori, Pamona (both ethnic groups from eastern Central Sulawesi), and North Sulawesi (mostly teachers), as well as two politically influential extended families of Wuasa village.

Starting with the colonial expansion, a process of differentiation occurred also in the field of religion. The suppression of the local religion went hand in hand with the introduction of Christianity, brought by the Dutch mission which, at that time, cooperated closely with the colonial administration. This cooperation arose from the shortage of administrative manpower of the Dutch. According to local history, the Protestant mission came into Wuasa in 1909 and the building of the first church started in 1910 at Watutau. This Church building is still in use today. This first proselytizing activity has culminated in the foundation of the Gereja Kristen Sulawesi Tengah (GKST), which dominated the religious life and the entire interpretation of Christianity of Central Sulawesi for a long time. The close relation with the administration continued after decolonization. The administrative wards of the village for example, the so called Rukun Tetangga (RT), have been formed according to the neighborhood Bible reading groups of the GKST church community. The weekly church activities of the different Bible reading groups – e.g. women, spouses, youth – were used by the village administration to spread information and government instructions. Thus, the role of the church came close to the aspirations embodied in the concept of "people's church" / *volkskerk* that was envisaged by the Dutch "Ethical Theologians" in the early 20th century. According to this idea, the people's church has to resemble and function as the *adat*/customary law of the traditional communities. That is, it fulfills the spiritual needs of the people as well as setting the norms of conduct in every aspect of life (Schrauwers, 2000: 43-47).

The GKST domination of the religious life of the indigenous population in Central Sulawesi and specifically in Wuasa was broken when the Pentekostal church came to Wuasa in 1955 and began to lay claim on part of the community. The Pentekostal church was introduced in Wuasa amidst a hostile attitude of the GKST church and the village administration. In effect, in its early existence the Pentekostal church had to endure intimidations and outright violence against its missionaries. As already mentioned, after the opening up of the Napu highland and the development of Wuasa as an administrative center of the sub-district, more independent churches have been established in Wuasa. There is now a total of six independent churches within Wuasa village with a population of 2638, of which 220 are Muslims. And in the immediate vicinity of Wuasa there are another three independent churches. The amalgamation of church and government administration became difficult when the socio-religious system of neighborhood groups cracked and splintered because of the growth of religious differentiation brought about by the new churches and the growth of the Muslim community. Today, the religious homogeneity is broken, and the perception of an inseparable community and its system of believe and norms is disturbed. A new development came to Wuasa village after the political reformation of 1998 and the fall of the Suharto regime. For decades the government party Golkar had dominated the political scene. The political opposition did not get a chance to express itself in an organized form. This condition changed dramatically with the approach of the general elections of 2004.[5] One year before the election, there were representatives of at least eight political parties in Wuasa: the former government party Golkar; the leading nationalist party PDI-P and its small challenger the Partisan Party / *Partai Pelopor*; three Christian based parties, the Indonesian Christian National Democratic Party / *Partai Kristen Nasional Demokrat Indonesia* (KRISNA-DEI), the Party of Peace and Prosperity/*Partai Damai Sejahtera* (PDS) and the Love the Nation Democratic Party / *Partai Demokrat Kasih Bangsa* (PDKB). There were at least two Islamic based parties present, imported by the immigrants from the south of Sulawesi, the large Party of Unity and Development/*Partai Persatuan dan Pembangunan* (PPP) and the new Crescent Moon Party/*Partai Bulan Bintang* (PBB).

The Transformation of the Natural Resource Management System

Before the commercialization of agriculture became dominant among the Pekurehua, the local adaptation system distinguished a multitude of tenure systems. The various tenure systems can be considered as a continuum, with individual based ownership of land at the one extreme and communal tenure at the other. The land within the settlement where the houses were built has always been and still is regarded as private property, *kintal*, of families that claimed it and passed it through inheritance from generation to generation. The permanent gardens fall into the category of private ownership too. However, a permanent garden would be returned to the communal sphere if it was

[5] The first election in 1999, directly after the Reformation of 1998, did not provide enough time for the organization of political parties. In Wuasa the election was accompanied by heavy intimidation on part of the village administration. Golkar won 80% of the votes, the rest went to the only oppositional party, the PDI-P.

neglected for a long time, and would than be available to be obtained by others. The shifting cultivation complexes, in contrast, were under the communal property tenure system. The traditional norm did not attach special rights to persons that pioneered the clearing of a patch of forest. Shifting cultivation land abandoned for fallow would turn automatically into communal land, free to every one to make use of. This may explain the local practice – in contrast with many shifting cultivation systems in other regions in Indonesia – of not planting tree crops in the shifting cultivation fields.

Since the introduction of coffee trees as cash crop in Napu in the Dutch colonial time, there has been a process of change in the local land tenure system. A stronger notion of private land rights has steadily encroached into the land that was previously defined as common property resource. Patches of coffee trees inside the forest became the object of inheritance, a phenomenon unknown before. The real change of the local land tenure system, however, was brought about by several factors: the introduction of the national land registration system in the 1960s and 1970s, the expansion of irrigated rice cultivation under strong government supervision and the immigration of land hungry farmers from south Sulawesi. The national land tenure system was forced upon the Pekurehua, as part of an overall integration of the community into the national economy and state administration system. In reality, however, the first two decades after the introduction did not greatly change the way people managed and used the land, which was always available to them. Until well into the 1980s shifting cultivation was still practiced under the principle of communal property. On the other hand the new national land tenure system fitted well to land categories that are furnished with strong property rights, as wet paddy fields, the *kintal* and the permanent gardens. Government programs to develop wet paddy fields in the 1970s were an important factor in pushing the growing notion of private land ownership. The development programs had two main objectives, to stop the practice of shifting cultivation and to establish sedentary agriculture. The rice fields were then registered under individual ownership. Fields in use were registered as dry land. People had also the right to register some reserve land. The land market brought about by the immigration of land-hungry farmers from the south, shortly after the opening of the road connection between Palu and the Napu valley in 1985 pushed the development of private ownership of land even faster. Besides the land shortage in their place of origin, the farmer migrants from the South were motivated by the high price of export cash crops such as coffee and especially cacao. The commoditization of land and the land registration have developed to such an extent that in 1996/1997 there was no land under communal property anymore. Table 1 shows that there existed a lively land market as early as 10 years after shifting cultivation ceased to be the dominant agriculture system, and with that ended as well the tenure principle of common access of land. The majority of land transfers involved dry land (96.5%), the category of land that was defined as common property resource in the past system of land tenure.

Year	Number of transaction involved	Land size involved in transactions (Hectares)		Total money involved** (Rp.)	Average price of land (Rp/ha)
		Total	Average		
1997	46	46.96	1.02	22,335,000	951,214
1998	112	107.24	0.95	232,405,000	2,282,105
2000	15	8.28	0.55	28,400,000	2,765,700
2001	15	6.95	0.46	32,500,000	4,388,489
2002	6	7.18	1.19	13,300,000	1,295,264
Totals	**194**	**176.61**	**0.91**	**328,940,000**	

- Land-transactions in categories of land: dry land 96.5%; wet paddy field 1.6%; land within the settlement 3.2%.
- Total agriculture land of Wuasa 1258 hectares, consisting of: 944 hectares dry land, 314 hectares wet paddy fields (Monografi Desa Wuasa, 1999)

Table 1. *Data on land transactions from 1997 to 2002 in Wuasa village**

*) *Analyzed from village administration data.*

**) *Information on the money value of the land transactions was not always available. The average value of land for each year was calculated in accordance.*

For the year 1997 data on the money value of transaction of 23.9 hectares land are missing; for the year 1998 the figure is 15.59 hectares; for the year 2001 the figure is 0.95 hectares.

At the outset, land transfers usually went one way from the indigenous people to the migrants. Very soon, however, land transfers went the other way too. In the span of 5 years, between 1997 to 2002, over half of the land transaction cases (109 land transfer cases from total 194 or 56.18%) were transfers from locals to immigrants, which involved 62.6% of land transferred. This left a substantial portion of other transactions, such as from migrants to migrants or from migrants to locals and between locals.[6] It showed that land transactions had already gone beyond a simple relation between the land hungry immigrants and the land rich local people. Today, an open land market is in operation.

The privatization of land rights resulted in the fragmentation of land, limiting the access to land to those who own land. Commoditization of land that goes hand in hand with its privatization resulted in a land distribution pattern that is not always in accord with the needs of the local people. The establishment of the Lore Lindu National Park (LLNP) set further limitations in access to land and other natural resources for the members of the community. The LLNP is composed of two conservation areas, the Lore Kalamanta National Park that was nominated as Biosphere Reserve under UNESCO's Man and Biosphere Programme (MAB) in 1977 and the Protection and Tourism Forest/*Hutan Lindung dan Hutan Wisata* of the Lindu Lake and two extension areas which encapsulate the Lindu Lake protection forest in its western, northern and eastern parts. The permanent border of the LLNP was marked with concrete poles in 1999, comprising 217,991.81 ha or 2.4% of the forested area of Sulawesi. A large part of the Lore Utara sub-district, where the village in study is located, falls within the national park, namely 90,525 ha with a park border of 99 km.[7] Along this border there are 14 indigenous villages and two transmigration villages. With the designation of the LLNP and with

---

[6] Cases of land transfer from migrant to migrant were substantial: in 1998 it involved 20% of the cases, in 2000 41% and in 2001 22.1%.

[7] Sub-District Lore Utara has been split into two Sub-Districts, Lore Utara and Lore Tengah.

the current park management concept, which banned all human activity inside the park, a large part of the sub-district's area became part of the park and thus legally became inaccessible for local people. In the case of Wuasa, 200 ha of the village land bordering the settlement, consisting of flat and formerly agricultural land, is lost – notwithstanding the loss of timber and non-timber forest products from what was formerly a common resource. This refers likewise to the other 116 villages surrounding the LLNP.

Over a span of 15 to 20 years the land tenure system of the Pekurehua has experienced a fundamental change. The originally diverse tenure system, with common property land tenure for the larger part of the land resources, changed into a uniform system of private land ownership. On top of that part of the agricultural land and the access to forest products were forfeited to the protectionist policy of the Park Authority. A list of land tax in the year 2003 from a centrally located section of Wuasa village may illustrate the distribution of land ownership in the community. From the 73 households 15 are immigrants. The total land involved in this tax register is 135.94 hectares.

Land ownership (hectares)	Number of household	% of households
0 – 0.5	15	20.5
> 0.5 – 1.5	21	28.7
> 1.5 – 2.5	22	30.1
> 2.5 – 4.0	8	10.9
> 4	7	9.6
	73	100

- Total land involved: 135.94 hectares (all categories of land)
- Average holding: 1.86 hectares
- Maximum ownership: 7.8 hectares
- Minimum ownership: 0.06 hectares

Table 2. *Land ownership in Dusun 2, Wuasa, sub-district Lore-Utara (according to land tax registration of 2003) Source: Register of tax on land and buildings for the year 2003 of Dusun II, Wuasa, sub-district Lore-Utrara.*

The tax register shows that the average land ownership is 1.86 hectares, which includes all categories of land. Almost half of the people registered (49.2%) owned land between 0 – 1.5 hectares. In terms of agriculture land only – irrigated as well as dry land – the figure is even less. The average land holding figure of 1.86 hectares does not deviate significantly from figures in the past. There are, however, two fundamental differences in the condition of the households in the past and at present, in terms of their economic position and in terms of their access to land. In the past the different categories of land were used to secure subsistence and to meet expectations bound to cultural and life cycle events. Such needs were defined and managed locally. At present, the need for subsistence is only a small part of what the land resource has to produce. There are other needs that have to be fulfilled, such as education, health, cloths and live style, house and land tax, and the provision of water and electricity. The definition of such needs is largely beyond the influence of local people. Another fundamental difference is the access of people to land. In the past, the whole Napu valley was a potential resource open to be used by each household, constrained only by the household work force, the character of the ecosystem and by the cultivation plans of others in the community. In other words, the resources of the Napu valley – the grass lands, forest, fallow land and ex-shifting cultivation fields, the wetlands near the river, irrigated paddy fields – as an integrated whole could be included in the household strategy. At present, each household is attached to a fix land holding, limiting the possibilities and reducing the flexibility of the household while confronted with continuously increasing obligations.

Data on individual land holdings do not reveal an unequal agrarian structure. On the contrary, they present a picture of marginalization of a local community, where agriculture does not provide a sufficient base for socio-economic development. However, this observation is only partly correct. Almost all local households have agricultural land, where they cultivate maize, vegetables, red beans, or potatoes, and many start to plant cacao in their gardens. Most local households have also irrigated rice fields. An important part of the community, however, lives only partly from agriculture. They spend part of their time as carpenters, construction worker, keeping a small store or food stall, etc. A fairly large group of the locals are government functionaries, engaging in agriculture in their free time. Most of the old elite families, the ones who are part of a line of ancestors reaching back to mythological figures, succeeded in preserving their social and economic status by occupying strategic positions in the civil service. From the moment modern education was introduced in Central Sulawesi, however, these old elite families have to share their influence and power to new prominent families that have climbed upwards through education and the Church.

The end result of a long process of external interventions is a community, horizontally and vertically differentiated through traditional as well modern categories. As already mentioned, the Wuasa community is already divided according to village of origin. Immigration created differentiation in both ethnicity and religion. The influx of different church congregations competing for the "lost souls" of the tribal communities, helped to differentiate further the Christian part of the community. Across these lines, and sometimes strengthening the already existing lines, are the electoral groups of political parties. The old elite families and the new prominent families make up the class that through the combination of education and their network of blood and marital ties dominate the political power in the region and through that gain access to the means for individual economic empowerment. These old elite families and the new prominent families became the upper class of the community. They are on the point of expanding their power and influence – through bureaucracy, education and the church – beyond the limits of the village and the sub-district to the district and even the provincial level. The migrants, especially the farmers from the South, largely dominate the agricultural sector, producing the bulk of marketable vegetables and export products such as cacao and coffee. We see here the development of a social class that can be characterized as an ethnic group and/or religious group, which dominates the economic sector largely because of their capability to get more from the land than the locals did. In this condition, transfer of land from locals to the migrants is a logic phenomenon. At the bottom of the social ladder, there are the majority of the locals that are mainly dependent on agriculture for their livelihood. People of this last category have to rely on their limited land as their sole economic resource, taking advantage of the local institutions for mutual help – such as the *mapalus* – in preparing land and for generating income.

Some complement the income from agriculture with petty commodity trading and services.

The fast process of social differentiation left little room for an essential process of internalization of new ideas and cultural elements and of acculturation. A condition is created where individuals live with multiple ideologies as alternatives, most of it without deep roots in the community, and with bits and pieces of the remaining indigenous culture and institutions. In this condition, collective action will be difficult to realize. In contrast, idiosyncratic behavior can go quite far without being confronted with public outcry.

CBFM and Local Reactions

The boundary of the National Park never received local recognition. Trespassing and encroachment became daily events. There are three sources of this perceived lack of legitimacy: First, it is the disputable status of the LLNP in the local perception. The borders of the LLNP were decreed without proper consultation and consideration of the rights and long term needs of the local community. At the start of the surveys for the boundary of the National Park, people were not fully aware of the consequence of the activities done by the surveyors. Only after the borders were defined and concrete poles were erected, people were sure that the surveyors had done the job in the easy way by avoiding the forest and taking instead the more easy terrain closer to the village. As already mentioned, the LLNP encloses approximately 200 ha of land perceived locally as common property and used as agriculture land. At the time the border of the National Park was marked, part of the land was already covered by individually owned coffee gardens and other fields with seasonal crops. Second, there are no comprehensive regulations on people's rights to extract timber and non-timber forest products essential for household use. Third, it is common knowledge that elements in the National Park Authority facilitated illegal logging initiated by large timber merchants and operated by local entrepreneurs who also employed individual chain saw owners from the village.

Despite the overall disagreement from the side of the local community, there was no open protest against official decisions on the boundary of the National Park. There were several reasons why the LLNP did not spark a community wide protest movement in Wuasa. In a certain sense, the park is a problem only for a part of the community. The LLNP has the most effect on people who depend on land complexes on the west side of the village and which became part of the National Park. Part of the land is covered with individually-owned coffee gardens. As a result, there is no common interest within the community as a whole to resist the encroachment of their agriculture land by the LLNP. There was no single dominant perception among members of the community on the LLNP, whether positive or negative. There were influential persons in the community who support the existence of the LLNP. There was also a local NGO that came to life focusing its activity on promoting the LLNP. This NGO was also active in reviving certain aspects of the local culture, such as local knowledge in natural resource management. One aspect that cannot be neglected as a factor that adds to people's favour for the LLNP is the national and international prominence that the Lore Lindu region and the community received once the LLNP was promulgated. This went hand in hand with a keen attention from different development and research programs of the national government as well as of international organizations. Furthermore, there is the fact that the community of Wuasa has become socially differentiated and quite complex, which countervails the development of a platform that could aggregate people's opinion on the LLNP and turn it into collective action.

It is important to know this background if one wants to understand the reaction of the local community towards the Community Agreement for Conservation program (KKM / *Kesepakatan Konservasi Masyarakat*) that was launched by the National Park Authority. The KKM program was a form of Community Based Forest Management concept, which was developed as part of the Integrated Conservation and Development Projects (ICDPs) which were meant to support the protection of several National Parks in Indonesia, among others the Siberut, Kerinci Sebelat, Gunung Leuser, Bunaken, Bogani, Wartabone/Domoga-Bone and the Lore Lindu National Park (LLNP). The common aim of these projects is "to ensure the conservation of biological diversity by reconciling the management of protected areas with social and economic needs of local people" (World Bank, 1996 cited in CSIADCP, Working Paper no. 2). The project that was assigned for the LLNP, the Central Sulawesi Integrated Development Community Project / CSIADCP, started in the middle of the 1990ties. In this project, the LLNP and its buffer zone of 60 adjacent villages were considered together as one unit, called the "Lore Lindu Bioregion". The basic assumption of this project is that "…biodiversity conservation depends almost entirely on people's values and the way in which they are put into effect through day-to-day decisions about the use of resources" (CSIADCP, Working Paper 2:2). Part of this assumption is the view that environmental degradation is related to the limitation of information, preventing people from acting in their own best, including practicing a sustainable natural resource management. On the other hand, social and economic distortion prevents people from turning their appropriate information into the right decisions. The CSIADCP project is designed to these limitations and distortions by responding in detail to the cultural, social and economic needs of the 60 villages in the buffer zone of the LLNP.

Part of the CSIADCP project is the development of a management plan for the LLNP, executed by The Nature Conservancy (TNC), an international, US-based environmental NGO. As part of the development process of this management plan, local NGOs were recruited for conducting research on certain fauna species and on different aspects of the village communities in and around LLNP. In a later phase TNC and some NGOs were involved intensely in developing the Community Conservation Agreement / *Kesepakatan Konservasi Masyarakat (KKM)*. In its first phase, the KKM agreement was closed with five villages at the border of the LLNP, Sedoa, Wuasa, Kaduwaa, Watutau and Betue. The concept of the KKM was based on the basic assumption of the undisputable status of the LLNP. This starting position exerts consequences for the NGOs involved. They had to embark on a two-pronged strategy of balancing the interest and rights of local people and the preservation of biodiversity. Within the large programs of CSIADCP and the development of park management under the coordination of TNC, solidarity with local community rights and interests had to stop at the border of the LLNP. In the following, the implementation of the KKM will be described and analyzed, based on two contrasting processes, that of the villages of Toro and Wuasa. This is to disclose the consequence of basing ones program on false assumptions on the community in case. Further, it will show the importance of internal dynamics in determining the outcome of external intervention, albeit the distribution of power between the intervening actors and the community subject to intervention. As stated by Long, (2000:13), all forms of external intervention are inevitably mediated and transformed by the actors and structures subject to the interventions.

The KKM agreement between the National Park Authority and the community of Toro was constructed within the context of the village's struggle to resist resettlement. In other words, the agreement was constructed with the active involvement of Toro leaders who had developed their specific visions of local natural resource management. This was directly linked to their struggle towards preservation of what they perceived as their ancestors' rights. In contrast to the KKM agreement in other villages, Toro appeared with its own tailored supporting institutions. In this context the *adat* council assumed a prominent role as the authority to supervise the management of natural resources through enforcing traditional forms of sanctions. Patrol squats were formed manned both by forest rangers of the National Park Authority and volunteers from the village. In developing a special institutional foundation for the community based management, Toro deviates from the formal concept where the supporting institutions follow the given government administrative system.

The process towards the establishment of the KKM agreement in Wuasa took a different form, which resulted in a different agreement. Although the village of Wuasa has lost quite a substantial part of its land and the access to timber and non-timber forest products to the LLNP, its existence was never in danger, such as in the case of Toro. Wuasa is located outside the LLNP and surrounded by quite extensive agricultural lands. The village as a whole never felt threatened. Therefore, problems related to the existence of LLNP were perceived more as individual issues. Problems with the LLNP were never taken up by the village as an institution, but were confronted at the household level. Conflicts were fought out between forest rangers with individual members of the community or with groups of individuals. Within this context, the external agents took TNC all the initiatives leading to the KKM agreement in this case. Activities such as participatory mapping, workshops to assess the threats and ecological services of the LLNP to the villages around the park, the forming of supporting organizations for the community conservation agreement, were all initiated by the TNC. Between December 2000 and March 2001 a process of participatory mapping of the resources and resource use of Wuasa was carried out under the initiative and coordination of TNC. This mapping activity was also used to identify historical or cultural sites and objects in the forested area. Despite the widespread complaints against the LLNP border, there was no mention in the document of land within the LLNP claimed by the village, nor of land and resources lost through the LLNP. The document only mentions the difficulty of finding the border signs because the majority of them were not anymore in their places, and the existence of coffee gardens inside the LLNP.

Close to the mapping activity, a large workshop was held in Wuasa called "Peoples Consultation for Conservation Planning", with participants from most villages of the sub-districts Lore Selatan, -Tengah and -Utara. In this three-day workshop, ecological and socio-economic functions of the National Park which directly affect adjacent villages, as well as threats to the National Park were evaluated through participatory processes. In this three-day workshop each aspect of the forest – such as forest ecological services, non-timber forest products, and occurrence of wild animals – was scored on a continuum from bad to excellent according to the perception of the participants. Thus, to the threats to the National Park, their magnitude and solutions were discussed. Some conclusions were drawn from an exercise, which convinced the participants that parts of the LLNP adjacent to the village are still in good shape. However, some threats to the National Park, which caused environmental problems for the villages, needed more resolute measures. One of the most critical problems identified was the effect of logging which, although still small in scale, already produced erosion that silted and clogged the Pembala river running through the village. At the previous rainy season this condition already brought flooding on a scale never experienced before. Interestingly, the most strategic solution that was suggested by the participants was the restitution of the ancestors' land, which was now occupied by the LLNP. The idea was that if parts of the margins of the LLNP were acknowledged as the ancestors' land, this move would encourage and intensify the community's responsibility for the security of the park. The legitimacy issue of the park area became a focus again when the participants discussed the critical problem of agricultural encroachment into the LLNP. It was emphasized by the participants that an important factor behind those encroachments was the shortage of agriculture land, a condition common to most villages on the border of the LLNP. In that relation participants suggested that the borders of the LLNP has to be renegotiated.

Amid demands of the participants to renegotiate the border of the Park, TNC emphasized the definitive and undisputable status of LLNP and its borders. The logical consequence of this position taken by the National Park is that the forest cover of the agreed KKM area has to be maintained. This second point means that the area cannot be converted into any other land use, not even into agroforestry systems. Estate crops such as coffee and cacao, which always are planted under shadow trees, are not allowed. The KKM area may be enriched with forest trees, which produce timber, nuts and fruits. However, the extraction of timber and certain non-timber forest products such as rattan from the KKM area is allowed only for private use, not for the market. Despite the lip service of participation and acknowledgement of local or ancestors' rights, the agreement was ultimately constructed according to the interest of the National Park, leaving many problems encountered by local communities unanswered.

The processes toward the KKM agreement in Toro and in Wuasa (and in most other villages) used participatory principles and methods. The detailed management system of the KKM in Wuasa, however, did not deviate from the restrictions put up by TNC and the National Park Authority. This did not consider complaints and suggestions for solutions that were expressed during the "participatory" workshops. From the perspective of TNC and the National Park Authority, the participatory processes were not seen as negotiations of concepts and solutions with the local community. From the outset it was obvious that the objective was mainly to convince the community of the indispensability of the LLNP and to involve people in transforming a finished concept into real organizations and regulations. As Mosse asserted, that "participatory planning" is more accurately viewed as the acquisition and manipulation of a new "planning knowledge" rather than the incorporation of "people's knowledge" by projects (Mosse, 2002: 22-23). Although the KKM agreement of Wuasa village was utterly unrealistic in its concept of land use and failed to meet the real needs of parts of the community, local perceptions of it were not uniformly negative. There were pro's and con's towards the KKM, and they were raised by different categories of people in the village. Those who perceived the KKM as positive were composed of at least a large part of the retired and active senior functionaries of the village and sub-district administration, the more educated members of the community, members of the community that owned and worked on land on the other part of the village far from the border of the LLNP, and young environmental activists. The perception of retired and active senior functionaries reflects in general the official opinion of the government on natural resource management, including the whole idea of a National Park and the KKM. Part of this category is a circle of highly educated senior members of the community who occupy important positions inside and outside the village administration. One of them holds the only master degree in the village and is also a member of the Pekurehua *adat* council at the sub-district level. Another person was the head of a secondary school in Poso predating the violent conflict. Part of this category consists of heads of the different Churches. Because of their formal position and social status they defended a legalistic standpoint and represented the urban, modernist and developmental point of view that considered their own community as backward, with low working ethos. In line with this idea, they refused the idea of shortage of land and blamed rather the low intensity of land use as cause of the deplored economic conditions. There were, however, important exceptions. Certain formal and informal leaders showed personal interest in the KKM area, among others because the obvious reason of their past activities in that area. In public these figures have to represent the formal standpoint, in reality they ignore the agreement.

Young environmental activists who were sponsored and trained by local and international NGOs from Palu who were active in advocating the LLNP, took a rather different position. These young activists from Wuasa, a couple of them had gone through university education, set up their own small NGO in close connection with NGOs in Palu. Although these youngsters support the KKM agreement, they positioned themselves quite independently *vis-à-vis* the village administration and the National Park Authority. In terms of ideas, this local NGO represents the combination of the typical urban environmental movement and the indigenous movement for cultural consciousness and human rights. As educated youngsters they perceive the social dimension as particularly relevant in the relationship between man and environment. That is why the NGO took Pekurehua as its official name, as to express its indigenous ethnic consciousness. It developed into an element that had to be taken into account in the local decision-making process – not only because of its capability to mobilize opinions among the people, but also because of certain qualities of its leader, such as knowledge in the decision-making process, environmental issues and the capability to mobilize opinions among NGOs in the provincial capital. With support from outside, this village NGO organized educational activities on environmental consciousness, public meetings on transitional justice or collective actions like in planting bamboo along the Pembala River.

There were four categories of people who did not agree with the KKM concept of collaborative forest management. The first category is composed of people who, since generations, used to farm at the western side of the village bordering the LLNP. When the land tenure system developed into a private property regime, people who cultivated land on that part of Wuasa perceived the land as theirs. The second category consists of youngsters who are still landless and live from the land of their parents. They heavily depend on inheritance because there is no reserve land any more. The third category is people who don't have traditional access to land. They come from other regions and don't possess any blood or marriage relation with the local population that would give them rights to land. Among these people are, besides farmer migrants, employees, teachers and members of the military and the police. The fourth category is comprised of people whose income depends on harvesting of forest products, in particular timber and rattan. For all these people, the KKM agreement means a real or at least potential threat, as it puts their economic security at risk.

Both, the representatives of pro and con alike, participated in meetings and workshops coordinated and facilitated by TNC with the objective of constructing the KKM agreement. Despite the disagreement of the whole community on the boundaries of the LLNP and on the unrealistic concept of the KKM land use, the agreement that came out of this process did not affect at all the positions taken by TNC and the National Park authority. It is therefore not surprising that one month after the signing of the KKM agreement, a group of people went into the KKM area and cleared about 40 ha of forest for agricultural purposes. Following this encroachment, a meeting was organized between the village administration, the *Lembaga Konservasi Desa* / Village Conservation Body and the National Park Authority, where it was decided that encroachment had to be stopped and the affected land had to be re-planted with forest trees. The already planted cacao or coffee plants, however, were left untouched. What is interesting in this incident is the fact that the encroachment was spearheaded by prominent leaders of the community, who also participated in the participatory process in developing the KKM agreement. These were not the ones that needed land most. In this contradictory condition, the drafting of the KKM agreement became a process in itself that had little correlation with reality. In the face of the ongoing activities of local people inside the designated KKM area, the locals assessed the stiff attitude of the National Park Authority and TNC with its predetermined conditions as a position remote from reality. For many people of Wuasa, outside and inside of formal institutions, pro or against the KKM, the KKM proposal that they helped drafting was perceived as a stepping stone to reclaim part of the LLNP area that was seen as an integral part of the community. There is, however, a fundamental difference between Toro and Wuasa in regard to the land regarded as ancestors' land. In Toro the KKM program became an integral part of their struggle for recognition of their village. A whole array of norms has been developed, including organizations to manage and enforce sanctions for encroachment into a forest zone that the village itself has defined. In the case of Wuasa, the reclaim of land was not related to a community wide agenda. With the signing of the KKM agreement, the role of the village community as a legitimizing factor ended. The formal organization – *Lembaga Konservasi Desa* / the Village Conservation Body – that was created as part of the KKM agreement to execute and control its implementation at the village level, did never really function. No socialization of the agreement among the members of the community was done. It was obvious that after the agreement there was no need on the side of the influential members of the community to implement it. At the village level there was no institution, modern or traditional, that was capable to exert pressure towards implementation of the agreement. Most of the strategic positions in the existing formal and informal organizations at the village level were filled by the same small circle of elite and prominent families that did not have interest in the implementation of the agreement. A situation was created where no authority can prevent people from breaking the agreement, and act on their own interest. This explains the large-scale encroachment of the KKM land shortly after the agreement was signed.

Some Deliberations on Participatory Approaches

In itself, the community-based management approach is an improvement compared to the development programs inspired by the modernization theory of 1950-1970 with their top-down structure of communication, centering on the idea of the transfer of knowledge and technology, management and financial aid to less developed parts of the society. However, programs of community based natural resource and environment management were dominantly based on too simple assumptions about the forest village communities that were involved in these programs (Leach, Mearns and Scoons, 1997; Johnson, 2004). The consequence is that both approaches show the same position in regard with the nature of rural/forest village communities: communities that are socially homogenous and imbued with conservationist values. In this perception, the formal and informal leaders of the community are believed to share the same interests with the whole community they represent. Based on this assumption, the participatory approach has been applied by involving these formal and informal leaders. In her foreword to the book by Agrawal and Gibson (2001: ix) Ostrom states "even if legislation or policy boasts a 'participatory' or 'community' label, it is rare that individuals from the community have had any say at all in the policy". Part of the blame for this condition, according to Ostrom, can be laid on the naive view most community based conservation (CBC) programs have of the "community". This paper tried to show how complex the structure of the community actually is, horizontal as well as vertical. Although Indonesia has decentralized its governmental system, the imprint of almost thirty years of centralistic authoritarian rule can still be felt. The village administration is the government administrative system at the lowest level, responsible to the next higher government administration level, representing rather the government's than the community's interest. Simplification of the nature of the local community has direct consequences for the participatory process. Uma Kothari (2002: 142) emphasizes the problem that lies in a simplification of the nature of power relations in local communities. Being ignorant of the complexity of social differentiation and the power relations within the local community, a community based participatory program could result in reinforcement of power and social control by certain social groups already in power and thus perpetuate the existing inequality rather than empowering the disempowered.

Observing the participatory approach from the implementing agent, Hall (1988) indicates four major modes of people's participation: 1) anti-participatory; 2) manipulative; 3) incremental; and 4) participatory (ibid. 93). According to Hall, most developmental participatory activities fall between the manipulative and the incremental modes. While the 'anti-participatory' mode precludes any popular participation, the 'manipulative' mode of participation is planned and controlled entirely for the ulterior motive of serving government objectives. The 'incremental' mode of participation is implemented on an ad hoc basis as a result of a half-hearted belief in the feasibility of popular participation or simply as a result of inefficient planning (ibid.: 93). In falling between the manipulative and incremental modes, the motives behind most of the participatory programs are 'instrumental' in character, such as gaining political support for the program or through taking advantage of local knowledge and institutions and thereby reducing costs of implementation (ibid.: 93-94; Cleaver, 2002: 37). Genuine participation, or more accurately the radical mode of participation, involves devolution of decision-making power to local institutions as part of basic social and economic reforms (Hall, ibid. 93). In this radical view, participation includes empowerment of the poor through class action, with transformation of structures of subordination through radical changes in law, property rights, and institutions (Hall, ibid: 95; Cleaver, ibid: 37).

The participatory approach applied in the development of the CBFM concept in LLNP can be characterized as being based on simplifications of the nature of rural communities and at the same time as a manipulative and incremental mode of people's participation. Although the Toro and Wuasa communities shared many common characteristics in terms of social structure, the outcome of the KKM process was profoundly different. In the case of Toro, the initiative in developing the KKM was entirely taken over by the community and exploited to safe the village as a whole, employing in the process the discourses of 'ancestors rights' and 'local environmental knowledge and wisdom'. In the case of Wuasa, there was no platform of common interest. The so-called representatives of the community took a compliant attitude in the process of developing the KKM, while at the same time conveyed their own agendas. The KKM became a stepping-stone for individual interests, spearheaded by the elite of the community. If a more radical participatory mode had been applied, the existing power structure would have been weakened, if not broken in favor of the community in general. A participatory approach in the context of a socially differentiated society or community has to abandon the idea of formal and informal leaders as representatives of the community. In a socially differentiated community, it is the powerful that will act as representatives of the community. A true participatory approach under such conditions would have to involve the smallest groups beyond the family such as neighborhoods, at least representatives of them. Only by this means, a counter balance against the power of the formal and informal leaders could be established.

References

Agrawal, A. and C. C. Gibson (2001): Introduction. The Role of Community in Natural Resource Conservation. In: Agrawal, A. and C. C. Gibson (Eds.): Communities and the Environment. Ethnicity, Gender, and the State in Community-Based Conservation. Rutgers.

Breman, J., P. Kloos, and A. Saith (1997): The Village in Asia Revisited. Oxford.

Cleaver, Frances (2002): Institutions, Agency and the Limitations of Participatory Approaches to Development. In: Cooke, B. and U. Kothari (Eds.): Participation – The New Tyranny? London.

CSIADCP TA NO 2518-INO, Draft Final Report. Working Paper 3: Social Analysis and Community Development.

Ellen, Roy and Holly Harris (2000): Introduction. In: Ellen, R., P. Parkes and A. Bicker (Eds.): Indigenous Environmental Knowledge and its Transformations. Critical Anthropological Perspectives. Amsterdam.

Flannery, Tim (1998): The Future Eaters. An Ecological History of The Australasian Lands and People. Reed New Holland.

Hall, A. (1988): Community Participation and Development Policy: A Sociological Perspective. In: Hall, A. and J. Midgley (Eds.): Development Policies: Sociological Perspectives. Manchester.

Hirsch, P. and C. Warren (1998): Introduction: Through the Environmental Looking Glass. The Politics of Resources and Resistance in Southeast Asia. In: Hirsch, P. and C. Warren (Eds.): The Politics of Environment in Southeast Asia. Resources and Resistance. London and New York.

Johnson, C. (2004): Uncommon Ground: The "Poverty of History" in Common Property Discourse. *Development and Change.* Vol.35/No.3/June 2004: 407-433

Kaudern, W. (1938): Ethnographical Studies in Celebes. Vol. I: Structures and Settlements in Central Celebes. Göteborg.

Kothari, U. (2002): Power, Knowledge and Social Control in Participatory Development. In Cooke, B. and U. Kothari (Eds.): Participation – The New Tyranny? 139-152. London.

Leach, M., Robin M. and I. Scoones (1997): Environmental Entitlements: A Framework for Understanding the Institutional Dynamics of Environmental Change. IDS Discussion Paper, 359.

Li, T. M. (1999): Marginality, Power and Production: Analyzing Upland Transformations. In: Li, T. M. (Ed.): Transforming The Indonesian Uplands: 1-44. Amsterdam.

Li, T. M. (2000): Locating Indigenous Environmental Knowledge in Indonesia. In: Ellen, R. Parkes, P. and A. Bicker (Eds.): Indigenous Environmental knowledge in its Transformation: 121-150. London.

Li, T. M. (2002): Engaging Simplifications: Community-Based Resource Management, Market Processes and State Agendas in Upland Southeast Asia. *World Development* Vol.3.No. 2: 265-283.

Long, N. (2000): Development Sociology. Actor Perspectives. London and New York.

Meinzen-Dick, R. and Anna Knox (2001): Collective Action, Property Rights, and Devolution of Natural Resource Management: A Conceptual Framework. In: Meinzen-Dick R., A. Knox, M. and D. Gregorio (Eds.): Collective Action, Property Rights and Devolution of Natural Resource Management. Proceedings of the International Conference 21 - 25 June, 1999.

Mosse, David (2002): "People's Knowledge", Participation and Patronage: Operations and Representations in Rural development. In: Cooke, B. and U.Kothari (Eds.): Participation – The New Tyranny? 16-35. London.

Osseweijer, M. (2000): We Wander in our Ancestors „Yard": Sea Cucumber Gathering in Aru, Eastern Indonesia. In: Ellen, R., Parkes, P. and A. Bicker (Eds.): Indigenous Environmental Knowledge and its Transformations. Critical Anthropological Perspectives: 55-78. London.

Schrauwers, A. (2000): Colonial "Reformation" in the Highlands of Central Sulawesi, Indenesia, 1892 – 1995. Toronto.

Sitorus, F. M.T. (2004): "Revolusi Cokelat" : Social Formation, Agrarian Structure, and Forest Margins in Upland Sulawesi, Indonesia. In: Gerold, G., Fremerey, M. and E. Guhardja (Eds.): Land Use, Nature Conservation and the Stability of Rainforest Margins in Southeast Asia: 105-118. Berlin.

Smith, E.A. and M. Wishnie (2000): Conservation and Subsistence in Small-Scale Societies. *Annual Review of Anthropology*, No. 29: 493-524.

Wertheim, W.F. (1956): Indonesian Society in Transition: a Study of Social Change. Bandung.

Western, D. and R. M. Wright (1995): The Background to community-Based Conservation. In: Lokakarya Promosi Pola Manajemen Sumberdaya Lokasional Berbasis Komunitas (Community-Based Forest System Management), Bandung, 10-12 Maret 1995. LATIN, WALHI, CARE International Indonesia, CARE Austria, dan Pemerintah Austria.

Chapter III

Discursive Strategies and Local Power in the Politics of Natural Resource Management: The Case of Toro Community

Moh. Shohibuddin

Abstract

In 2000, the Park Authority of the Lore Lindu National Park issued a so-called "conservation agreement", in which it acknowledged the "customary area" of the community of Toro located within the boundaries of the Park. Through this recognition, Toro regained rights of access to parts of the forest resource that were previously prohibited. With this act the Park Authority recognizes explicitly the "traditional wisdom" of customary resource utilization. Since that, the people of Toro have been actively engaged in the mobilization of support in order to strengthen and affirm their claims over their territory. This mobilization took the form of an articulation and reproduction of customary identity and institutions, especially those related to the organization of natural resource appropriation. Thus, the revitalization of Toro's cultural identity is more than a simply return to ancestral customs. Rather, it is an articulation of "cultural politics" which involves a process of "cultural reproduction". This process allowed Toro cultural identity to be reinterpreted and redefined, local institutions to be reformulated and reconfigured, and the allocation of local power and resources to be re-negotiated and re-shaped. This paper investigates the impacts of this cultural policy articulation on the community conservation agreement as well as on intra- and inter-community relations.

Introduction

Influenced by the state's decentralization policy and the ensuing rise of indigenous rights movements all over Indonesia, the area around the Lore Lindu National Park has experienced an important transformation of its landscape status. This process takes place at a complex national and international conjuncture.

At the national level, the conjuncture occurred in the form of the establishment of the Lore Lindu National Park (LLNP) through the Decree of the Minister of Forestry No. 593/Kpts-II/93, dated on October 5th, 1993. The demarcation of the Park, covering an area of 229.000 ha, results from the fusion of three formerly distinct nature reserves: the *Lore Kalamanta Wildlife Sanctuary* (established in 1973), the *Sungai Sopu and Gumbasa Wildlife Sanctuary* (established in 1981), and the *Danau Lindu Recreational and Protection Forest* (established in 1978).[1] The Park was also declared as a World Heritage Site by the UNESCO, due to its richness in bio-diversity and its high endemism. The management of the Park is carried out by the *Balai Taman Nasional Lore Lindu* (BTNLL, Authority of Lore Lindu National Park); an regional administrative office under the supervison of the Ministry of Forestry. According to State Regulation No. 68/1998, the management of Indonesia's National Parks is carried out through a zonation system.[2] The interests of local communities bordering the Park – despite their social, economic and cultural dependence on the forest and despite their long history of forest interaction – tend to be overlooked in the zoning system.

Conservation policies are in line with the global commitment to save the remaining tropical forest. In addition to the state's involvement through multilateral agreements, it requires the involvement of various international NGOs and donor organizations in order to operate conservation programs effectively and in order to fulfill the global objectives. *The Nature Conservancy*, *Conservation International* and *CARE International* are among those international NGO's conducting conservation activities in the LLNP area. Within this framework, since late 1997 the community of Toro mobilized collective action in order

---

1 Through the exclusion of the Besoa and Danau Lindu enclaves in 1999, the area covered by the Lore Lindu National Park was reduced by 11,000 hectares (Decree of Minister of Forestry and Estate Crops No. 464/Kpts-II/1999). The official size of the remaining area of the Park is 217.991,18 ha.

2 According to the zonation system, the park is divided into eight zones: (1) core zone, (2) wilderness zone, (3) utilization zone, (4) intensive utilization zone, (5) traditional utilization zone, (6) rehabilitation zone, (7) social-historical zone, and (8) recreational zone.

to revitalize its cultural identity and to empower its traditional institutions which had been deprived of their functions by Soeharto's New Order regime (1965-98). Toro's effort links up with the rise of numerous ethnic movements prior to the fall of the Soeharto regime, the political expression of which culminated at the Indonesian Customary Community Congress (*Kongres Masyarakat Adat Nusantara*), held in Jakarta in 1999.[3]

As the national political atmosphere opened during the initiation of the political "reformation" in 1999, the community of Toro responded to this by systematically organizing its claims over its customary territories. The arguments for Toro's demands evolved primarily around issues of cultural identity, territorial integration and social and political autonomy. The major initiative to achieve this end was the revitalization of "local wisdom" (e.g. customary law, indigenous environmental knowledge and traditional practices in regard to natural resource management) as well as the participatory mapping of the customary area.

In July 2000, as a result of the concerted effort of Toro's leaders and the Sulawesi-based NGO *Yayasan Tanah Merdeka* (YTM, the "Free Land Foundation") and good will of the rather "progressive" Head of the BTNLL, the Park Authority finally recognized Toro's territorial claims inside the Park. Following formal recognition, the residents undertook further efforts in order to strengthen and affirm Toro's claims. These efforts were mainly articulated in terms of Toro's customary institutions, especially those related to the organization of natural resource appropriation. Thus, the revitalization of Toro's local resource institutions represents less a return to ancestral customs than a process of cultural-political articulation. In this process, the community reinterpreted and redefined its cultural identity and re-negotiated and reshaped the allocation of local power and resources.

The village of Toro is part of several communities belonging to the *Moma* ethnic group, which is generally believed to be the indigenous group of the Kulawi valley. The articulation of cultural politics referred to above represents just one of several discursive strategies which people use in reclaiming territories inside the Park. However,

---

[3] The political assertions and other congress results can be found on the following website: http:/dte.gn.apc.org/AMAN/index.html

with the recognition of Toro's ancestral rights by the National Park Authority, the discursive and social practices mobilized by Toro became the "exemplary model" of the "traditional ecological wisdom" (*kearifan tradisional pengelolalan sumberdaya alam*) of Kulawi culture. Toro's newly acquired position challenged the regional customary and administrative authorities of Bolapapu, the capital of Kulawi sub-district and the center of Kulawi's pre-colonial royal courts.

Therefore, this paper will discuss the contesting representations of culture and space and their strategic use by different actors. It will highlight the politics of natural resource management both in terms of inter-community relations (e.g. among Toro community and other surrounding communities) as well as intra-community relations (e.g. among different actors within Toro community itself).

Theoretical Orientation

The perspective adopted by the New Order government perceives local communities as retarded and alienated entities, whose knowledge reflects an intrinsically inefficient mentality, antithetical to development and modernization. Therefore, it was the declared task of the Indonesian government and its development experts to adapt them to the global and national mainstream culture (Dove 1985, Peerson 1988). On the other extreme, there is the opposite perspective (adopted by indigenous rights activists and most NGO's), which considers local knowledge as the panacea for all problems, and as a holistic pattern of understanding that provides an alternative world view to the destructive reductionism of western science. Local communities, perceived as living in ecological balance with nature, are regarded as "environmental experts" and keepers of equitable and sustainable resource management systems (Nygren 1999). The explanation illustrates the "comeback" of an old polarity that divides knowledge into two diametrically opposed archetypes: "western knowledge" is considered to be reductionist and theoretical in nature, while non-western knowledge is considered to be holistic and practical. Both knowledge systems are totally segmented and occupy different cultural spaces with only little exchange between them (Nygren 1999).

In the following discussion of the mobilization of collective action in Toro, our adopted approach goes beyond these opposing perceptions. Instead, it views local knowledge and its underlying cultural identity in terms of a *performance* which develops in the context of people's practical engagement with the environment, and as a part of human destiny embedded in nature and engaged in situated, practical acts. Within such engagements, local knowledge is considered to be the result of those practical actions, namely as "embodied capacities in the performance of tasks in social contexts shaped by particular cultural logics" (Escobar 1999). According to Hobart's characterization (Hobart 1993), local knowledge is a practical, situated activity, constituted by a past, but changing, history of practices.

In short, local knowledge operates through a compound of real practices in a particular context and is not based on a shared, context-free, abstract knowledge system. Instead of being permanently established in the past, local knowledge complies basically with continuous change in history, culture and power. This means that any analysis of local knowledge must take account of the existing cultural plurality as well as the concrete social relations, through which knowledge and identity are produced and reproduced. With its emphasis on practice and agency, the *reformative* approach towards local knowledge allows us to realize a more active conception of culture, by acknowledging that human actors shape knowledge and identity through social practices within specific relations of power. Meanwhile, the actualization of local knowledge at a certain conjuncture represents a strategically "political work of articulation" (Li 2000), as its discourses relate to social and economic powers.

Ngata Toro: Social Structure and Traditional Resource Appropriation

Community Establishment and Social Structure

The village of Toro is situated on the western side of the Lore Lindu National Park, approximately 15 km south of the sub-district capital Bolapapu and 86 km south from Palu, the provincial capital of Central Sulawesi. Geographically, Toro protrudes into the National Park area and is connected to its neighboring villages by a narrow, meandering trail. This geographical setting turns Toro into an enclave area, directly bordering the Lore Lindu National Park. The native population of Toro belongs to the Moma group, which is the major ethnic category in the sub-district of Kulawi. The people of Toro however, differ from other Moma-speaking villages according to their specific cultural ecology and mythology of origin. Hence, from a cultural point of view, the Toro community represents a distinct sub-culture, especially in regard to its isolated geographical situation and historical background.

According to local myth, the founders of Toro originated from a settlement called Malino, located some 40 km northeast from present Toro village. Once upon a time, invisible spirits (*bunian*) attacked the native village of Malino and killed all people, except for a number of families who managed somehow to survive. After a long odyssey, these remaining families finally arrived in the Kulawi valley and decided to settle there. These expatriates then called themselves and their new village "Toro", which means "the remainders". Thus, the name of the village refers to their existence as the remaining survivors of the spirit attack.[4]

From a cultural perspective, the confirmation of this myth of origin (*ethnogony*) is part of the group's identification. Although the myth is based on stories that are usually perceived to be impossible in the "real world", it is not merely a story without meaning. In traditional societies (and no less so in modern ones), myths constitute the main elements of the formation of a "*social imaginary*" (Arkoun 1994), which provides a group with identity and which gives meaning to its history. Therefore, the mythical statements about the event's origin offer a significant meaning towards the confirmation of community identity. The same applies to the multi-ethnic community of Toro in the sense that social integrity can be achieved and the social order can be reproduced among the community members.

The multi-ethnic character of Toro relates closely to the demographic history of the area in general. During the pre-colonial period, the formation of inter-ethnic villages was mainly caused by inter-tribal

[4] Ngera (1989) provides a comprehensive narration on this myth of origin, The analysis is provided by Shohibuddin (2003).

marriages. However, the cultural landscape was drastically changed during the 1950s, due to large-scale immigration. This migration stream involved two distinct ethnic groups, e.g. the *Rampi* people from South Sulawesi and the *Uma* people from the western part of the Kulawi valley. All of these early migrants obtained their land from the village chief. The migrants built their own settlement on the hill slope area, which forms the present village entrance. Later, their settlement was relocated to the edge of the road to Kulawi, which is currently known as Hamlet V.

Hence, according to the demographic history, the current composition of Toro inhabitants consists of three major ethnic groups, namely the Moma ethnic group as the native villagers, and the Rampi and Uma groups as "newcomers". In addition, there is also a small number of villagers of other ethnic backgrounds who came to Toro via spontaneous migration. From the total village population (2070 individuals), the Moma hold the major share, followed by the Rampi and Uma groups. The ethnic diversity of Toro correlates more or less with the establishment of hamlets of ethnically homogenous populations. As mentioned above, this pattern relates to the migration waves themselves, in the sense that each phase involved only a particular ethnic group, so that the allocation of land was centralized in different locations for each ethnic group. To the present day, the native villagers (Moma) are concentrated in the "native hamlets", e.g. Hamlet I (Bola), Hamlet II (Bulu-Lempe) and Hamlet III (Kinta), as well as in some parts of Hamlet IV (Lengkoue), whereas the Rampi people of the first migration wave (1952-1953) established a settlement in what is now Hamlet V (Roupa) and parts of Hamlet IV. Most of migrants who arrived within the second wave (1959-1960) reside in Hamlet IV. A rest of both groups, along with a number of new Rampi families from neighboring villages, were provided with land in the vicinity of Hamlet VII (Bulukuku). The Uma people on the other hand, lived from the beginning in the southeastern part of the village, which then developed into a new hamlet, namely Hamlet VI (Nente Baru).

In terms of social structure, the community of Toro – like other villages in the Kulawi valley – traditionally consisted of three hierarchical social ranks, namely *Maradika* (the noble society); *Todea* (the common people), and *Batua* (slaves). Noble status of the first rank is acquired via kinship relations with the founders and the first inhabitants of Toro. This corresponds to a well-known pattern that commonly prevails in pre-state communities "who are the first are the ones who get". The noble Maradika were also considered as *primus inter pares* with personal charisma, leadership qualities and compassion. As a result, the Maradika held control over all sources of political and economic power, including the redistribution mechanism within society. For example, the principle of traditional reciprocity in social interactions places the Maradika as beneficiaries of presents (*pepue*) from the commoners in every harvest period. As compensation, these presents are "cosmically" believed to bring "improvement of the strength to live" (*tinuwui*) to the ones who deliver them. This is because the Maradika are believed to be blessed with divine power that can be transferred to common people.

The second social rank are the Todea, or "commoners", consisting of free people who are not bound by fixed obligations, albeit they have no exclusive privileges over other groups. As common people, Todea possess their own belongings and cultivate their own land, albeit limited in size. The lowest social rank, Batua, is a stratum in society that functions primarily as servants. People of this rank have no privileges; they are exclusively obliged to serve the stratum of Batua. First, Batua originated from members of the tribe itself (*autochtonous* or native) after they lost their former position by breaking the customary law in a way that could not be rectified in accordance with customary means (rituals, fines, etc.). In these cases, a Maradika had to interfere by taking over the execution of the customary sanction (otherwise, the offender could be killed or driven out of the village). As a consequence, offenders of customary law could become slaves of the Maradika. Second, Batua could also originate from other tribes in case they were arrested during inter-tribal warfare and finally turned into the status of slaves. The only right in the hands of a Batua is to make an effort to "position him or herself" (*montolo woto*), which means the right to pay for freedom. This is done by paying back the cost of the fine that was paid by their guardian Maradika, or alternatively, by carrying out a series of certain rituals (Garang 1985). The three social ranks can be easily identified by dress, speech and behavior, or by the types of involvement in customary activities. Although slavery has been discontinued since the arrival of the Dutch, the basic differences are still identifiable through family trees, participation in customary ceremonies, or through the amount of bride price (*mahar*) that has to be paid in customary nuptials. In line with the hierarchical structure of society, leaders in Kulawi society can only be recruited from the first social rank, the Maradika. During the pre-colonial period, this traditional leadership system was composed of two leadership institutions, namely the *Totua Ngata* and the *Maradika.*[5]

Totua Ngata refers to a council of village elders who collectively run the leadership of the village. In practice, this council administers all village affairs, both internal (*irara ngata*), such as farming issues, land clearings, customary ceremonies etc. as well as village external affairs (*hawaliku ngata*), such as inter-village cooperation and federations, the declaration of wars etc. Besides, the council serves also as the local committee of justice that deals with cases of violation of customary law and the arbitration and reconciliation of conflicts among society members. One structural component of the council are the so-called *Tina Ngata*, who are female figures who are asked for their consent in each decision the Totua Ngata are about to make. These women even had some sort of veto right, especially in matters of conflict resolution. This role corresponds with women's cultural status as *pabolia ada*, e.g. the ones who guard and maintain village custom.[6] The second traditional leadership institution, the Maradika, is composed of a person elected by the Totua Ngata to become the chief of the community in question.

[5] Thus, the term "Maradika" conveys two meanings: (1) a term describing noble social rank as explained above, and (2) a term for a position as chief of a tribe.

[6] A woman's position as *pabolia ada* relates to her noble status, which determines the value of her brideprice in nuptials, as well as all privileges she possesses. Social status of a child born from a marriage is determined by the noble status of its mother and whether the value of her brideprice reflects her social status or not. This is the reason why women are called the ones who guard and maintain the custom (*pabolia ada*), because it is them who ensure the continuity of the hierarchical social order.

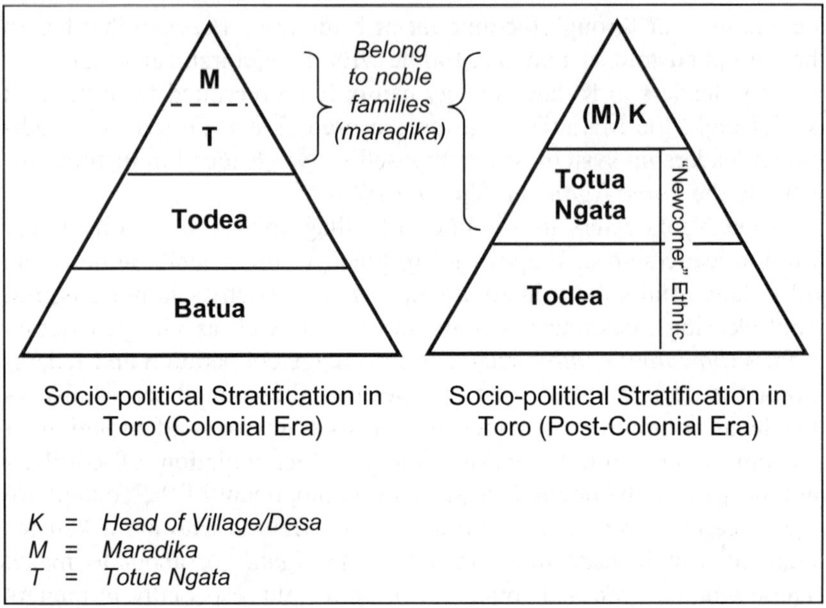

Figure 1. *Tradtional social structure in Toro*

One important criterion in the election of Maradika, in addition to his lineage membership and leadership capacity, is wealth and the degree of control over economic assets. This relates to the Maradika's role as guardian (*asylum*) for those who break customary law but who are not capable of fulfilling the sanctions imposed on them. In such situations, the Maradika is expected to take over the responsibilities of those people by paying the fine (*givu*) to the party that suffered the loss. Although a Maradika is not directly engaged in routine administrative tasks (which is carried out by the Totua Ngata), he is nevertheless regarded as a symbol of community unity. Therefore, his presence in the Totua Ngata deliberation assembly is always expected, and his opinion may become an influential source of consideration in every decision-making process.

Since the establishment of the Dutch colonial administration, fundamental changes of the traditional leadership functions have taken place. The need to exert stricter control and coordination, urged the colonial authority to proclaim the Maradika as the chief of the village, expected to carry out daily routine administrative functions. Although this policy did not immediately alter the existing local leadership structure (since every decision was always made through a deliberation process), it weakened the position of Totua Ngata in the long run, as their role became increasingly limited to judicial and customary ceremonies without any entitlements to take care of governmental affairs.

This situation continued after the establishment of the post-independence government, especially since the "village" (*desa*) became the smallest official administrative unit through Government Enactment No. 15/1979. According to the enactment, the chief of the village is elected directly by the people at the same time he is an integral part of the bureaucracy and, as such, accountable to the head of the sub-district.

The Cycles of Shifting Cultivation

Representing a political demand articulated around arguments of "indigeneity", the traditional regime of natural resource appropriation is transformed into a strategically means in Toro's claims over its specific unity of "identity" and "territory". The section below will be a discussion of the regime of natural resource appropriation in Toro as it was practiced before its was modified in order to fit the wider transformation process that took place in the research area. Actually, this traditional regime of natural resource appropriation did not differ from those operating in other neighboring communities. Land management was based on a rotating shifting cultivation system, involving regular cooperation between households, especially during the peak seasons of clearing. The labor-intensive collaboration was coordinated by and derived its cultural significance from traditional leaders who decided which tract of forest is to be cleared, when it is cleared, which group members are involved, how the land will be distributed and how the land will be cultivated in the future.

In Toro's rotating cultivation system, individual families obtained a piece of land by *mopangale* or opening *pangale*. *Pangale* is a category of primary forest, defined by topographical, geographical and soil condition considerations, which may be converted into agricultural land. Thus, *pangale* includes forest situated on flat terrain, forest close to the settlement area, and several kinds of fertile land. The clearing of forest was done in groups (*mome ala pale*), with the cleared land distributed among individual members according to their individual level of effort. The cultivation system was based on the following cycle: A piece of forest cleared is called *popangalea*; the person who cleared it for the first time has the right to own the land. This ownership prevails continually and will be inherited by his descendants, even if the land is abandoned and grows again into forest. This means that the land does not turn into an *open access* area, since its ownership status remains untouched. A cleared piece of dry land in productive state is called *bone*, usually cultivated with rice, corn or other kind of seasonal crops. After several planting cycles, when the land's fertility decreases, it is called *balingkea*. In case it is still possible, *balingkea* may be cultivated again with vegetables or corn for one more planting season, in which case it is called *mobalingkea*. Alternatively, it may be used as grazing land. *Balingkea* that has been left fallow for one or more years is called *oma*. Depending on how long the fallow period lasts, *oma* can be divided into three categories, namely *oma ngkuku* (if it has been left fallow for 1-2 years), *oma ngura* (if it has been left fallow for 3-5 years), and *oma ntua* (if it has been left fallow for 5-25 years). The last category is also known as *pahawa pongko*. In terms of vegetation *pahawa pongko* already resembles *pangale*. Tall trees are common, thus it needs *pongko* or a "ladder" to fell them in order to reach the upper parts of the stem. In short, the traditional mode of land cultivation can be depicted as diagram in figure 2.

In addition to the categories listed above, home gardens (*pampa*) may also be found. *Pampa* are a piece of land nearby the settlement area which is continuously cultivated throughout the year. The land is usually planted with vegetables for daily subsistence consumption. Since it is continuously cultivated, land-lease practices occur only rarely on *pampa*. Practices of land-lease and benefit sharing prevail on rice fields of the *bone* or *balingkea* types. However, since the introduction of commercial crops like coffee, clove and cacao, many dry land plots that were used to be cultivated with agricultural subsistence crops before, were converted into cash crop plantations, including the *bone, balingkea* and *oma* categories. This shift indicated an increasing number of land under the category of *pampa*, at the same time it meant a noticeable decrease of agricultural crop cultivation activities on dry lands. It further reduced dry land accessibility through tenancy relations, since most of the dry lands had turned into *pampa* without the possibility of being leased.

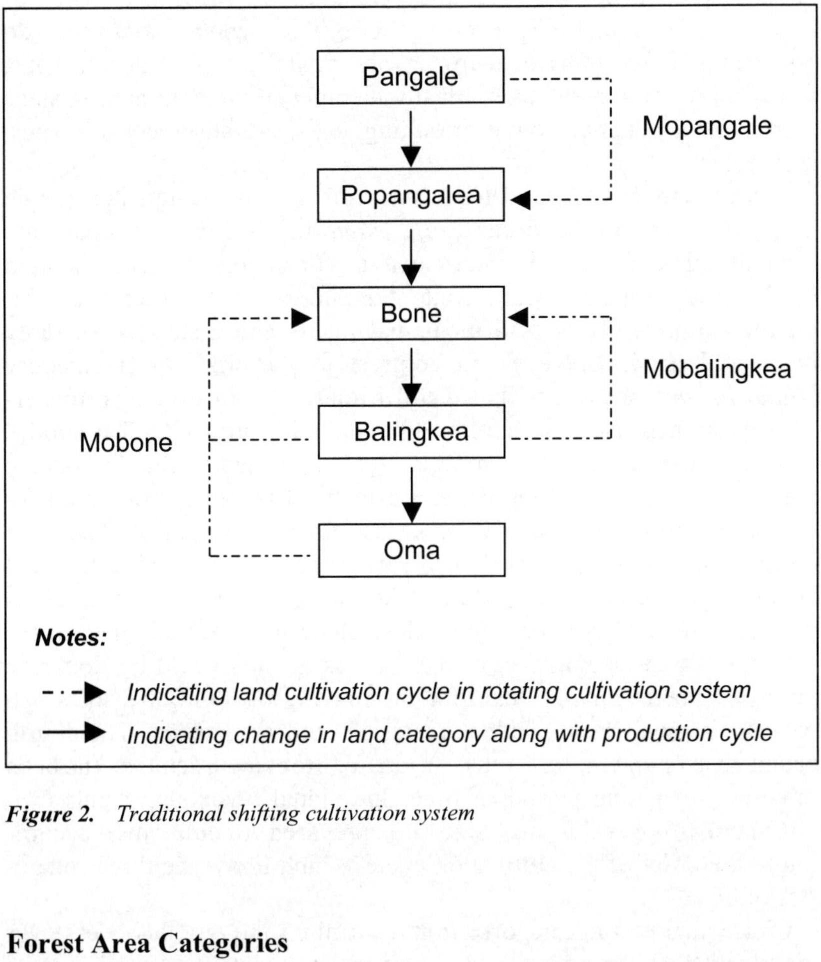

Figure 2. *Traditional shifting cultivation system*

Forest Area Categories

While the cycles of the shifting cultivation production system created different land categories, Toro society also recognized a traditional delineation of the forested area. The categorization scheme employed mainly four vegetation characteristics and types of access as its major differentiating criterion: The first category is called *wana ngkiki* and refers to forested areas on top of mountains, usually dominated by grasses, mosses and shrubs. Such areas are difficult to reach and limited human activities take place there. However, these areas are highly valued by Toro people, since they are assumed to be the source of fresh air (*winara*). Being almost inaccessible, *wana ngkiki* areas do not acknowledge individual ownership rights (*dodoha*) at all. Thus, forest products may be taken by any member of the community, since there is no productive activity resulting in the establishment of ownership rights.

The second category, called *wana*, refers to the untouched jungle area. According to the cultural rule, w*ana* may not be converted into agricultural land. This is because it is considered to function as a buffer zone for the soil water content. Besides, it also serves as a habitat for animals, plants, and medicinal plants and materials for daily household need. However, in contrast to the first forest category (*wana ngkiki*), *wana* entails several resources with exclusive ownership rights because they have been continually utilized. This applies first of all to the *damar* (resin) tree, which is owned by the first person who discovered it and tapped its resin the first time. Ownership of *damar* is permanent and will be passed on by inheritance. The third category is *pangale*. As explained earlier, *pangale* is a primary forest area likely to be converted into dry farmland, whereas its flat terrain is usually converted into dry rice fields. Since it is located close to the settlement area, people may also extract rattan and wood for domestic purposes, forest *pandan* leafs for the making of traditional mats and baskets, as well as medicinal materials and fragrances. The fourth category is *oma*, e.g. forest that originated from a garden that has been left idle for a long period of time. Individual ownership rights (*dodoha*) prevail over this area, since it is prepared for cultivation according to the order of the cultivation cycle, which forms the three categories of *oma*.

Consequently, the categorization determines the type of access over lands and forest products. One category may not be cleared, but its products may be utilized. Another forest category may be cleared, but its ownership has not been decided yet. In addition, there is also a category and sorts of forest products that have an established ownership, so that other people may not access them without the owner's consent. The reproduction of these socio-cultural aspects and technoeconomical institutions served as the cultural means by which Toro society articulated its "traditional wisdom" within the context of the contemporary struggle over territory and local resources.

Forms of Cultural Policy Articulation

As a cultural identity-based political claim, the cultural politics articulated by Toro community will be conditioned, hampered, and at the same time enabled by a given set of local socio-cultural resources, which give "cultural obstruction" to individuals in organizing their experiences and perceptions, deciding actions and choosing among available alternatives. They also function as the foundation on which the process of the creation of symbolic worlds gains its starting point, so that it results in new cultural identities and manifestations. On one level this process involves an effort to redefine local cultural values and meanings; on another level it becomes a reconfiguration of existing social institutions, which include local organizations as well as social actors in a broader sense. All of these constitute a "cultural reproduction" process conducted in an effort of strategically positioning the socio-cultural system of Toro itself. The following section highlights the actual forms of cultural reproduction in Toro and their manifestation through collective action in asserting Toro's cultural and political claim over the local territory and its natural resources.

Return to Ngata: A Claim for Indigeneity and Cultural Autonomy

The validity of a cultural-political claim is primarily determined by how far the existence and autonomy of that cultural identity (that serves as the basis for the claim) can be demonstrated. One of the most important challenges faced by local communities in Indonesia in asserting their ethnic and cultural identity was the implementation of a uniform pattern of village administration in the New Order era which acknowledged no indigenous communities in Indonesia.[7]

[7] With the observation that no community in Indonesia is more or less "native" than the others, official policies during the New Order were rather "ideologically" oriented than empirically. Therefore, "all native inhabitants of Indonesian archipelago

Within the past decade, the establishment of a main agenda to regain ethnic identity and "native" local government structures characterized the rise of ethno-cultural movements in Indonesia. In general, the term "native" refers to a system that existed since the pre-colonial period, although in reality many aspects of these local identities and institutions have been susceptible to standardization by the colonial government. From the beginning of their struggle, Toro people emphasized the prominent role of the "native" governmental structure as a constitutive part of their cultural identity. The term for this structure, *Ngata*, refers to an institution that has been claimed as an autonomous community unit in socio-cultural, political and economic aspects. Hence, "return to Ngata" becomes the main project, stressing that the community was a unique unit of ethnic and ecological identity, from which an indivisible unity of "culture-territory" originated. This provided the basis for an ethno-cultural approach, which promoted the political assertion over territory, community, autonomy and the right to have their own vision about future developments. Interestingly, during this process, issues of gender, which are an important aspect within the construction of cultural identities in general, became the most prominent aspect within the agenda of the ethno-cultural movement.

In reality, this process has brought about an internal struggle among Toro's community members, as well as in their interaction with neighboring village communities. This process has created an internal controversy about the question in how far and in which aspects the original *Ngata* identity could be recovered under contemporary conditions. This issue is closely related to the issue of traditional leadership institutions and the question of reconciliation with the present situation, dominated by formal governmental institutions that cannot simply be eliminated. The controversy reached its climax with the occurrence of increasing disintegration among Toro's key actors. Some of them wanted the *Ngata* identity to be used consistently, including all governmental and administrational affairs with external parties. They emphasized their point with the Decentralization Law (UU 22/1999), which allows local governments to make use of any term other than

---

are indigenous people, otherwise no one is indigenous people" (Kusumaatmadja 1993, cited in Li 2002).

"village" (*desa*) in order to foster local identities. Most opinion leaders, however, did not approve of such a radical approach, since they believed that a too frontal kind of strategy would inevitably evoke resistance from higher government officials, especially among the bureaucratic echelon of Kulawi sub-district.

Another controversy occurred within the wider context of interaction, namely the one relating to the other villages in the Kulawi sub-district, which possess an ethnically and culturally similar identity. For example, a number of customary figures from other villages opposed the effort to return to the native *Ngata* identity. For them, the original "community identity" was enshrined in the concept of *Bola* instead of *Ngata*. The integration of this term in several village names, such as *Bolapapu* and *Boladangko* supported their argument. Additionally, they also mentioned that from a cultural point of view, Toro was located in a marginal area, because it was Bolapapu that became the capital of the Kulawi Kingdom and which served as the cultural center of Kulawi. Indirectly, this statement affirmed their rejection of the claims to cultural revitalization made by Toro's leaders.

"Local Wisdom": From a Traditional Pattern of Production into a Conservation Model

The *claim* of Toro's indigeneity was further articulated by reproducing various local ecological practices and bits of knowledge as a type of "local wisdom" related to natural resource management, comparable to the modern concept of conservation. Various aspects of the traditional ecological knowledge, which have been revitalized, encompassed numerous taxonomies of flora and fauna, as well as general ethno-biological knowledge regarding the local habitat. This included, for instance, the identification of natural resource types which reflected the knowledge held about the existing bio-diversity, such as wood, rattan, medicinal plant species, etc. All aspects of this ecological knowledge were articulated in terms of a proof of the richness of local traditions shaped by generations of interaction with the local ecosystem. This issue was also stated in terms of a communal intellectual right that should be protected from foreign bio-prospectors, who always attempted to commercialize it.

Several *traditional* ecological practices in natural resource management were also revitalized. Some of these aspects included ritual procedures in opening land, various phases in the production process and in utilizing the various types of forest resources. The latter refers to traditional procedures in extracting timber, rattan, incense and fragrance, *damar* resin, traditional medicinal materials, forest *pandan*, as well as traditional mining. In addition to these rituals, several traditional preconditions that provide restrictions in natural resource extraction in order to allow the ecosystem to recover were also revitalized. Some of the well-known restrictions include the obligation to perform *ra ombo* (applying a rotation system) in rattan extraction, both in regard to space rotation as well as to time cycles. The latter (time cycle restriction) is also determined by the prohibition of floating rattan and forest *pandan* leaves on rivers or the prohibition of passing through the rice fields during the ripening period. These restrictions put indirect control over the extraction of rattan, so that it can only be extracted outside of the rice cultivation period.[8] Another kind of restriction being revitalized is the prohibition of carrying out any kind of exploitation activities in areas that fall under the category of *taolo*, that is watershed areas, water sources and sloping areas that are prone to landslides, places surrounding ficus trees and sites that are considered to be sacred (*keramat*). All of these restrictions have a culturally informed taboo system, which manifests itself in terms of prohibition and abstinence and which is believed to have a good impact on ecosystem recovery.

It is interesting to note that these forms of traditional knowledge and ecological practices – within the framework of the present historical project – have been reproduced as detailed points of technical requirements in the regulation of natural resource utilization. These changes represent basically a transformation of traditional practices and knowledge from a production pattern, the practice of which was embedded in a concrete adaptation necessity, into a standardized and

---

[8] At present, this regulation is in fact difficult to apply, as there is a change in the rice sowing cycle, which is now taking place throughout the year. Since the introduction of new rice varieties with shorter planting periods, farmers can now harvest rice twice a year. Consequently, there is no working time left outside the rice farming period.

"essentialized" corpus in order to support a modern version of human-nature interactions. Thus, the knowledge and ecological practices which existed in a specific context and which were formed through an experiential involvement in daily production routines have been extended and anchored to a totally different setting. This has led to an abstraction and universalization process over cultural aspects, bearing the risk of becoming detached from the real context into which they were born and being pulled out of their cultural bond.

Not less important in this context was the revitalization of traditional categories of forest, such as *wana ngkiki, wana, pangale* and *oma*, as explained in the previous sections. These categories relate to a rotating cultivation pattern and a specific land use rotation mechanism. It is these traditional institutions that were articulated in form of a zonation scheme, quite similar to modern principles of conservation management. This articulation has reproduced what was initially known as a traditional land use typology in regard to habitat uniqueness and the physical contour of the ecosystem. It is obvious from this point of view that the reproduction of traditional knowledge and ecological practices by articulating them in terms of a modern technical agenda of conservation was only made possible by the development of cultural codes and practices that could be recognized. It is clear that the "epicenter" that drove such cultural reproduction came from external sources. By doing this within the existing framework of cultural codes and practices however, it has provided the foundation for this historical project to draw on notions of traditional wisdom that were rooted within its own tradition.

Interpretation and Re-creation of Cultural Concepts

In addition to its relation to the natural resource management aspect, the cultural politics articulation of the people of Toro has also developed more complex cultural concepts in order to provide the proof that their assertion is justified by their own culture system. With this regard, an effort has been made to further develop philosophical conceptions regarding traditional harmony in human interaction with nature and other human beings as the basis for the technical project of conservation.

The cultural concept of *katuvua*, for instance, is a concept in Kulawi culture that illustrates an ideal value about the harmonious relationship between the traditional human being and its surrounding environment (*mapahilolonga katuvua*). This way of life is nothing else but a reflection of a mythological space in traditional life; a time in which the borderlines between man and environment, or between subject and object had not been clearly defined. This is what Peursen (1988) called the "socio-mythical space" in which humans do not possess their own identity and can only achieve their identity within a given relation that exists within the socio-mythical space.

Although this cultural conception represents more a reflection of a universal worldview, it nevertheless provides the stock of a linguistic code – with a real cultural reference – to be projected as a cultural nomenclature in the technical agenda of natural resource management. This allows the re-conceptualization of this cultural code in the framework of the current historical project. Therefore, in the effort of positioning oneself within the existing discursive context, such a conception was articulated by Toro in terms of a "community vision" in the context of the so-called "conservation agreement" jointly developed with the National Park Authority. The agreement was formally articulated during a meeting in October 2002 using the following expression: "*Hintuvu Ngata Molingku Katuvua"*, which is translated as: "The Fulfillment of Ngata Togetherness for a Harmonious and Sustainable Living with Nature". This formulation obviously reflected the technical agenda of conservation in a modern connotation.

In reality, this re-conceptualization was the crucial point in determining how far the claim over community uniqueness in relation to natural resource management could obtain cultural justification. Such a justification makes the claim culturally "meaningful" as it appears to be rooted in its own tradition. On the other hand, it served as a distinct mode of articulation in the wider interaction with conservationist and ecological discourses, which basically became a political as well as an economic issue. The latter referred especially to the relation with the hegemonic discourse of conservation voiced by the National Park Authority and various international and local environmental NGOs.

Customary Law: From Maintenance of Social Relations into a Legal Instrument

What was said with regard to the re-conceptualization of the cosmic world-view into a technical agenda of conservation, occurred also in the transformation of the customary judicial system, having some fundamental implications. Within the framework of the conservationist technical agenda, the transformed customary judicature had turned into an abstract and legalistic law procedure rather than performing its concrete social function in maintaining and renewing social relations. This transformation applies especially to violations of the natural resources utilization rules. It works on two levels. First, on the level of judicial principles, the technical agenda demanded the adoption of a legalistic element as the basis for the acknowledgement of customary institutions. This altered the initial character of the conception of customary violation (*tiboiki ada*), which formerly served as a basis for the operation of customary justice itself. The *tiboiki ada* conception put basically more emphasis on the collective impact of a customary violation – with its implication on the reconfirmation of individual conformity and social integration – that is, by linking the impact of such customary violation to the fate of all community members, which would be cosmically disturbed. This collective implication faded out when the sanction of customary law, in the cases of violation against natural resource management rules, was individualized only to the actor (in accordance with the legal procedure allegation) without being linked to a category of violation with cosmic impact towards the entire community.[9]

Second, on the level of customary judicial purpose, it has brought about a process in which unwritten law, the implementation of which was practical in nature, tended to undergo a process of "positivization" and "codification". Hence, the concrete function of reconciliation and social reintegration that was emphasized in the practice of customary justice, has been altered through new elements of abstrac-

[9] With regard to the impact on morality it brings about, customary law violation is classified into three categories, namely: (1) *longko ngata*, violation that is considered causing harassment towards community honor; (2) *rumu ngata*, that is violation that causes disgrace to the community; and (3) *rungku ngata,* violation that causes the community to express a sense of repugnance.

tion and legality for the sake of fulfilling demands of certainty and law enforcement.

This tendency has become an inevitable dilemma, especially because the efforts to revitalize customary institutions must include the documentation of customary law practices themselves (Eyzaguire 2001). In fact, if this customary law possessed an inherently local and dispersed character and gained its vitality through its integration into daily livelihood practices, its documentation and isolation from the dynamic flow of life seems to be contradictory (Agrawal 1995).

Huaka: From a Traditional Utilization Area into an Exclusive Territory

One prominent feature in the cultural dynamics of Toro community is the redefinition of "conservation" as "territory plus culture". This definition regards territory as an ecological and cultural unit, constructed through the cultural and daily practices of a local community (Escobar 1998). Thus, a territory becomes a real articulation between the types of socio-economic utilization, as well as cultural meaning and the natural environment it covers.

There are two concepts used by Toro people to explain their living space and traditional management area, as they are formed through their interaction with nature, namely *huaka* and *dodoha.* The first concept refers to a communal area, the claim of which is based on factors of historical origin and cultural meaning rather than on factors of economic utilization. Nevertheless, this area can be intensively used for economic utilization and converted into *dodoha*, e.g. a limited form of ownership of an extended family or individual. The latter covers the *oma* and *balingkea* categories, namely areas that have formerly been utilized as farmland and that are currently left uncultivated as well as gardens and rice fields that are still intensively cultivated.

The *huaka* concept reflects basically the "mental map", the boundaries of which cannot be clearly defined. Moreover, it is possible that overlapping utilization by other communities takes place such as hunting activities, extraction of medicinal plants etc. This is because the traditional land tenure system that prevails in this area acknowledges the "horizontal principle", which makes a clear distinction between land tenancy and plants growing on the land (that can be owned by another person, e.g. *damar* trees).

The Participatory Mapping activity facilitated by *Yayasan Tanah Merdeka*, abbr. YTM ("The Free Land Foundation") attempted to create a physical interpretation of this mental map. The physical existence of this map served as an important tool during the negotiation process with the National Park authority. In fact, the acknowledgment granted by the LLNP was based on the argument that the zonation system applied in the National Park was compatible with the traditional land use system as it was successfully illustrated through the participatory mapping activity.

Reconfiguration of Local Leadership Institutions

As mentioned previously, the arrangements of local leadership institutions have undergone various modifications. These were in turn associated with conflicting interests between local actors. This effort was initially carried out during a village meeting on May 7, 2001, which produced several agreements. Some of them are: (1) expanding the membership of the Customary Council by including community member representatives in addition to the "Elders" (*Totua Ngata*); (2) establishing a village government that is identical to the function of *galara* (arbiter) under the control of *Totua Ngata*; and (3) establishing a deliberation institution that is called *Lembaga Musyawarah Ngata Toro*, abbr. LMN (Ngata Toro Deliberation Council). The agreement resulted in a local governmental structure that was arranged in such a way that it reflected a sharing of legislative and executive authority resembling the one existing in the national government of Republic of Indonesia (see figure 3). Thus, this institutional arrangement was actually done by seeking historical reference over the original forms of pre-colonial government in the region. The leadership institutions in question turned out to be a controversial and sensitive issue among Toro opinion leaders themselves, as it was strongly related to the question of how far such an innovation could or could not be given cultural justification according to the prevailing cultural system and its codes. The controversy over this issue has finally led to a severe disintegration among Toro figures. At last, in a village meeting held in October 2002, a more pragmatic approach to resolve the institutional arrangement issue was adopted in order to avoid an endless dispute regarding the actual native form of Toro's institutions, which could hardly be achieved in its original form under prevailing conditions.

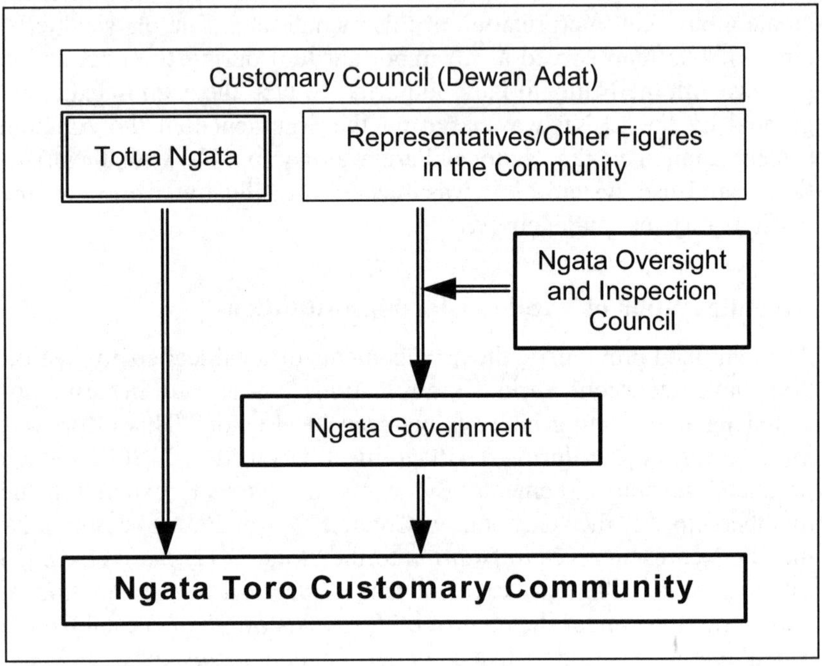

Figure 3. *Structure of relation between local institutions in Toro*

The approach adopted is based on four institutions, namely *Totua Ngata* (Village Government), the *Lembaga Musyawarah Ngata* (Customary Council), the *Badan Perwakilan Desa* (BPD, Village Representatives Council) and last but nor least the *Organisasi Perempuan Adat Ngata Toro* (OPANT, Ngata Toro Customary Women Organization).[10] The adoption of these four institutions was made possible by maintaining the existence of "secular" leadership institutions (Village

[10] The Ngata Toro Customary Women Organization (*OPANT*) was formed in August 2001; it was supposed to serve as an organizational translation of the traditional institution of the "Mother of the Village" (*Tina Ngata*).

Government and BPD), yet reducing most of their authority. Thus, both of them could be integrated into the other two institutions, that is, the traditional institution of *Totua Ngata* and the neo-traditional institution of *OPANT*, complemented by a clear allocation of tasks and authorities in order to avoid overlaps in responsibility (figure 4). There were two interesting aspects that could be observed during the process of institutional adjustment. First, gender, being an important dimension in identity construction, became a prominent aspect in the institutional arrangement agenda, giving significant impact to the existing leadership structure by ensuring women's access to *all* decision-making forums. This was achieved by transforming the functional institution of *Tina Ngata* (the "Mother of the Village"; a female *totua ngata*) into an autonomous and independent modern organization, namely *OPANT*. The adopted transformations resulted in a fundamental change, since *OPANT,* as a modern organization (at least in principle) based its recruitment and succession of leadership on a direct vote by its members and not on positional status in an "ascribed" manner. It is different from the former *Tina Ngata*, which was rather functional in nature and involved only women from noble society.

The emphasis on gender aspects relates also to the second aspect, the representation in the course of the process of institutional arrangement. This issue has in fact become the main concern since the process of exploring "local wisdom" started, a fact proven by the effort to establish a deliberation institution and to expand the membership of the Customary Council. However, since the customary rule defined that a *Totua Ngata* Council member could only be recruited from the noble families, the issue of representation was not aimed at this traditional leadership institution. Toro instead chose another way via the introduction of a new institution *(OPANT)* and modifications of the Customary Council and the modern institutions (Village Government and BPD). It has been approved for all of these institutions to hold an equivalent level of position. It also insisted that all of four institutions must held in high esteem the spirit of togetherness, deliberation and consensus as reflected in the value of *hintuwu libu ngata*.[11]

[11] The term *hintuwu libu ngata* refers to an agreement that results from consultation and discussion among the community members and is comparable to the principles of *musyawarah* and *mufakat* in Java.

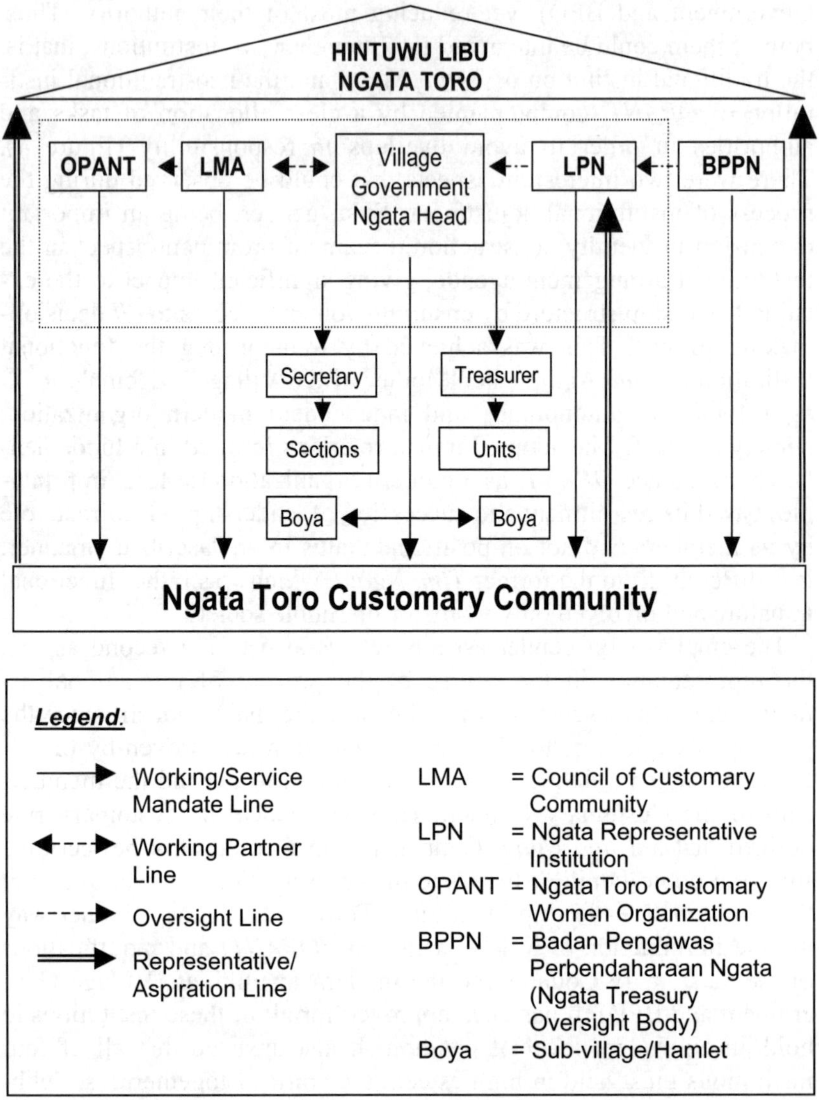

Figure 4. *Structure of authority and control between Toro institutions*

Besides formulating the structure of relations between the four institutions, in the end the effort to arrange local leadership institutions required also a similar effort in regard with the Customary Council itself which was then turned into *Lembaga Masyarakat Adat* (LMA, Council of the Customary Community). This has become an inevitable consequence, as the institution's authority was expanded to control also natural resource management and the enforcement of regulations, and not, as in the past, taking care of customary problems and daily rituals only.

This effort has produced four categories of LMA roles. First, the *hintuwu* roles govern interactions between community members such resolving conflict and dispute among them as well as performing customary ceremonies and life cycle rites. Second, the *katuvua* roles govern community members in their interaction with the surrounding environment, especially in the utilization and management of natural resources. Third, the *gagu ada* roles, govern the management of customary and cultural gears, which comprise the traditional meeting building (*lobo*), customary dress, customary ceremony equipment, and the traditional art and games equipment. The fourth category consists of functions related to issues of customary law enforcement and control, carried out by the customary security officers (*tondo ngata*). The *Totua Ngata* coordinated all of these functions. This arrangement of local institutions has succeeded in bringing about synergistic effects on the existing four institutions through the agreement that all of them would hold equivalent positions and would jointly run governmental tasks in Toro. Moreover, the effort has also modified the *Totua Ngata* into an institution with clearer and more detailed functions.

This arrangement effort seemed to be the most crucial issue in the course of Toro's internal dynamics. This is understandable, since the effort has precisely hit the core part of the community's social structure as it relates to traditional leadership institutions and the social ranking system on which they are based. It is not surprising that the effort succeeded only after an intense struggle, which took almost two years (from 2000 to 2002). Therefore, what initially was intended as a political project in order to place Toro's community identity in a strategic position within the context of the existing discursive environment, developed into an intense cultural project about adequate leadership and community representation.

Significance and Impact of a "Cultural Politics Articulation"

Towards Community Conservation Agreement

The process of cultural reproduction in Toro resulted in socio-cultural dynamics that were of high significance and impact for the local landscape. Within the context of the conservation agenda, the process of cultural reproduction made possible the incorporation of a technical conservation agenda into local knowledge and the institutional system. This has enabled the technical agenda, which was developed under the framework of the Community Conservation Agreement (CCA), to incorporate cultural nomenclature and local practices of natural resource management. Moreover, it was also implemented through the existing local institutional mechanisms and authority structures of Toro.

The incorporation of a conservation agenda into local knowledge and the institutional system makes a big difference between the CCA in Toro and the CCA-patterns in other communities surrounding the National Park with regard to the formulation of the agreement itself, the level of villagers' understanding, as well as their depth of knowledge about the conservation agreement (cf. Birner and Mappatoba 2002). All of these dynamics have facilitated a devolution process in natural resource management, which currently becomes the main trend in the making of natural resource policies (Birner and Wittmer 2000). Through this process, the authority, rights and obligations were transferred from the government (in this case the LLNP) to the local community of Toro, which then responded by revitalizing and reproducing its local cultural and institutional system, so that it could perform the new role that it has been granted. This process was relatively successful in making the local institutional system efficient enough to support the conservation project run by the government in this area, which was proved by the establishment of clear regulations on natural resources utilization, as well as their strict enforcement mechanism.

The Impact on Toro's Social Structure

Meanwhile, at a wider level of the social process, the effort of cultural reproduction within the context of natural resource management has led to a structural transformation which had significant impacts on the local social system. This could happen because the process of cultural reproduction had reached the deepest core of the existing social structure: the leadership system and social ranking system on which it is based.

On the internal level, the process drove the social structure of Toro in opposing directions: On one hand, it led to a de-elitization of the local social structure, due to the adoption of non-traditional leadership institutions as envisaged in the local governmental structure. This indicated the abandonment of other options, i.e. to fully resurrect native institutional forms of the pre-colonial period, albeit the goal to revitalize these native institutions was actually the initial obsession. This decision resulted in a fundamental change, since it offered a new channel for the involvement of non-noble community members in decision making. In a social system in which most cultural institutions were traditionally controlled by the nobles, the institutional reconfiguration opened the way towards a more diverse institutional participation of Toro's non-noble community members, especially through the newly emerged OPANT and BPD-institutions. As modern organizations, these two new institutions have, at least in principle, based their recruitment and leadership succession system on the direct election by their members.

It is important to point out that in other Kulawi villages, where the adopted leadership structure involved only the Village Government and the *Totua Ngata*, institutional access was also limited because in practice both of them were considered as the continuation of former noble functions. Paradoxically, a strengthening of the traditional hierarchical structure took place as well. This could happen because the process of cultural reproduction that occurred at that time was based completely on the articulation of local identity and the socio-cultural system. Such a reproduction process could in fact never be neutral, since the power relations that ensured its operation were also reproduced during the process. Bourdieu (cited from Jenkins 1992: 105) described this process as the "function of social reproduction of the cultural reproduction process", that is when the process, quoting Jenkins, is

> *"..... tending to reproduce the uneven distribution of cultural capital among the groups or classes which inhabit the social space in question, hence* ***reproducing social structure****." "..... tends to reproduce the uneven distribution of cultural capital among various groups or classes that occupy social space in question, hence* ***reproduces social structure****." [bolds added]*

In the articulation of Toro's traditional wisdom, the reproduction of the former hierarchical social structure occurred actually in conjunction with the transformation of this structure into a more open form. These contrasting processes might happen, as the de-elitization of the former social structure did not occur through a total transformation of the hierarchical traditional leadership institutions, but through a parallelization and synchronization with new institutions which were based on intrinsically rational principles. As a consequence, this "institutional duality" has created a situation where modern institutions could exist simultaneously with the strengthened former hierarchical structure.

The strengthening of the hierarchical structure was intrinsically linked to notions of "traditional wisdom". The process proceeded within the context of a cultural claim over territory and community identity, resulting a construction of *the other*, including the intra-community relations of Toro people themselves. This construction has placed the long established immigrants from Rampi in the Bada valley, about 300 km to the South of Kulawi, several households of Chinese descent as well as the households of more or less "regional migrants" without relations to the leading nobles as categories of people whose identity and culture gained less consideration in the process of revitalization. Instead, the process of cultural reproduction was in fact based on the cultural identity of Toro's native villagers, namely the members of *Kulawi Moma* culture.

A similar negligence also occurred towards members of the native ethnic group that had no lineage connection to the local noble society. Their position, which was already culturally marginalized in the traditional system, became even more insignificant when the system was reproduced under the current circumstances. Although the traditional hierarchical system has formally been cancelled, it nevertheless persists in many cultural practices, for example through the dowry system in customary nuptials. Revitalizing this cultural system and its articulation in natural resource management has brought about an expansion of new spaces, in which the old values and hierarchical orders were actualized.

One manifestation of the cultural reproduction that confirmed this hierachical system was the articulation of the various forest area categories as an equivalent form of the modern zonation system. This confirmation has placed the forest area surrounding the village as *oma* as compatible with the "Traditional Utilization Zone" in the National Park zonation system. On one hand, the revitalization of forest zones stabilized the local land use system. On the other hand, this confirmation has also reproduced the land ownership structure, as it was defined under the traditional system. This is because land inside the area which was culturally categorized as *oma*, belonged to a land category that according to local history was under the ownership of noble families. Hence, the stability of this area could be guaranteed, since not all people were allowed to open land situated in this area. Clearing of land might only be done, if a person has been granted permission from a noble family member who owned the forest plot in question. At the same time, such a restriction has in fact limited the participation of any community members in monitoring and protecting the forest area, since most of them did not have a sense of ownership and responsibility over it.[12]

Impact on Inter-Community Relations

In addition to its impact on intra-community relations, the process described led also to more competitive inter-community relations, especially in regard to the arrangement of village territory. This had to do with the transformation of space that occurred in this area since the beginning of the participatory mapping process. In principle, participatory mapping was an attempt to "make a map with the involvement

[12] In recent times, several cases happend where plots of land being cultivated for years by new-coming ethnic groups were snatched by "native" people. The reason behind this is that the land is included in *oma* which belongs to certain families.

of all parties located or living in a particular area through equivalent roles and actions". Thus, the resulted map could meet a standard, effective and optimal form.

> *"Standard means being in accordance with the principles and technology of map construction; effective means comprehensible by the individual who is used to employ the (modern) map as well as by native villagers; optimal means that map utilization dynamics developing in line with the (local) situation and condition. This is made possible by the involvement of many village parties or individuals from the planning stage until the creation and design of the further use of the participatory map" (ATOK, an NGO activist from the participatory mapping team of the Sulawesi-based NGO* Yayasan Tanah Merdeka *(Free Land Foundation).*

As explained previously, the participatory mapping activity has resulted in a fundamental transformation of the spatial status of the area concerned. Through the mapping activity, the natural landscape that experienced an active human interference by cultural meanings and techno-economic activities (e.g. mental map and traditional utilization area) has been transformed into a modern map that confirmed a sense about territory and the sovereignty over it. This allowed the map to be used as a negotiation tool with the Lore Lindu National Park Authority, resulting in the acknowledgment of Toro's customary area inside the National Park.

However, the confirmation of the map in the modern connotation transformed the categories of "management space" and "mental map" that were open in nature and that allowed overlapping ownership rights (according to the prevailing horizontal tenure principle) into a "territorial map" in the modern sense, which was rather exclusive and restrictive. Within the context of inter-community relations, the last denotation has led to a competitive struggle in defining the territorial status of the surrounding forests.

Although during the mapping process it was stated that the claimed territorial boundaries were based on authentic evidence according to the "language of wisdom", in fact each community made use of different criteria and standards in determining territorial boundaries. This caused continuous boundary conflicts between Toro (which has made the map) and other neighboring villages. The conflict over area boundary was even fought out with the village of Katu, albeit it did also customary area mapping facilitated by YTM.

In this case, the customary area map claimed by Katu comprised some areas located in the old settlement of Malino. The entire complex of the settlement has been included in Toro's customary area map, since the location was considered to be Toro's place of origin. Hence, it was treated respectfully and regarded as a sacred site. This overlap in mapping has led to an open conflict when it was discovered that Katu people placed boundary marks surrounding the old village site. They even began cultivating cacao plants in the related area.

The conflict was finally resolved through a mediation process organized by YTM. In the meeting held on November 7, 2002, both parties put forward their argument for the determination of their area boundaries. It was then discovered that the determinations were based on different criteria. Toro's claim was mainly based on historical arguments, e.g. that Malino was the place of origin of the founders of Toro and that the people of Toro still maintained a spiritual relationship with the historical site. The people of Katu confirmed the historical claim of Toro. However, they believed that the founders of Toro left the area because they escaped from the attack of a certain Bunian, thus the area was deserted. Moreover, the community of Katu based its claim on the factual argument that in fact it was the people of Katu who did more activities in this location and who protected it from external exploitation.

The meeting reached the final resolution that on one hand confirmed the historical claim of Toro over the Malino area, while on the other hand it granted the management right of this area to the community of Katu. The agreement states four points of resolutions, as described in the following box.

Points of Mutual Agreement between Toro People and Katu People

1. **It is agreed that**: Malino belongs to the native people of Toro and functions as the boundary between the customary areas of Toro and Katu.

   The boundary will be marked using:
   - Bolo Watu plants (small tree species, used as marker)
   - Areca plants
   - Wala Tana (small tree species, used as marker)

2. **It is agreed that**: The right to supervision over the Toro area in Malino and its surrounding is given to the community of Katu.

3. **It is agreed that**: The right to manage Toro's customary area that has been and is currently managed for coffee and damar trees by the people of Katu stays in the hands of the community of Katu. It is also agreed that the coffee plantation managed by Katu people may not be expanded.

4. **It is agreed that:** Any large-scale management of forest products, namely rattan, for commercial purpose will be carried out upon deliberation between the people of Toro and the people of Katu.

The case of conflict over the area boundary as explained above was actually no more than a superficial expression, especially since both villages have concurrently conducted customary area mapping, thus the overlaps of the area boundary became very obvious. The conflict could be easily resolved as well, since the two disagreeing villages were equally facilitated by YTM, so that a mediation for resolving the problem could be accomplished. However, there may still be many other problems to come, which are out of sight and may emerge at any time, the resolution of which will be much more difficult to find.[13]

---

[13] One of them was a claim from Mataue community, a village in the northern direction of Toro, that its customary area boundary overlaps with the area that is currently claimed by Toro. The argument underlying this claim was the ownership of a number of damar trees inside the area. Toro nevertheless denied the claim using the historical argument that damar resin gained its commercial value only after the Dutch entered the area, whereas Toro's claim was based on a preceding historical evidence, namely the existence of Malino long before the colonial period.

Conclusion

The process of articulating cultural politics with regard to natural resource management was a "cultural reproduction" that involved a fundamental reinterpretation and redefinition of local cultural identity. Although this process was born under a specific power constellation and articulated as a political project in an attempt to negotiate interests, it was not merely a "cultural mobilization" process, in which a set of practices, knowledge, values and organizational forms of a cultural system were arbitrarily reassembled and modified in order to fulfill demands from external forces.

In fact, the process was proceeding through the involvement of a local cultural identity in the re-definition and re-actualization in order to respond to the ever-changing horizon of meaning and expectation along with changes in the social, economic and political context that covers it. In other words, it has by itself become a cultural project, the dynamics and metamorphosis of which were determined by the logic of the culture itself. Therefore, the local cultural identity served as a cultural means that facilitated and impeded the process of its articulation in the course of a current historical project. It became a *cultural means that allowed* the articulation of a cultural claim over territory and community. At the same time, it became also *a cultural means that restricted* the manifestation of such a cultural claim in a form that was acceptable in terms of meaning and institutional frame to the culture itself.

The dynamics induced by the cultural transformation process had a significant social impact. On one hand, the political project, being framed under the scheme of a conservation agenda and a claim to autonomy in natural resource management, was successfully embedded in the local cultural value system and institutional setting. This allowed a devolution process to grow in which authorities, rights and obligations in the control of natural resources and their management were devolved from state institutions downward to the community level.

On the other hand, as a cultural project driven by the hermeneutics of authenticity, the cultural reproduction was made possible only to the extent that it was acceptable to the cultural and institutional structure of the cultural system itself. This process has in fact resulted in a structural transformation of the community of Toro, as it is indicated by two opposing streams. On one hand, it created a more diverse and open local institutional structure compared to the native structure before. Its achievement rests on the modification of local leadership institutions by adopting new, modern organizations (OPANT and BPD), and by suspending the idea of a complete revitalization of traditional institutions in their original forms. In this way it has provided an alternative institutional channel to social groups that were culturally not represented in traditional institutions, namely the groups belonged to new-coming ethnic groups and non-noble society.

On the other hand, because the cultural reproduction process was based on the cultural identity of Kulawi Moma, the reproduction of the old hierarchical social structure that underlies the cultural identity took place as well. This "function of social reproduction of the cultural reproduction process" has placed the new-coming ethnic groups and the non-noble native ethnic groups in a category of neglected groups (*the other*) in the ongoing cultural dynamics. Thus, the dynamics have basically reflected reproduction of merely one subculture in Toro community.

A similar process occurred also in regard to inter-community relations which then developed into a more competitive form following Toro's claims over its customary area. Because this claim, which was confirmed through participatory mapping, had basically realized a spatial transformation of the landscape status in the area, that is from what used to be categorized as a mental map and a traditional management area (which was open and overlapping in nature) into a category of territorial unit in a modern connotation (which was rather exclusive and restrictive).

Within the context of inter-community relations between Toro and neighboring villages, such a "spatial status transformation" has factually brought about relations that involve a more competitive negotiation of interests. This is mainly due to the struggle to determine the territorial status of each community, the boundaries of which (and their criteria of determination) are conflicting.

References

Acciaioli, G. L. (2001): Memberdayakan Kembali Kesenian Totua: Revitalisasi Adat Masyarakat To Lindu di Sulawesi Tengah (Empowering the Art of the Elders: Customary Revitalization among the To Lindu People in Central Sulawesi). *Jurnal Antopologi Indonesia*, 65: 60-83.

Agrawal, A. (1995): Indigenous and Scientific Knowledge: Some Critical Comments. *Indigenous Knowledge and Development Monitor*, 3, 3: 3-6.

Aragon, L. V. (2000): Fields of Lord: Animism, Christian Minorities, and State Development in Indonesia. Honolulu.

Atok, K. (1998): Palasar Palaya, Sebuah Filosofi Pemetaan Partisipatif. (Palasar Palaya: The Philosophy of Participatory Mapping). *Kabar JKPP (Jaringan Kerja Pemetaan Partisipatif)*. Palu.

Arkoun, M. (1994): Nalar Islami dan Nalar Modern: Berbagai Tantangan dan Jalan Baru (Islamic Ratio and Modern Ratio: Challeges and New Directions). Jakarta.

Birner, R. and M. Mappatoba (2002): Community Agreement on Conservation in Central Sulawesi: A Coase Solution to Externalities or a Case of Empowered Deliberative Democracy? Storma Discussion Paper No. 3. Bogor.

Birner, Regina, and H. Wittmer. (2000): Converting Social Capital into Political Capital: How Do Local Communities Gain Political Influence? A Theoretical Approach and Empirical Evidence from Thailand and Colombia. Presented at "Constituting the Commons: Crafting Sustainable Commons in the New Millenium", the Eighth Conference of the International Association for the Study of Common Property, Bloomington, Indiana, USA, May 31 - June 4.

Bourdieu, P. (1977): Outline of Theory of Practice. Cambridge.

Darma, F. and W. Satya. (2001): Ngata Toro: Antara Mitos dan Realitas. Studi Refleksi Gerakan Pemetaan Partisipatif (Ngata Toro: Between Mythos and Reality. A Study on the Movement of Particpatory Mapping). Palu.

Dove, M. (1985): Peranan Kebudayaan Tradisional Indonesia dalam Modernisasi (The Role of Traditional Indonesian Culture in Modernization). Jakarta.

Ellen. R.F. (1998): Comments. *Current Anthropology*, 19, 2. Respond on Paul Sillitoe's thesis, "The Development of Indigenous Knowledge: A New Applied Anthropology" in the same publication: 238-239.

Escobar, A. (1998): Whose Knowledge, Whose Nature? Biodiversity, Conservation, and the Political Ecology of Social Movements. *Journal of Political Ecology*, 5: 53-82.

Escobar, A. (1999): After Nature: Steps to an Anti-essentialist Political Ecology. *Current Anthropology*, 40, 1: 1-16.

Eyzaguirre, P. B. (2001): Global Recognition of Indigenous Knowledge: Is This the Latest Phase of 'Globalization'? *Indigenous Knowledge and Development Monitor*, 9, 2.

Foucault, M. (1973): The Order of Things: An Archeology of the Human Sciences. New York.

Garang, J. (1985): Dunia Kulawi: Masyarakat, Budaya dan Gereja di Sulawesi Tengah (The World of Kulawi: Society, Culture and Church in Central Sulawesi). Edisi Khusus Majalah Peninjau, 10, 2 dan 11, 1.

Goh Beng Lan. (2002): Cultural Hibridity in Southeast Asia: Locating What's Local and Specific as also Comparative and Global. *Jurnal Antropologi Indonesia*, 67: 69-78.

Gold, A. G. (1995): Magical Landscapes and Moral Order: New Readings in Religion and Ecology. *Religious Studies Review*, 21, 2.

Guba, E. G. and Y. S. Lincoln. (1994): Competing Paradigms in Qualitative Research. In: Denzin, N. K. and Y. S. Lincoln (Eds): Handbook of Qualitative Research. London.

Hefner, R. W. (1999): Geger Tengger: Perubahan Sosial dan Perkelahian Politik (Social Change and Political Struggles in Tengger). Yogyakarta.

Hobart, M. (1993): Introduction: The Growth of Ignorance? In: Hobart, M. (Ed.): An Anthropological Critique of Development. The Growth of Ignorance: 1-30. New York.

Jackson, P. (1994): Maps of Meaning: An Introduction to Cultural Geography. London.

Jenkins, R. (1992): Pierre Bourdieu. London.

Kaudern, W. (1941): The Noble Families or Maradika of Koelawi, Central Celebes. Goteborg.

Kusumaatmadja, S. (1993): The Human Dimensions of Sustainable Development. In: Proceedings, Seminar on the Human Dimensions of Environmentally Sound Development. Jakarta: Walhi.

Laban, B. Y. (2000): Speech in the Meeting and Congress of Central Sulawesi Customary Community, May 16, 2000.

Lagimpu, A. (2002a): Revitalisasi Kelembagaan Tradisional Masyarakat Adat (Revitalization of Customary Organizations). Paper presented in the Seminar and Workshop of Ngata Toro Customary Community, October 25-30, 2002.

Lagimpu, Andreas. (2002b): Kearifan Lokal dan Pemanfaatan Sumberdaya Alam (Local Wisdom and Resource Use). Paper presented in the Seminar and Workshop of Ngata Toro Customary Community, October 25-30, 2002.

Li, T. M. (2000): Articulating Indigenous Identity in Indonesia: Resource Politics and the Tribal Slot. *Comparative Studies in Society and History*, 42: 149-179.

Li, T. M. (2002): Masyarakat Adat dan Masalah Pengakuan (Custoamry Communities and the Problem of Recognition). *Wacana*, 3, 11.

Li, T. M. (2003): Situating Resource Struggle: Concepts for Empirical Analysis. Forthcoming in *Economic and Political Weekly.*

Lumba, F. (2002a): Tahapan-tahapan Perjuangan Ngata Toro (Step-by-step Struggles in Nagat Toro). Note for the Inter-Village Discussion regarding Local Community-Based Natural Resource Management, Lobo Ngata Toro, March 30, 2002.

Lumba, F. (2002b): Pengalaman Perjuangan Ngata Toro (Experiences from the Struggles of Ngata Toro). Note for the Inter-Village Discussion regarding Local Community-Based Natural Resource Management, Lobo Ngata Toro, March 30, 2002.

Lumba, F. (2002c): Pergulatan Teologis dan Praksis Seorang Pendeta dalam Rangka Perjuangan Agraria dan Hak-hak Masyarakat Adat (Theological and Practical Struggles in the Context of Agrarian Struggles and Indigenous Rights). Note for Discussion in LPA Awam Green, Palu, November 8, 2002.

Mattulada, A. (tt.): Sejarah Kebudayaan "To Kaili" (A Cultural History of the Kaili People). Palu.

Miles, M. B. dan A. M. Huberman (1992): Analisis Data Kualitatif: Buku Sumber tentang Metode – metode Baru (Qualitative Data Analysis: A Basic Book on New Methods). Jakarta.

Munawar-Rachman, B. (1997): Kesetaraan Gender dalam Islam, Persoalan Ketegangan Hermeneutis (Equality of Sex in Islam, Hermeneutic Problems). *Majalah Filsafat Driyarkara*, 23, 2.

Myer, L. (1988): Biodiversity Conservation and Indigenous Knowledge: Rethinking the Role of Anthropology. *Indigenous Knowledge and Development Monitor*, 6, 1.

Ngera, P. (1989): Desaku (My Village). Handwritten Paper, Unpublished, 123 pages.

Nygren, A. (1999): Local Knowledge in the Environment-Development Discourse: From Dichotomies to Situated Knowledges. *Critique of Anthropology*, 19, 3: 267-288.

Persoon, G. (1988): Isolated Groups or Indigenous Peoples: Indonesia and the International Discourse. *Bijdragen tot de Taal-, Land- en Volkenkunde*, 154-II: 281-304.

Peursen, C.V. (1988): Strategi Kebudayaan (Cultural Strategies). Yogyakarta.

Radam, N. H. (2001): Religi Orang Bukit: Suatu Lukisan Struktur dan Fungsi dalam Kehidupan Sosio-Ekonomi (Hilltribe Religion: A Picture on the Structure and Function in the Socioeconomic Sphere of Life). Yogyakarta.

Rizal, R. (2002): Kilas-Balik Perjuangan Organisasi Perempuan Adat Ngata Toro (OPANT). (A Review on the Customary Women Organization in Ngata Toro). Paper part of the activity in Women Political Participation Empowerment in the Era of District Autonomy, collaboration with OPANT and The Asia Foundation.

Sangaji, A. (2000a): PLTA Lore Lindu: Orang Lindu Menolak Pindah (The Lore Lindu Dam Project: Lidu People Refuse to Move). Yogyakarta.

Sangaji, A. (2000b): Orang Katu di Taman Nasional Lore Lindu: Perjalanan Panjang menuju Rekognisi (The Katu People in the Lore Lindu National Park: The Long March to Recognition). Paper presented in the International Symposium and Workshop on "Entering the 21st Century: Regional Autonomy, Local Culture and Nation Culture". August 1-5, 2000. Makassar.

Sangaji, A. (2001): Moratorium Taman Nasional Lore Lindu (Moratorium for the Lore Lindu National Park). Paper presented in the Public Debate on Lore Lindu National Park Moratorium. July 5, 2001. Palu.

Setiawan, B. (Ed.) (2000): *Perjuangan Demokrasi dan Masyarakat Sipil: Reposisi dan Peran Ornop/LSM di Indone*sia (Democratic Struggle and Civil Society: Repositioning and the Role of NGOs in Indonesia). Jakarta.

Shohibuddin, Moh. (2003): *Artikulasi Kearifan Tradisional dalam Pengelolaan Sumberdaya Alam Sebagai Proses Reproduksi Budaya: Studi Komunitas Toro di Pinggiran Kawasan Taman Nasional Lore Lindu, Sulawesi Tengah* (The Articulation of Traditional Wisdom in Natural Resource Use as a Process of Cultural Reproduction: A Community Study at the Forest Margin of the Lore Lindu National Park). Master Thesis of the Study Program in Rural Sociology, Institut Pertanian Bogor.

Shohibuddin, Moh. (2005): The Ethical Dimension in Revitalizing the Identity of Ngata for Claiming Territory and Local Resource: The Struggle for Village Autonomy among the Toro Community in the Adjacent of Lore Lindu National Park. Paper presented at "International Seminar VI on Dynamics of Local Politics: Ethics, Politics and Democracy," organized by Percik and Fourd Foundation, Salatiga Central Java, August 02-05.

Sillitoe, P. (1998a): The Development of Indigenous Knowledge: A New Applied Anthropology. *Current Anthropology*, 19, 2: 223-235.

Sillitoe, P. (1998b): Defining Indigenous Knowledge: The Knowledge Continuum. *Indigenous Knowledge and Development Monitor*, 6, 3, December.

Towaha, C.H. (2002): Proses Perkara Perkawinan Poiwa Pakasi Dan Ponona Monoho Yang Diurus Oleh Totua Ngata Toro Pada Tanggal 8 Agustus 2002 S/D 19 Agustus 2002 Di Rumah Ketua: Boya III Kinta Ngata Toro (Customary Marriage Clothes and Attributes as they are Managed by the Community Elders). Unpublished.

Tsing, A. (1999): Notes on Culture and Natural Resource Management. Working Paper, Berkeley Workshop on Environmental Politics. Berkeley.

Van Marrewijk, A. (1998): Indigenous Knowledge: The Proof is in the Eating of the Pudding. *Indigenous Knowledge and Development Monitor*, 6, 3.

Yayasan Hukum Bantaya (2001): Masalah Agraria di Sulawesi In: Tim Lapera (Ed): Prinsip-prinsip Reforma Agraria: Jalan Penghidupan dan Kemakmuran Rakyat (The Principles of Agrarian Reform: The Road to Living and Peoples Welfare). Yogyakarta.

Yin, R. K. (1997): Studi Kasus: Desain dan Metode (Case Study: Design and Methods). Jakarta.

Chapter IV

Customary Law Communities, Property Relations, and the Management of Common-Pool-Resources: Evidence from Toro Village, Central Sulawesi, Indonesia

Günter Burkard

"It is not enough to live in the forest. One must have a stable village that can be identified and funded. One must have a distinctive culture worth studying and saving. And one must have a strong, visible leadership to articulate community concerns in ways that these agencies can understand" (Tsing 1999: 185).

Abstract

This paper analyses some structural problems related to the construction of a "customary law community" inside the Lore Lindu National Park in Central Sulawesi, Indonesia. The resource management institutions of the village of Toro are generally perceived as being inherently sustainable, sufficiently fulfilling the Park Authority`s demand for forest conservation. However, rather than being indigenous creations of the ancestors, Toro's institutions and principles are based on the state-defined notion of "community controlled land", the existence of which is denied by the regional government and by neighboring villages. Controlled by those who have access to certain bodies of relevant knowledge, the process of adequate "community representation" is characterized by a high degree of participatory exclusion. Whereas village leaders justify their claims to a "right of avail" (*hak ulayat*) in terms of local wisdom, common people perceive their claims to private property (*hak milik*) as crucial to socio-economic security. Different notions of "community" and different representations of property relations, which are embedded in different politico-legal discourses and regional histories, hamper the formation of "village alliances" in law enforcement and monitoring. These, in turn, constitute a necessary precondition for the sustainable protection of the forest resource.

Introduction

This article analyses the problems related to the construction of a "customary law community" in the Lore Lindu National Park (LLNP), Central Sulawesi, Indonesia. In the broadest sense, this paper is meant to be an investigation into the practices, through which local communities (or their leaders) seek to enhance their influence within a regional politico-legal discourse on communal identity, property relations and forest conservation. In July 2000, the community of Toro, located in the Kulawi valley, 86 km from the provincial capital of Palu, was granted a high degree of autonomy in the protection and utilization of its surrounding forest resources by the regional Park Authority (*Balai Taman Nasional*). The area to be managed independently by the community comprises no less than 22,950 ha of secondary and primary forest, successfully re-claimed by the community leaders as "customary village land" (*tanah adat*). 18,000 ha of the area affected are located within the National Park. After successful nomination by *Care International*, Toro gained even international acclaim in winning the UNDP-based "Equator Award" in April 2004. The official nomination form[1] distinguishes three periods of Toro's initiatives for acknowledgment and self-determination: (1) The *Foundation Phase* (1993 - June 2000) is characterized by the exploration of customary law (*adat*) and local institutions, documented in written form and "enacted" in a systematic land use map with active support of "The Free Land Foundation" (*Yayasan Tanah Merdeka*), a local indigenous rights advocacy NGO. (2) The *Acknowledgment Phase* (July 2000 - Oct. 2001) is governed by the struggle to reclaim traditional village lands (*tanah adat*) and the acknowledgement of Toro's customary law and local management schemes as adequate means of protection by the Park Authority.[2] This phase includes also the foundation of local institutions and the revitalization of traditional concepts of Leadership, above all the gender-related "customary women group" and the "mother of the village" as the traditional bearer of

---

1 See www.careindonesia.or.id and www.equatorinitaiative.net/files/2004-0136.

2 Acknowledgement took the form of a so-called "Conservation Agreement" which is only recognized by the Park Authority, but not by the Forestry Department. Thus, its "legal reliability" must be rendered as low.

indigenous environmental knowledge. (3) During the *Dissemination Phase* (Nov. 2001 - present), Toro is actively engaged in facilitating neighboring villages to follow Toro's example and in building strategic partnerships with national and international organizations. The issues referred to above are more than characterizations of different periods of time. In fact, they reflect some of the most crucial contextual aspects in community-based resource management. The first and second phase relate to the processes of "community construction" within the wider framework of the Indonesian agrarian and forestry statutes and global concerns with *indigenous societies*, *local knowledge*, *women empowerment* and *local organization*. The third phase links up with the practical problems as they relate to the making of inter-village alliances in regard to rule creation, enforcement and monitoring.

The theoretical framework adopted in this paper consists of five key concepts: *common pool resources*, *practice*, *social actor*, *local knowledge* and *legal pluralism*. Their inter-relatedness will be fleshed out within the framework of *discourse*, which functions as an umbrella concept or an 'adhesive' between the five concepts.[3]

Irrespective of the *de jure* status they enjoy, **common pool resources** (CPR's) share the feature that "exclusion" is difficult, which means that they can be kept from potential users only at a high cost while they are "subtractable" in the sense that they can be depleted with individuals and groups using them more or less randomly (Gibson et al. 2000: 6). Upon examining the literature on the "problems of the commons" published within the last decade, two outstanding themes of investigation can be identified. The first revolves around the adequate institutional level for managing such resources. There is an increasing awareness that neither a pure "higher-level management" by central governments, nor a pure "local-level management" by individual communities works sufficiently well by itself. Instead, successful management schemes need to be supported by *cross-scale interac-*

[3] Following Michel Foucault's reasoning as outlined in "'The Order of Discourse" (1971) and "The Archaeology of Knowledge" (1972), *discourse* denotes a form of practice, that produces specific fields of culturally and historically located meanings or a specific problematic by establishing meaningful relations between concepts and reality.

tions, which link institutions horizontally across space as well as vertically across different levels of organization (Berkes 2002: 293). The second theme can be described as "diachronic analysis": On the one hand there is the notion of the profit-maximizing individual whose calculus will inevitably overexploit and deplete any common pool resource. The other extreme is the community image of the "moral economy of the commons" with its emphasis on the sustainability and resilience of customary institutions and indigenous livelihood patterns (McCay 2002: 380). This view tends to consider customary communities as isolated, bounded "micro-cultures", more or less closed systems which had evolved to be functionally stable. The one-sided focus on the "design principles" of local institutions tends to ignore the role of outside institutions and discourses that – albeit at varying rates – shape and influence them. Local communities however, do not act uniformly with these larger influences, but bring their own ideas, histories and interests into this interplay.

Arguing that viable common institutions depend to a large extend on the creation of "encompassing organizations", McCay (ibid: 381) calls for a stronger awareness of the social and political "embeddedness" of common pool resources in order to reconcile the discrepancies between theories of individual agency and structure-based approaches. In our conception, the idea of embeddedness is inseparably linked to the notions of "practice" and "social actor".

Practice theory, as developed by Bourdieu (1977) and discussed by Ortner (1984, 1989) focuses on the culturally informed practices and dispositions (*habitus*) of actors. In this sense, **practice** is always action in relation to structure. "Practice" derives from structure, it reproduces structure and it has the capacity to transform structure.

> *"Thus it is structure that is doubly practised: it is both lived in, in the sense of being a public world of ordered forms, and embodied, in the sense of being an enduring framework of dispositions that are stamped on actors' beings" (Ortner 1989: 13).*

Practice is intrinsically linked to notions of asymmetry and domination. Rather than the free floating, profit-maximizing agent adopted by neoclassic economic theory, practice theory sees the individual actor as being "heavily constrained by both internalized cultural parameters and external material and social limits" (Ortner 1984: 14). This is certainly true, but any given system consists simultaneously of a variety of at times conflicting discourses, interests, and subsystems within which an individual chooses and acts. The system being open, the individual actor is exposed to new models of behavior, new possibilities of choice. In such situations, questions of motivation and strategic action become central. Further, as Hefner (1990) reminds us, actors' preferences are not formulated in "solitary introspection", but are informed by a process of positioning oneself through identification with significant others and their evaluation of what is appropriate and desirable. This "social referencing" is heavily influenced by the relations between the different sub-groups within a given society, as well as their moral and political histories (Hefner 1990: 21).

In recognizing that it is "social referencing" as well as socio-cultural differentiation which shapes the practices of structurally positioned actors, the individual actor is transformed into a **social actor** (Long 1992), whose attitudes and actions are always embedded in larger attachments or "commitments". Such commitments however, transcend the scope of the individual calculus attributed to the free agent of utilitarian interest theories:

> *"The view of the person as a clear-headed maximizer over clearly defined preferences must give way to the image of a more complicated and less certain actor, attempting to sort out what is worth doing and what sort of person to be" (McPherson 1983: 111, cited in Hefner 1990: 20).*

Expanding the concept of *social actor*, the notion of "social organization" is better perceived as a "set of practices" (Nuijten 1992: 204) than in terms of persistent structural entities. Different patterns of *organization* emerge from interactions and negotiations between different actors. In this sense, "negotiation" refers first of all to the "negotiation of meanings, not solely the maneuvering of individuals within agreed rules. It is a social process which draws upon cultural traditions while in the process is transforming them" (Li 1995: 510). Turning now to our point of entry, the management of common pool resources, it seems less that an individual calculus can explain a commons system, rather, the individual calculus must be explained from the viewpoint of the socially, culturally and politically embedded commons (see Peters 1987).

Following an actor-oriented perspective on *knowledge* as developed by Norman Long (1992); **local knowledge** is perceived as emerging from social interactions between specific actors and their "life worlds". Working outwards from this basis, local knowledge is not constructed by a single (local) "knowledge system", but is produced by *knowledge interfaces* which constitute the critical points of intersection between the actor's diverse life worlds. Different live worlds however, need negotiations within *knowledge encounters* which "involve the struggles between actors who aim to enroll others in their 'projects', getting them to accept particular frames of meaning, winning them over to their points of view" (N. Long 1992: 27).[4] In the course of this process, certain discursive statements will be reaffirmed, whereas others will be reformulated or transformed. This is why certain discourses are more persistent in time than others. Because knowledge processes are integrated into social processes, "discursive analysis" must always take issues of power and inequality into account (A. Long 1992: 168). However, it is not only that those in power are engaged in knowledge production. There may also be a reverse situation in which the ownership of valued knowledge itself constructs an asymmetric relational structure. As Foucault (in Gordon 1980) reminds us, discourse can be both, an instrument and an effect of power. Under specific circumstances and at certain points of time, certain bodies of knowledge come to be privileged over others, that is to say the "value" which is attached to specific sectors of knowledge will rise. Because not all members of a society or community have equal access to the privileged or valued domains of knowledge, those who control valued knowledge are becoming *experts*; and it is their expertise which conveys them authority, status and power. In other words, those who have access to valued knowledge control the process of knowledge production. In regard to the notion of "local knowl-

---

4 It should be noted however, that because knowledge is embedded in individual life worlds and multiple realities, the process of "winning over" is not a complete one in the sense that certain perspectives presented by actors engaged in negotiations are accepted in totality. It is rather that "meaningful bits of information" are integrated into a unique individual life world, a unique individual reality "which may differ from the 'reality' of the original knowledge provider" (Fremerey 2005: 256).

edge", it is important to keep in mind that what constitutes "valued knowledge" in a given local community is often less defined by the community itself, but is shaped by regional dominant forces and influences (NGO activists, environmentalists, political agendas etc.). The validity accrued to local knowledge has thus less to do with its "indigeneity", but rests with the authority which is accorded to it by significant outsiders. Thus, the authority attached to local knowledge must be analyzed in the context of the wider discourses to which it is linked and from which it derives its legitimacy.

Thus, as noted by Mosse (2001: 17), local knowledge should be conceived *relationally*, as a product of social practices rather than in terms of a fixed commodity with the subsequent question arising, what actually counts as evidence of local knowledge (or customary practice) within a given socio-historical setting. In the Indonesian context, local knowledge is primarily portrayed in terms of local beliefs and customs (*adat*). Within the last decade, the term *adat* has experienced an unprecedented valorization within the Indonesian environmental discourse and struggle for self-determination. Originally an *objective* term to denote the way of life of village communities, encapsulating customary law, ritual practices, cultural dispositions and social relations, *adat* has been transformed into a *rhetorical* term in the struggle of local communities to reclaim their rights over natural resources and territories. The concept of *adat*, then, needs deconstruction so that it is seen as a socially constructed and negotiated process, not simply the actualization of a given system of norms, rules and values. As far as the Indonesian reality is concerned, the decisive indicator for the existence of a customary law community (*masyarakat hukum adat*) however remains its recognition as such by the Indonesian government. In other words: communities must be able to articulate their identity in terms of an "adequate representation" which fit these kinds of characteristics recognized by government agencies. Thus, there is the paradoxical situation that knowledgeable actors may use the national law in order to prove the existence of customary law (Campbell 2001: 114). Because the 'knowledgeable' are usually those who are experienced in dealing with higher authorities as well as non-government organizations (those who are familiar with the relevant laws and discourses), it is generally the village leaders who are in possession of the relevant knowledge. In such cases, the "production of knowledge and the exercise of administrative power intertwine, and each begins to enhance the other" (Allen 2003: 70).

In this paper, the notion of **legal pluralism** is confined to the issue of property rights. In its broadest sense, *legal pluralism* refers to the coexistence of multiple legal orders within a certain sphere of social life, whereas *property rights* regulate who is allowed to make use of a certain resource and under what conditions. Given the fact that there are always multiple legal/normative frameworks at the same time, varying "bundles" of property rights may coexist and interact simultaneously (Meinzen-Dick and Pradhan 2001: 11). As indicated in the last paragraph, *adat* has already become part of a plural legal reality. The notion of legal pluralism however, is not exhausted with an identification of different legal sources and their varying representations of rights (e.g. "local law" versus "state law" / "forestry law" versus "agrarian law"). This stresses the intrinsically *social nature* of property rights in the sense that they represent not only a relationship between actors and things, but also a social relationship. However, this insight is anything but new. In his anthropological classic *Crime and Punishment in Savage Society*, Malinowski (1926, cited in Dietz et al. 2002: 8) was quite aware of the complex nature of any property regime:

> *"Ownership, therefore, can be defined neither by such words as "communism" nor "individualism", nor by reference to "joint-stock company" system or "personal enterprise", but by the concrete facts and conditions of use. It is the sum of duties, privileges, and mutualities which bind the joint owners to the object and to each other" (Malinowski 1926: 21).*

As Li (1995: 502) points out, the quality and formation of property rights must be understood in terms of the particular normative social/cultural whole, the locally held "vision of community" in which they are embedded. It is important to note here that 'discursive fields' always consist of a number of "competing and contradictory discourses with varying degrees of power to give meaning and organize social institutions and processes" (Weedon 1987: 35). This means that within the discursive field of "property rights", individual actors and communities may make use of different "normative repertoires" or contradictory "property discourses" in order to legitimate their respective claims.

This article is organized as follows: Section one discusses the cultural localization and legal situation of "customary law communities" in contemporary Indonesia. Section two analyses the "revitalization" of the major customary community features in Toro, "traditional leadership" and the spatial mapping of "customary land use" (against the background of alternative counter-interpretations) as they are informed by an overall conservation discourse and the Indonesian forestry and agrarian statutes. Section three analyses the contradictions between an adequate "representations of community" on the one hand and "community security" on the other, pointing out that conservation concerns and equity concerns are not always compatible, but rather conflict with each other. Further, it compares Toro with its neighbor villages in regard to some key socio-economic features that are generally linked to land-use patterns and deforestation in the research region. Section four looks into the "legitimacy" of the revitalization process in the light of the different property discourses existing within the research region. It further highlights the major practical problems related to the making of "village alliances" in community-based forest management as they relate to the contradictory normative frameworks adopted by different villages. These frameworks, in turn, are based on different phases and interpretations of local history. Thus, the perception of local knowledge and custom as a "fixed commodity" is not only inadequate because of its hybrid history, but also because this history itself is contested by the actors themselves.

Indigenous Societies and Natural Resource Management in Indonesia's Forest Frontier

So-called "indigenous peoples" and their locally held knowledge systems have long been a major focus of anthropological research. Whereas the original, academically informed interest in indigenous knowledge was primarily aimed at discovering its structure and systematic, the last two decades witnessed an important shift in the evaluation of local knowledge as it became increasingly crucial to the implementation of conservation and development projects. The often remarkable effectiveness of traditional livelihood patterns in the sustainable use of natural resources has led to a valorization of local perceptions which gained not only international legitimacy, but were even endowed with the aura of "ecological wisdom" (Posey 2000: 35). By linking it to issues of bio-diversity preservation and by transforming it into "wisdom", local knowledge was put at the center of the environmental discourse. On the other hand, the fact that many indigenous peoples live in areas of high biological diversity is often used strategically: In portraying their land-use systems as sustainable modes of resource management, indigenous people present themselves as ideal stewards of the environment. Since the devolution of greater power to local communities has become a major imperative in natural resource management, an increasing number of program agendas aim at the elimination of gender-related discriminations and the full empowerment of women in development and conservation planning. It is generally believed that a stronger integration of women in decision-making will raise the effectiveness of sustainable development strategies.

Within the last two decades, a significant number of developing countries have undergone a paradigm shift in natural resource management from state-centered, top-down management towards more decentralized management schemes, in which local communities are given wider responsibilities (Shackleton et al. 2002). Indonesia has been no exception. Since late-1998, after 32 years of centralist rule under President Suhartos "New Order" (*Orde Baru*) regime, the country experienced a process of rapid fiscal and administrative decentralization. Whereas in comparison with other nations, political decentralization came rather late, it was preceded by a vivid discussion on the empowerment of local communities and indigenous rights (IR) since at least the early 90ties. The Indonesian term for indigenous peoples, *masyarakat adat* was created in 1993 when representatives of different ethnic groups and members of the Indonesian Environmental Forum WALHI (*Wahana Lingkungan Hidup*) formed an alliance called the "Network for the Defense of the Rights of Indigenous Societies" (*Jaringan Pembelaan Hak-Hak Masyarakat Adat*) in Tana Toraja, South Sulawesi. In 1999, with the foundation of the nationwide "Indonesia's Indigenous Peoples' Organization" AMAN (*Aliansi Masyarakat Adat Nusantara*) in Jakarta, the indigenous people of Indonesia once and for all articulated their specific demands on the political agenda (Sangaji 2001).

The probably most important outcome of the foundation of AMAN was the "Joint Decree on the Resolution of Traditional Rights Conflicts" (PP 5/1999) of the Minister of Agriculture and the Head of the Agrarian Board (*Badan Pertanahan Nasional*, abbreviated BPN). In this decree the BPN, which has the legal authority to acknowledge traditional land rights, declared rights over customary village land as non-transferable (Fay and Sirait 2002: 141). The BPN's authority however, is not recognized in areas designated as "forest land". The problem with this is that more often than not, "customary forests" are secondary forests which are embedded in cyclical agricultural systems and as such form an integral part of mixed farming strategies (see Burkard, "security", this volume). Given the highly sectoral character of the Indonesian legislation, there is neither the common vision, nor a common single law regulating natural resource and protected area management in a conclusive manner. In contrast to the BPN's commitment to devolution, the new forest law (UU 41/1999) preserved many of the centralist features of its predecessor (UU 5/1967). Within UU 41/1999, community forest is not a distinct category from private forest and state forest, but represents a subaltern category within the national forest area (Art. 1, §6). Consequently, Act No. 41/1999 recognizes only the right to forest management (*pengelolaan*) according to customary principles, but not the right to open forest for conversion into agricultural land.[5]

Despite the unclear distribution of authority within the government, Agrarian Law (BAL; UU5/1960) and Forestry Law (UU 41/99) share similar views on the peculiarities which define customary communities: Traditional land rights are respected only as long as customary claims do not conflict with national priorities (UU 5/1960 Art. 3, UU 41/1999 Art. 67). The devolution of rights related to the management of natural resources remains confined to so-called customary law communities (*masyarakat hukum adat*), leaving no chances for communities which either do not qualify as customary law community (e.g. multi-ethnic villages or transmigration sites) or which lack the

[5] Traditional management practices must not conflict with the state defined function of the forest as conservation or protection forest (Art.37).

necessary indicators.[6] The required indicators listed in the Forest Law (UU 41/1999, elucidation of Art. 67) are: a legally identifiable social organization (*rechtsgemeenschap*), a formal customary law institution, a clearly defined territory regulated by customary law, the presence of traditional *adat* authorities and the collection of forest products to meet subsistence needs. Thus, devolution is channeled through a social organization, which mirrors the features of bureaucratic structures, which are strictly defined by the state. In emphasizing the management capabilities of *adat* institutions, other criteria such as socio-economic security and social stability are not recognized as decisive factors for devolution.[7]

Deeply ingrained in the Indonesian legislation, especially the BAL (Art. 5, 16, 65), and an indispensable part of the customary law society is the notion of *ulayat*, referring to a special relationship between the local community and its "customary territory" (*tanah adat*). The term preferred by Dutch colonial scholars, *beschikkingsrecht*, is usually paraphrased with "right of avail" (Burns 1989: 9).[8] In its broadest sense, *ulayat* refers to community-controlled land. Whereas group members have a right to make use of virgin land within the community's territory, outsiders can only do so after obtaining permission from the community. The regulation of access remains in the hands of a traditional authority who may be either an individual elder or a council of elders. The collective nature of *ulayat* does not mean that

---

6 Indonesian law does not appreciate the term *masyarakat adat*. Instead, the law prefers the term *masyarakat hukum adat*, literally: "customary law society" (UU 5/1960, UU 41/1999, PP 5/1999). This difference is more than semantic: Whereas the term "masyarakat adat" is preferred by environmentalists and local rights activists in order to pronounce the "indigenousness" of local people and their practices, the term "masyarakat hukum adat" refers to the features of institutional formality which are required in order to qualify for official recognition.

7 The most important law on protected area management, the 1990 Act on the "Conservation of Biodiversity and Natural Ecosystems" (UU5/1990) offers almost no chances for community participation and empowerment. General statements dominate the act, such as that people should be involved in nature conservation and that the government must raise the conservation awareness among the local population (Art.37, §2). Similarly, the "Environmental Management Act" of 1997 (UU 23/1997) makes no reference at all to the local knowledge held by indigenous peoples.

8 Burns (1989) prefers the term ‚right of allocation' over 'right of avail', because it better fits its intend that land cannot be alienated in perpetuity.

individuals cannot hold resources for their own needs. Important is rather the fact that individual rights are defined by their relationship with the rights of the community to which the individual right holder belongs and from which his individual rights are derived. In general the community recognizes an individual's "preferential right" to a certain piece of land at the same time he is bound by customary restrictions imposed by the community (Evers 1995: 3). The most common restriction in this regard is that land cannot be sold. Perceived in terms of a "delicate balance of rights and restrictions" (Evers), in the *hak ulayat*-conception individual and community rights are complementary and mutually constitutive. In contrast to the "right of ownership" (*hak milik*), which allows for alienation and which can be registered by land titles (e.g. certificates), the "right of allocation" (*ulayat*) is an unwritten "entitlement" based on community consent.

The most critical feature attributed to customary law communities however, is that they must be persistent in time, since generations and without interruptions (UU 5/1960 Art.2, UU 41/1999 Art. 67, PP 5/1999 Art. 2). Thus, according to the law, there cannot be an "invention" or "adaptation" of *adat*, nor can there be a "revitalization" of "lost" customary principles. Here, the legal situation becomes a paradox: It was no one other than the Indonesian government itself which deprived forest dependent communities of their self-governing capability and sanctioning patterns by imposing a uniform, bureaucratic structure in the countryside (UU 5/1979) and by concentrating the full management authority over forest resources in the hands of the state (UU 5/1967). On one hand, local communities were withdrawn from their traditional role in forest protection by government regulation (PP 28/1985, Art.10), but on the other hand, they are expected to prove their embeddedness and integration with nature in order to qualify for recognition as customary law community. As Campbell (2002) points out:

> *By depicting a romanticized version of* ***adat*** *as a glorious living tradition of harmony with nature that is fully operative in forest dependent communities, it is easier for government critics to push their equally simplistic view that most customary systems (as static self-perpetuating operating systems) have already broken down" (ibid: 115).*

Whereas laws regulating natural resource use in direct manners offer only little "room for manoeuvre" for local communities to manage their resources autonomously, there exist a number of laws regulating resource use in indirect manners which seem more dedicated to the devolution of management rights. A Parliament Resolution from 2001 (*TAP No. IX, 2001*) states that the "management, ownership status and use of land must be ordered anew with a just land reform that acknowledges more clearly the ownership rights of the common people" (Art.5, §1). The same paragraph introduces the concept of "transitional justice" as a means to establish social stability during the transition from centralist state to regional autonomy. The *Law on National Development* (UU 25/2000, Art.10) points out that "regulations about resource use must include access of *adat* communities as well as (other) local communities" and the new *Law on Human Rights* (UU39/1999, Art.6, §1, 2) declares local rights over resources as a part of the basic human rights with community land rights being portrayed as "natural rights". Already drafted in the Suharto Era, the *Law on Population Development and Family Welfare* (UU 10/1992) proclaims the "right to the beneficial use of territory that constitutes a traditional customary inheritance" (Art.6). Albeit these laws contain important statements, there exist no implementing regulations to enforce their basic provisions.

From the point of view of local communities, however, the "state" is regarded as an assemblage of various agencies and institutions (the Forest Department, the Park Management, organizations acting on behalf of the government etc.), not in terms of a monolithic central structure. Given the overall legal and cultural localization of "indigenous people" and "customary communities" in Indonesia, the fulfillment of the state-defined features of a "customary law community" (as stated in the Forestry and Agrarian Laws) remain the single possibility for communities to regain control over their ancestral territories.

The Construction of a Customary Law Community: Why is Toro Different?

Similar to other enclaves inside the National Park, such as the village of Katu in the Besoa valley and the villages surrounding Lake Lindu (documented in Li 2000), Toro villagers were threatened with resettlement from their village since 1997 (Sunito 2005: 151). Efforts to resist resettlement ended with the formal recognition of Toros ancestral lands when a "village conservation agreement" (*Kesepakatan Konservasi Masyarakat*) was signed with the National Park Authority in July 2000. Influenced by an NGO-attached village leader, Toro decision makers realized that the success of their resistance towards resettlement depended on their capacity to proof the authenticity of their ancestors' rights and their ability to construct a viable and enforceable customary "counter-culture" (Zerner), which fulfilled the Park Authority's demand for adequate preservation. This effort however cannot be analyzed in isolation from the wider national discourse on development and indigenous rights. Resettlement is nothing uncommon in Indonesia where thousands of farm households are relocated from the overpopulated island of Java via the national transmigration scheme. As pointed out by Li (2000: 134), in case local communities present themselves as "ordinary farmers", their livelihood can – at least theoretically – be recreated elsewhere like that of thousands of other transmigrates. This means that in resisting resettlement, Toro leaders had to manage a fivefold task:

1) Their local culture had to be presented as unique, sufficiently different from other villages in order to match with outsiders images of "indigenous societies",

2) Their livelihoods had to be presented in a way which proofs their attachment to place, in order to convince outsiders that their specific form of existence is tied to this place,

3) The representation of traditional forms of social organization must match with the nationally defined features of a customary law community,

4) The traditional mode of land use must be presented as being sufficiently sustainable to sustain the integrity of the National Park and should – if not mirror – at least be not in conflict with the national forest zoning scheme,

5) Toro must win the hearts of NGO's in order to make the "Toro case" public and to mobilize the necessary support.

I think this is what Tsing (1999: 163) meant with the necessity to create a "tribal situation" in contrast to a mere "village situation". Thus, more than other villages near the Park, Toro represented the "kind of community that environmentalists and green developers might choose for co-operation, learning, and alliance" (ibid: 161). Support was organized via a Sulawesi-based NGO called "The Free land Foundation" (*Yayasan Tanah Merdeka;* abbr. *YTM*) which had already long established connections with part of the village leaders. In contrast to the two other leading NGO's in Central Sulawesi, "The Nature Conservancy" (TNC) and "Care International", YTM's activities are not framed within the government's policies of development and village-state-relations, but are often formulated in opposition to the state. Less concerned with issues of "development" (as in the case of "Care") and "conservation" (as in the case of "TNC"), YTM is regionally known as the specialist for "empowerment". In the case of Toro, "empowerment" meant nothing else than the actualization of the tasks outlined above.

The adequate representation of the *Customary Law Community of Toro* manifested itself in two major features. First, the notion of "traditional leadership", including the establishment of a "Mother of the Village" (*Tina Ngata*) as the traditional carrier of environmental wisdom and a "Customary Law Council" (*Lembaga Adat*) having the full jurisdiction over Toro's ancestral territories. The second feature concerns the spatial mapping of Toro's customary lands and traditional modes of land use. Within the course of time, these basic "features" have been supplemented by several minor features, including the important aspect of a "low external input sustainable agriculture" (LEISA). The three features listed figured prominently in the proposal for the "Equator Award" as the most pronounced distinctive elements of "Toro identity". According to the proposal, the rather successful preservation of the forests in Toro is directly linked to these "customary features". Whereas most NGO-activists and some bureaucrats and researchers share this view, the following discussion is aimed at unveiling these direct relationships as merely "myths".

Traditional Leadership and the Customary Council

Traditional leadership positions have been framed into two concepts, the *Tina Ngata* (literally: the "Mother of the Village") as the female carrier of "local environmental wisdom" (*kearifan lokal*) and the *Totua Ngata* as the custodian of customary law. Between the two "customary leaders", it is first of all the *Tina Ngata* who won regional and international attention and acclaim and is meanwhile perceived as a sort of "cultural core" of the customary law. Thus, the journal *Down to Earth* (No. 63, 2004)[9] reports empathically of the prominent role of the *Tina Ngata* in the sustainable resource use of Toro. The emergence of the *Tina Ngata* is intrinsically linked to the foundation of the "Customary Women Organization" *OPANT* (*Organisasi Perempuan Adat Ngata Toro*) in May 2001. Legitimating its existence with the prominent role that women played in village decision making in the past, *OPANT* was founded as a substitute for the state-organized groups of the "Family Welfare Program" (*Pendidikan Kesejahteraan Keluarga*, abbr. "PKK") which focused primarily on issues of household hygiene, child care, nutrition and birth control. The PKK was abolished because of its exclusively "domestic" character which was perceived as being antithetical to the local tradition in which women played important public roles as "keepers of the *adat*" (*Pabolia Ada*), decision makers (*Pangali Baha*) and carriers of environmental knowledge (*Tina Ngata*). Whereas the "ecological function" of the *Tina Ngata* lies in the determination of the right "time quality" for forest clearing according to astrological principles, the *Totua Ngata* knows about the ecological suitability of forest locations for agricultural conversion (distance from river streams, steepness and soil erosion, fertility etc). Both functions relate to the "rights of avail" of traditional authorities in the allocation and distribution of community controlled land. *Tina Ngata* and *Totua Ngata* function ex officio as leaders of *OPANT* and the Customary Council (*Lembaga Adat*) respectively.

Whereas the "official" Toro version argues that the *Lembaga Adat* was the central authority (*otoritas tunggal*) in regulating social life

[9] See: http://dte.gn.apc.org/news.htm

and natural resource use since times immemorial[10], this version is heavily contested by two different "folk versions". Following the first version, the *Lembaga Adat* was founded in 1939, when several religious leaders, aristocrats and elders met in order to restore the social norms, which had been deteriorating with the arrival of the Dutch. In the second version, the origins of the *Lembaga Adat* can be traced to the regular meetings that were held in the "communal meeting house" (*lobo*) during the Japanese occupation. It is important to note that in both versions the *Lembaga Adat* appears by no means as an invention of the ancestors, but is rather perceived as the outcome of reactions against outer circumstances or informal meetings. Further, the issue of local resource management plays no role at all in the foundation of the council. According to the "Rules for Resource Management" drafted in 2002 every planned land clearing must be reported to the *Lembaga Adat*, which gives its formal consent after the *Tina Ngata* and *Totua Ngata* have been consulted. The point to be made is that both "traditional functions" are linked to formal organizations and display rather formal than traditional characteristics. Thus, the *Tina Ngata* herself held the former position as head of the PKK. Compared to the *Lembaga Adat*, the role of the *Tina Ngata* is even more contested. In the common version in the cultural centre of the Kulawi valley, the town of Bolapapu, the term *Tina Ngata* relates to the cultural and social role of Bolapapu, which is also the administrative district centre, as the "Mother of the Villages".[11] This version is popular among the Islamic descendants of the Kulawi kings who deny the existence of *Tina Ngata* as a person. The *Tina Ngata* of Toro is merely perceived as an, albeit clever, invention of Toro in order to gain outside attention.

Some elders in Bolapapu on the other hand accept the personal aspect of the *Tina Ngata*, though with three limitations: First, the phenomenon *Tina Ngata* is confined to the core village of Bolapapu. Second, the *Tina Ngata* is by no means a continuous institution; rather the

---

[10] Lembaga Masyarakat Adat Toro: Kearifan Masyarakat Adat Ngata Toro dalam Pola Interaksi Pemilikan dan Pengelolaan Sumber Daya Alam (Customary Society of Toro: The Wisdom of the Customary Community of Toro in the Interaction of Property and Natural Resource Management) 2002. Unpublished Manuscript, for quotations see: www.beritabumi.org.id.

[11] Indonesian language does not differentiate between singular and plural; both interpretations are possible.

Tina Ngata emerges sporadically because she is in possession of a relevant expertise (Suryo Adiwibowo, personal communication). Last but not least, in our representative survey on *adat*-conceptions, many commoners in the village related the term *Tina Ngata* to the wife of the village head who normally carries the bulk of the PKK-work load and is thus jokingly called "Mother of the Village". This is not to say that the personal notion of *Tina Ngata* as an ancestral institution adopted by Toro leaders is as such "illegitimate". What is to be stressed, however, is that Toro leaders present their specific conception to the outside world with quite some success; at the same time their legitimacy to do so is still under debate inside the village.

The main topic within the regional "Toro discourse" however concerns the relationship of traditional authorities and formal organizations. Among most elders in adjacent villages as well as among Toro residents themselves there is consensus that in the past the *Totua Ngata* was not a formalized position. The same holds true for the individuals who gathered to discuss a certain problem. Far from being a fixed set of people, the individuals who gathered depended on the specific problem that had to be solved. Thus, the *Totua Ngata* was an informal position, which was rather determined by "individual" or "situational" expertise, than by formal rules of succession. Informants pointed out that in terms of gender *Totua Ngata* is a neutral term and that women have not been excluded in decision making in case they were in possession of the relevant expertise. In this sense, the original meaning of *Totua Ngata* relates rather to the way people were "organizing their affairs" than to organizational entities and gender issues. Only within the "revitalization process", have positions of *ascribed* status been formalized and converted into an *achieved* status with the role of *Tina Ngata* being transferred to the former leader of the PKK and the role of *Totua Ngata* being transferred to the head of the *Lembaga Adat*, both of them already in powerful administrative positions before the revitalization process itself started.

The discourse around issues of "organizing" versus "formal organization" however is not exhausted with the traditional role of *Totua Ngata* and their position in village life, but meanwhile includes the role and "appropriate" understanding of the Customary Women Organization *OPANT* itself. *OPANT* (formally established in 2001) has its origins in an informal movement of some younger women, searching for ways to secure a stronger participation of women in village decisions. Given her capacity and experience as the former leader of the PKK, the *Tina Ngata* was chosen to represent their concerns to the village administration and outside organizations. With the formation of *OPANT* however, the women saw that the leader increasingly captured the benefits of their struggles. Due to in-transparent management practices in the case of funds for a nursery school, several key position holders (secretary, cashier and the specialist for traditional arts) withdrew their involvement in *OPANT* with the effect that at the time when *OPANT* (represented by its leader) won international acclaim in form of the "Equator Award", *OPANT's* activities had already come to a standstill in Toro. A former OPANT-activist pointed out that

> *"Formerly, in the PKK there was not much money...but the channels were clear and everybody knew about the state subsidies... so and so much for the village, so and so much for the hamlets and so and so much for every neighbourhood group. ... So it was with the information.... One became ten, ten became one hundred and a hundred became thousand. ... Now, with Opant "she" makes proposals to all kinds of organizations in the name of Opant, but we never know anything... There are no clear channels for information and funds. When we asked about the funds for the nursery school, we received Rp. 50.000 each and that was all. Before we become "cheated"* ***(dikasih bodoh)*** *I think it is better to withdraw".*

Meanwhile, the women call for the re-establishment of PKK. The lesson to be learned from the *OPANT* case is that with the establishment of a formal organization, which mirrors bureaucratic structures, a structurally positioned leader could capture the organization's concerns for her private "empowerment" in presenting *OPANT*'s activities to the public and making its concerns intelligible to outsiders. Thus, a genuine movement of "organizing women's affairs" ended up with a formal organization characterized by a high degree of "participatory exclusion". Because of the emphasis on "empowerment" for its own sake, neither *YTM* (which facilitated the foundation of *OPANT*), nor *Care International* (which made the proposal for the "Equator Award") realized that state-organizations (PKK) are fulfilling valuable functions that will have to be performed by the "empowered" organizations if they are eliminated. Thus, when the Department of Health distributed medicines in order to kill mosquito eggs in water basins, Toro households lost out because the medicines were distributed via the PKK-branches in every village. One must be careful with the idea that the formation of "own" organizations is always in favour of local communities. As observed by Ahrne (1994: 36), "when people act on behalf of organizations their abilities and scopes are enhanced; they get extraindividual strength as well as goals and motives." Once a new organization is established, its leaders may well capitalize the name and principles of the organization for their own benefits.

In regard to "local knowledge" related to the appropriate planting time, which Toro portrays as an "exclusive knowledge" embodied in the *Tina Ngata*, our survey on "customary practices" revealed that – far from being exclusive – this aspect of local knowledge is commonly shared knowledge, almost evenly distributed among the community members. 69% (n = 40) of the women in the representative household sample determined the planting time by themselves, using astrology, moon phases, age of seedlings or traditional calendars. The others, who followed the example of their fellows, based their decisions on the instructions of the extension service (n = 4), the working plan of the reciprocal working groups (n = 5) or modelled their behaviour on the activities of elder women who are perceived as being knowledgeable in this question (n = 3).[12] The point to be made is that by monopolizing and appropriating commonly shared knowledge, certain individuals who are `knowledgeable` about what kind of knowledge "counts" in the eyes of significant outsiders may well enhance their power and influence within a regional discourse on empowerment and customary practices. It is within this discourse, where "knowledge and power intertwine and where each enhances the other" (Allen 2003: 70). Our discussion on the failure of *OPANT* revealed that as in most approaches to local organization in development, NGO's often resemble to state agencies in their lack of "understanding of how people strategically organize themselves", a lack which is

[12] Knowledge about the determination of planting times is neither exclusive to a certain person, nor exclusive to a certain community. A household survey of Storma-subproject A4 revealed that most farmers determine planting times by themselves in all 12 villages under study (Nunung Nuryantono, personal communication).

mainly due to "the persistence of notions of formal bureaucratic rationality" (Nuijten 1992: 189). Without a formal organization (*OPANT*) existing, Toro's struggles for women empowerment could not have been presented to the public easily. In the case of Toro however, the focus on "organization" as a *persistent entity ingrained with bureaucratic rationality* led not to the empowerment of a clear defined section of village society (women), but rather enhanced the position of parts of the village elite who put themselves on top of the organization and who are well equipped to capture the benefits of what were formerly uncoordinated meetings of women "organizing themselves". The person who fulfils the role of *Tina Ngata* is not a free-floating, profit-maximizing individual "engaged in endless creativity" (Ortner 1989), but a structurally positioned actor using certain aspects of locally held knowledge *strategically* in her encounters with outsiders, in making her concerns intelligible to the life-worlds of others. On the other hand, the women who feel deprived of their struggles are not passive recipients of a process shaped by forces outside their control. Under the given circumstances, withdrawing oneself from engagement is more than an effective "*weapon of the weak*" (Scott 1985). In successfully bringing *OPANT*-activities to a standstill and in calling for re-establishment of the PKK, the actions (or "practices") of Toro women – because they are related to the existing structure – may well have the capacity to transform that structure in the long run. Of course, this process will involve other negotiations in sorting out "what is worth doing and what sort of person to be" (McPherson 1983: 111).

The Spatial Mapping of Customary Territories and Traditional Land Use and the Feature of a "Low External Input Sustainable Agriculture" (LEISA)

As common in other communities in the research region, Toro villagers distinguish between several types of forest according to the age of the vegetation: (1) *wana ngiki* refers to the forested landscape deep inside the forest, usually on hills and out of reach for people's activity, (2) *wana* refers the border areas between used an unused forest areas and functions primarily as a reservoir for hunting activities, (3) *pangale* are the old secondary forests which have been left idle for more than 25 years and which have become almost "similar" to virgin forest, (4) *omah* are forested landscapes with a vegetation age ranging from 5 to maximum 25 years and which are actively used by the village population, predominantly for "agro-forestry" systems. In the so-called "Participatory Map of Land Use and Local Wisdom" (*Pemetaan Partisipatif dan Penggalian Kearifan Lokal*) clear categories and clearly distinguishable territories represent the several kinds of forest. In portraying the different kinds of vegetation as clearly differentiated areas inside the forest, the cartographic representation of Toro's customary forests does not only fulfil the Park Authority's demand for adequate environmental protection; it also mirrors the national zoning scheme as outlined in the Indonesian Conservation Act (UU 5/1990, Art.34, §3). It further allows for the construction of a more or less congruent overlap of resource boundaries with legal boundaries.

In the official Toro documents, this zoning scheme is represented as the manifestation of an indigenous conception called *katuwua* which orders man's interaction with the forest. The *katuwua*-concept is formulated in relation to a complementary term, *hintuwu*. The *hintuwu*-principle is portrayed as those elements of *adat* that order the social life of the community, whereas the *katuwua* relates to those elements that order the community's relationship with the natural environment. The original meaning of *katuwua* however seems much less exclusive and specified as the Toro version suggests. All over Central Sulawesi, including the city of Palu, the term is usually known in the meaning of "plants" or something that "sprouts up". Etymologically, the term consists of two parts: in the Kaili and Kulawi languages, *ka* means "something", whereas *tuwua* means "to grow".[13] Thus, *katuwua* represents a universally shared term in the region. Denoting the "life force" of plants, in its original meaning, *katuwua* exists irrespective of humans and is not at all linked to the community's interaction with its surrounding resources.

Following the official Toro version, it is an integral part of the local customary law that *omah*-forests may be used by private households

---

[13] Examples of *katuwua* given by our respondents are that one should not trample on harvested maize cobs or that one should not draw rattan along the wet rice fields when the rice begins to ripen in order to avoid that the soul of the rice gets lost.

on the *ulayat*-principle as long as the utilization of the secondary forests does not conflict with the restrictions imposed by the village leaders based on their "local wisdom". Use rights of *omah*-forests can be transferred from one generation to the next. The *pangale*-forests on the other hand are portrayed as "community lands" where no "right of avail" applies and which cannot be converted for agricultural or agroforest purposes. The common village property is called *huaka,* whereas private use rights are labelled as *dodoha.*[14] Thus, different types of forest relate to different property rights. Our survey on customary practices however revealed that the villagers do not perceive the term *huaka* as denoting a common village property. Instead, *huaka* is considered as a "family pool" which derives its ownership status from the rights of the first clearer. In the daily conversation *huaka* and *dodoha* appear as two sides of the same thing: whereas forested lands are often referred to as "this forest is *huaka* of family X", the term *dodoha* relates to individual property which has been formally inherited by an established inheritance procedure, for example: "this wet rice plot or this machete is my *dodoha*". In the folk version, *huaka* is less linked to common village property, but refers to the common family-pool, which has not yet been divided among the co-heirs. Interestingly, in explaining *huaka*, informants refer to both, *omah*-forests as well as *pangale*-forests. In this version (which is the dominant version throughout the research region), *omah* and *pangale*-forests are no independent legal categories, but are perceived as being embedded in the local tenure system. Regulations do not apply in terms of vegetation (forest cover which can be converted against forest cover that needs to be preserved), but in terms of ownership. Utilization of forest is restricted by the fact whether it belongs to a certain "family pool" or not, irrespective of the age of the vegetation. Interestingly, in the local language *Bahasa Kulawi*, the dominant way of land acquisition in the past, the opening of primary forest, is called *mo pangale*. In the *mo pangale*-conception the opening of virgin stands

[14] Lembaga Masyarakat Adat Toro: Kearifan Masyarakat Adat Ngata Toro dalam Pola Interaksi Pemilikan dan Pengelolaan Sumber Daya Alam (2002). (Customary Society of Toro: The Wisdom of the Customary Community of Toro in the Interaction of Property and Natural Resource Management) 2002. Unpublished Manuscript, for quotations see: www.beritabumi.org.id.

was succeeded by long fallow periods. The vegetation was cut at least two times before the plot was used for the cultivation of upland rice. More often than not, the clearing of forest (*mo pangale*) functioned as a mechanism to secure "reserve lands" for future generations. Following this line, the original term *pangale* seems more related to issues of expansion than to issues of preservation. The first clearer is called *po pangalea*, whereas the plot being cleared is called *po pangaleanya*. The argument made by some villagers is that the suffix "nya" (a possessive pronoun in Indonesian language) points out rather to the rights of a first clearer than to the "rights of avail" claimed by Toro leaders. If private forests are transformed into a community forest, rights are transferred from individual owners to the community with the question arising who in the community makes the decisions and who is actually empowered by this shift of ownership status? Faced with different situations, leaders and villagers draw upon totally different property discourses in legitimating their claims.[15]

It should be noted that – like in many other communities – Toro's dominant form of land use in the past was of the shifting cultivation type. The "territorialization" of customary law however, as "enacted" in the cartographic material, gives the impression as if the settlement and its surrounding resources have been temporally and spatially stable throughout generations. The customary territories, represented in form of fixed, simply identifiable zones of legal and biological diversity, appear as clearly distinguishable areas which make monitoring and rule enforcement an easy task (see also Zerner 1994: 1082). On the one hand, with the help of the *YTM*-sponsored "customary land use mapping" (informed by the national statutes of the Conservation Law) Toro could prove its capability to fulfil the Park Authority's demands for adequate protection. On the other hand, the villagers' interpretations of individual ownership rights that surround the *mo pangale* and *po pangalea* conceptions have become almost neutralized in the course of the process. Given the vivid property discourses existing in the village, the spatial mapping of traditional land use patterns appears less in terms of a realistic representation of a glorious past

[15] After the "Conservation Agreement" with the Park Authority was signed, people's understanding of the agreement was that they could now open their *huaka*-family pools in the forest, leading to dozens of households clearing land in the forest.

characterized by local wisdom, but rather in terms of a "politicized social and cultural construct" (Rodman 1992: 640) in order to meet the needs imposed by outside agencies.

The second aspect of the sustainable character of Toro's land use system, the feature of a "low external input sustainable agriculture" (LEISA), was already promoted by Toro in its facilitating efforts in neighbor villages before the proposal for the "Equator Award" was drafted.[16] This term relates to the rather low degree to which Toro villagers make use of chemical inputs, especially fertilizers. In our survey, 31% (n=18) of the villagers made sporadically use of such fertilizers, the use being confined to wet rice cultivation. No use of chemical fertilizers is documented in the cultivation of non-rice food crops, such as red beans, maize and cassava. It is tempting to attribute the very limited use of fertilizers to the feature of *indigenous environmental knowledge* (IEK). Informants however pointed out that – given the fact that Toro receives almost no government subsidies – the private investments of fertilizers are too high compared to the limited increase in yields, which can be expected. This finding is shared by plot experiments that were conducted by *Storma* researchers in neighboring Napu valley: N-based fertilisers can raise maize yields, but the increase is not significant. Given costs of fertiliser and labor as well as the low returns for maize make it just not worthwhile to use fertiliser as long as yields do not drop below a certain minimum (see Anthofer et al. 2003). Therefore it is rather "risk management" than IEK, which is responsible for the non-use of fertiliser in the dry land sector. In case input costs are hypothetically excluded, a different picture emerges: In our survey, 57 out of 60 respondents stated that they would use fertiliser if it was delivered by the government or other organizations. Thus, Toro farmers are by no means "nature-bound tribesmen", but "rational peasants" who act within the limited scope of action provided by the government and their own village administration.

The representation of *adat* through the "Green Lens" (Zerner 1994) is not surprising under the given circumstances of an intensive eco-

[16] Masyarakat Adat Ngata Toro: Diskusi Kampung Bolapapu (Customary Society of Toro: Discussions in Bolapapu Village) 2003: 6. (Unpublished manuscript, vailable from the author).

logical discourse promoted by the National Park Authority and environmental organizations. The struggle about who has the relevant knowledge to use the common-pool-resource forest margin in a sustainable manner is also a question about who has the legitimate right to control this resource. Within this process, *adat* is transformed into a powerful legal "counter-culture" (Zerner), which has to position itself as a viable alternative to the state's control over resources. Paradoxically, this "counter-culture" has to make heavy use of the state's national statutes itself – even at the cost of locally held perceptions (as exemplified in the *mo pangale*-principle) – if it strives for official acknowledgement. This, in turn, depends on 'knowledgeable' leaders who are aware of the discursive fields surrounding their communities, who are familiar with the requirements provided by the law and who are capable of using "the national law in order to prove the existence of customary law" (Campbell 2001: 114). This includes the knowledge about the "right" property discourses to draw upon in legitimating claims.

Customary Community, Socio-economic Security and the Stability of the Forest Margin

Within the Indonesian legislation, the official recognition as *customary law community* depends on the existence of a "right of allocation" (*ulayat*), a major feature of which is that patterns of land use are restricted by limitations imposed by the community. The notion of alienable rights as embodied in the concept of private property (*huk milik*) is not compatible with the features of a "customary law community". This means that there is no room for private land certificates within the framework of "customary law community". Once a farmer owns a certificate, his plot can no longer be disturbed by anybody. It is safe, and rules on land use (like those written down in the "Conservation Agreement"), can hardly be applied. Thus, Toro leaders try to avoid the issuing of certificates for two reasons: (1) In case a farmer owns a certificate, he can sell the plot at any time to anybody. The village internal "ban on land sales" (which is part of the customary law of community struggle) would thus not be effective. (2) During the process of certification, bordering plots must be identified. If the ownership status of neighboring plots is not clear and identification turns out as rather complicated, BPN officials are very fast in declaring neighboring plots as "state land" (*tanah negara*) which may lead to troubles between people obtaining the certificate and others who have been displaced from their lands via the declaration as "state land". Further, if a substantial amount of Toro forests would become state land or documented "private property" (*hak milik*), the construction of *huaka* as common village property and the struggle for recognition as "customary law community" would definitely fail. The notion of the *Lembaga Adat* as the sole authority having full jurisdiction over Toro's territories could not be upheld.

Interestingly, both parties involved, the villagers who feel themselves deprived of their private rights in their lands as well as the village administration that strives to maintain the "customary image", legitimize their claims with "intergenerational access". Farmers argue that they cannot intensify their plots in an effective manner. The argument of the village leaders on the other hand points out to the disastrous situation in several neighboring villages where the common practice to sell land to outsiders (migrants) has led to a severe shortage of land. Certificates, however, are not only an important means to enhance plot security; they are even more essential if villagers want to borrow money from the bank. Under the Indonesian credit scheme, it is primarily land certificates that function as collateral. Thus, Toro villagers are severely disadvantaged in obtaining credits compared to their fellows in neighboring villages.[17]

There is another rationale behind the ban on land sales. Security in terms of intergenerational access can only be sustained if land sales to outsiders are limited. But the role of the village head (*Kepala Ngata*) and the customary council (*Lembaga Adat*) in controlling the land market should not be underestimated as a source of power. Whereas other villages rely on neighbor's agreements of borders in land clearings and transactions and are thus recognizing individual ownership

---

[17] Research of Storma sub-project A 4 revealed that the number of Toro residents who obtained credits from the People's Bank (*Bank Rakyat Indonesia*) in the county capital of Bolapapu is significantly lower compared to the residents of other villages (Nunung Nuryantono, personal communication).

rights, in Toro all activities related to land must be managed via the *Kepala Ngata* and the *Lembaga Adat*, thus fulfilling the state's prerequisites of *ulayat*. It is not only the sanctioning capacity of the *Lembaga Adat* that is enhanced (as in other villages), but control over land. Only by such means can Toro leaders avoid an influx of land-hungry Buginese migrants, whose rather intensive kind of land use would not only convert the enriched secondary agro-forests of Toro into modern cacao plantations (in neighbor villages Buginese migrants managed land certificates quite quickly on their own) and undermine the "customary image", but in its effectiveness in terms of product value the Buginese system could present a dangerous counter-model to the "sustainable" model as envisaged by the leaders of Toro.

However, Toro must endow its people with some other form of "land proof" in order to obtain credit. This is the so-called "transfer letter" (*surat penyerahan*) which is usually issued in case of inheritance. It is seldom used for purchase because according to the law (in difference to the "land transaction document", the *acta jual-beli*) one cannot use it for renting out, mortgaging, sharecropping, leasing and exchange. In a certain sense land documented by this kind of letter cannot be used for all kinds of transactions peasants make in order to survive. At least it can be used for – albeit very small – credits from the bank. The option of rather small credits (compared to credits available with certificates) does not compensate for the loss of security if plots cannot be leased, rented out or exchanged. The policies of Toro is that in case somebody needs credit, the village administration issues a *surat penyerahan* in which it states that so and so much "customary land" (*Huaka Adat Ngata Toro*) has been transferred for usage to the receiver by the village authorities. Toro leaders issue these letters for two purposes: (1) In difference to certificates or transaction documents, the "transfer letter" must be approved and signed by the village head which once again enhances his control over the land market. (2) The official form for the "transfer letter" (issued by the regional county administration of Donggala) lists several types of village land including the term *milik adat* ("customary property"). Thus, irrespective of whether a plot has been purchased, inherited or was acquired by individual clearing (*mo pangale*), in the "transfer letter" all plots appear formally as *milik adat* (wet rice fields – usually perceived as *dodoha* – included).

There is a serious inconsistency in regard to "customary territories" between the provincial government and the county administration. Whereas a provincial regulation from 1993 (SK 529.2/8158) denies the existence of "customary lands" (*tanah adat*) in Central Sulawesi, the county of Donggala recognizes its existence inside its territory. There exist not only multiple legal/normative frameworks inside the village, the coexistence of multiple legal orders can also be found within the state apparatus itself. In such a situation, both, villagers and leaders can draw upon "state law" in legitimating their claims. The lesson is that the notion of "customary law community" and priorities of socio-economic security among the village population are not always compatible.

However, there remains an unsolved question. As pointed out above, the forest margin in Toro is indeed more stable in comparison to most other communities in the research area. There is a widespread assumption shared by NGO's and local bureaucrats that the relatively stable forest margin in Toro is directly linked to the process of "customary revitalization". An alternative explanation lies in a systematic comparison of the socio-economic characteristics of Toro with other villages in isolating the factors that count for deforestation in the research region in general. According to our findings, wet rice cultivation *can* reduce the pressure on the forest cover. Given the fact that in comparison to dry land agriculture, wet rice production is characterized by a higher absorption of work force and lower incomes, it leaves less time and surplus that could otherwise be invested in new land clearings. Thus, Martens et al (2004: 188) found that the "forest-saving" effect of yield increasing technologies in the lowlands of the forest margin is stronger if new technologies are also labor intensive. In our representative sample of 144 households in four villages, there exists a clearly established negative correlation between wet rice cultivation and land clearings. The area of wet rice worked by an individual household relates negative to the area of forest that has been cleared within the last five years ($r = -3.2$; significant on the 0.01 level). However, the number of owned adult cacao trees relates positively to forest clearing ($r = 0.41$, significant on the 0.01 level). Given a weak enforcement of rules and monitoring, a significant increase in cash income during the economic crisis (1997/1998) put cacao farmers in an advantageous position to extend their holdings as compared to wet rice farmers and mixed farmers. Table 1. presents a systematic comparison between Toro and two other villages in regard to these three "context variables":

	Watumaeta	Rompo	Toro
Average areal of wet rice fields (*sawah*) operated	**0.15**	**0.4**	**0.5**
Percentage of households working wet rice fields	**28.2**	**64.5**	**82.1**
Percentage of households opening land within the last 5 years	**80.4**	**38.4**	**20.5**
Average number of adult cacao trees per household	**522**	**108**	**136**

Table 1. *Characteristics of three selected villages in regard to factors related to forest clearing*

If we compare Watumaeta with Toro, there is a clear established dominance of wet rice agriculture over cacao production. The average number of adult cacao trees per household is only ¼ of those in Watumaeta. In terms of these socio-economic "context variables", Rompo is comparable to Toro. Here, the forest margin is also relatively stable, though with no *adat* revitalization. Thus, the comparatively stable forest margin in Toro seems rather to be linked to several circumstances in the wider socio-economic environment than to the revitalization of customary principles. These, instead, seem to have been constructed *around* the given situation in order to resist resettlement, to broaden the chances for official recognition as "customary law community" and to mobilize outside support.

Toro, its Neighbors and the State – The Regional Discourse on "Kulawi Identity"[18]

As Agrawal (2000) points out, the preliminary focus of social science has been on the internal dynamics of groups, but external dynamics are important as well. Councils engaged in resource management are often based on village boundaries, with each village making its own rules. In their effort to preserve natural resources, small villages will have greater difficulties in raising the necessary funds than larger villages. As it is well known from examples in India, groups of adequate size can often be more successful than small groups (Agrawal and Yadama 1997). First, because they find it easier to raise the necessary funds, second because they can also be more effective in dealing with higher authorities. The argument made is that small councils are disadvantaged in their efforts to generate sufficient human and financial resources to monitor and enforce local rules. This raises the question as to the formation of *village alliances* in natural resource management. The success of such an endeavor however depends to a large extend on the compatibility of rules and sanctions related to the forest. This demands that the villages concerned draw upon the same legal discourses in legitimatizing their rules. It remains an established fact that a village cannot be successful in preserving its environment alone. Therefore, with the help of YTM, Toro has started an intense process of "facilitating" in other villages since the end of 2001. The facilitating contents are mainly defined by "women empowerment" and "forest protection". This process however seems to be hampered by several intervening factors. In contrast to Toro, whose striving for self-determination is partly framed in opposition to the regional administration and which focuses on the sustainability of its traditional agricultural techniques (LEISA), its neighbor villages framed their engagements with the government largely within the state's discourse of development. Toro has dropped central institutions in the village that

[18] Part of the information concerning Bolapapu was uncovered by Momi Huwono in his Master Thesis "Forest Resource Management: Battlefield of Local Knowledge among Local Leaders. A Case Study of Local Leadership in the Kulawi Community of Bolapapu, Central Sulawesi; Bogor (MSc Thesis, Graduate School, Bogor Agricultural University (unpublished master thesis).

have been replaced by new organizations, as e.g., the PKK which was replaced by *OPANT*. In addition, the village operative assemblage LKMD (*Lembaga Kesejahteraan Masyarakat Desa*), which assists the village head in the implementation of government programs, was dropped totally. Central functions of the village parliament (*Badan Perwakilan Desa*) have been transferred to the *Lembaga Adat*.

Government institutions (provincial government, BPD, Development Agency) have less a problem with the re-invention of *adat* as such; the fear is that via the "facilitating process" the virus of "institutional heresy" will spread as well which could make regional policies messy and difficult. Under the given circumstances, the villages have been rather reluctant towards Toro's efforts. Further, during the Suharto-Era the village elders have been heavily involved in applying state regulations with the effect that traditional power and political power have been highly intertwined. Despite political decentralization and the new legislative framework, the ingrained perception that any strive for self-determination is automatically directed against the state still survives in the minds of local leaders. Accordingly, investigation, redefinition and revitalization of *adat* are perceived as being in opposition to the government. The fear on behalf of the villages is that if they drop central village institutions like Toro did, their fate can become like the one of the enclave village of Katu, which was punished for "overacting" by a total withdrawal of state subsidies. This fear is especially pronounced in the field of "women empowerment". The paradox is that if women would be empowered by a new group (like *OPANT*), villagers at the same time could find themselves less empowered to control their own agrarian development if agricultural subsidies are cancelled. Thus, the village head of Sungku withdrew the idea of founding an *OPANT* because he feared trouble with state agencies. In the village of Bolapapu on the other hand, women stated that *OPANT* in Toro serves only the needs of certain people and that they prefer to create an "*adat* branch" under the roof of PKK (Lakmi Savitri, personal communication). The village head of Matawe explained that he rejects the involvement of NGO's in village affairs. The women on the other conceded their disappointment with the PKK and explained their plan to create a women's organization, which will exist beside the PKK. Thus, Toro did not convince other villages to follow its example, but rather, every village searched for its own organizational solution, which is quite different from Toro. In the end, the failure is also a matter of strategy. Because Toro made its arrangements with YTM and the Park Authority without sufficient involvement of neighboring villages, but entered other villages with "ready for use" concepts afterwards, it followed a typical "top-down-approach". Thus other villages could not identify a "win-win"-situation if they co-operate closely with Toro.

As the relationship of Toro with its closest neighbor, Sungku shows, the foundation of a "customary law community" and the grant of conservation agreement are not always in favour of neighboring villages. Whereas Sungku people living in Toro can still cultivate their fallow lands in Sungku, Toro people living in Sungku cannot do so in Toro. Toro established a more or less congruent overlap between resource boundaries and legal boundaries in declaring *omah* as open for conversion under certain conditions (*ulayat*-criterion), but forbidding the conversion of *pangale*. Neither in Bolapapu, nor in Sungku or Matawe such a concept exists. Fallow plots are *huaka* (family property), irrespective of the age of the vegetation. The Toro concept of legal boundaries and principles of access are not compatible with other villages, which follow an antithetical legal discourse. As mentioned above, Sungku, Bolapapu and Matawe reject the existence of "customary village lands" in their community. Therefore, the establishment of village alliances proved not practicable in this part of Kulawi. That may be the reason they only marginally addressed the forest issue in the facilitating process.

In the eyes of Bolapapu leaders, Toro exports a "heterodox" version of *adat*, whereas the "orthodox" version in its purest form is only to be found in the village of Bolapapu. Toro is accused as a "tradition dealer" (*perdagang adat*) selling a hybrid version of customary principles to outsiders, thereby destroying *adat.* Within this conflict, Bolapapu and Toro draw upon two contested Kulawi "histories". Bolapapu leaders argue that Toro is actually a new village, founded in 1926 by immigrants from the neighboring villages of Bolapapu, Sungku and Lonca (a fact which is proved by our household data but which is not mentioned in the materials published by Toro) and that the "traditional court" (*pengadilan adat*) was located in Bolapapu, the centre of the Kulawi kings. The supremacy of Bolapapu in matters of "customary law" is still recognized by other neighboring villages that regularly consult Bolapapu in case they themselves cannot solve a conflict within or between villages. In 2004, descendants of the Islamic *Tomampe* branch of Kulawi rulers founded an organization called "The Tina Ngata Group of Kulawi" (*Kelompok Tina Ngata Ngulawi*) in order to restore the "real" meaning of the *Tina Ngata*-concept as "The Mother of the Villages". Supported by the regional government, a descendant of the Christian *Djiloy* branch founded the "Kulawi-wide Customary Council" (*Majelis Adat Kulawi*) as an organizational forum to counter-balance Toro's predominance in *adat* representation. A definite weakness of Bolapapu's counter-action is that it is not co-ordinated by a single movement or village organization, but is highly dispersed along the familial and religious lines of the local aristocracy.

However, Toro argues that Bolapapu's claims refer to the <u>direct</u> past and points out that the rise of Bolapapu as the administrative centre of Kulawi and the instalment of the "Kulawi kings" is actually a product of Dutch "indirect rule" and that in the <u>distant</u> past, before the arrival of the Dutch, the relationship between the villages in the Kulawi valley was not at all a hierarchical one (Suryo Adiwibowo, personal information). Further, the original resource management in Kulawi was quite well-organized by *hak ulayat*, before it deteriorated with the introduction of private property by the Dutch in their effort to facilitate the collection of taxes! These representations of local history however should not be judged in terms of "ethnographic accuracy". But within this "battlefield of knowledge between local leaders" (Huwono 2005), social actors do not only draw upon a hybrid history, but a history which is heavily contested.

Interestingly, the issue of natural resource management is not addressed in either of the Bolapapu organizations. Lacking the necessary knowledge and conceptions, the response of Bolapapu's established leaders to the "new discourse" probably came too late. The most pronounced village-internal opposition and advocates for Toro's movement are the educated members of the younger part of Bolapapu's population. Whereas the established *adat* leaders regard the interpretation of *adat* as an exclusive "affair of the elders" and argue that according to local custom younger people have no significant say in village affairs, the "youth group" calls for a new "social contract" (*kontrak sosial*), including the preservation of Bolapapu's resources for future generations (Huwono 2005). There exists a central conflict about the meaning of *adat* in current times: should it be preserved as locally established practice or should it be transformed in order to match the needs of current times (e.g. stronger participation, nature conservation). A subsequent question then arises: what kind of persons and what kind of community do we want to be? It is actually "Kulawi identity" which is at stake. Whereas the former head of the village made a strategic alliance with Bolapapu's elders in accusing the "youth movement" of being a disguised Toro-related NGO which wants to exploit *adat* for individual interests, the election of a new village head in March 2004 marks probably a caesura in Kulawi history. Being young and the first leader of non-aristocratic descent, the head shows an empathetic attitude towards the "youth movement" and makes no secret of his sympathy for Toro's policies – a view which is shared by the new head of the village parliament (BPD) in the neighboring village of Matawe. Whereas the existence of *tanah adat* in Kulawi is denied by the elders (a point of view which is in line with the provincial government), the new leaders in Bolapapu and Matawe made the preservation of "ancestral lands" for future generations become a cornerstone in their rhetoric. Thus, the "system" has become more open and which direction the future mode of Kulawi social organization and resource management will take and which discourse will prove more robust and persistent in time, should largely depend on the capability of the various actors "to enroll others in their projects" and "winning them over to their points of view" (N. Long 1992: 27). "Power" in the Kulawi valley cannot be grasped by Max Weber's conceptions of "charismatic" and "bureaucratic" forms of leadership (Weber 1922:).

> *"Power is not to be taken to be a phenomenon of one individual's consolidated and homogenous domination over others... power must be analysed as something which circulates... power is employed and exercised through a net-like organization (Foucault, in Gordon 1980: 98).*

Like knowledge, "power" is embedded in social practices. In this sense, its degree should largely depend on the capability to draw upon the appropriate discourses in giving meaning to and organizing social institutions and processes. In the long run, however, the outcomes of these accommodations and negotiations may well be presented in the form of cultural statements or in terms of *adat* (see A. Long 1992: 165).

Conclusion

The foregoing pages aimed at revealing the importance and interrelatedness of five major theoretical conceptions, *common-pool-resources, practice, social actor, local knowledge* and *legal pluralism* in regard to local resource management with special reference to Central Sulawesi. The notion of *discourse* served as an "umbrella concept" which binds the five dimensions together. In regard to the management of the common pool resource *forest margin* it was shown that social actors construct "local knowledge" within regionally dominant discourses by drawing on different legal/normative frameworks. "Individual calculus" is not a suitable concept to explain conditions of the commons. In the case of the Kulawi valley, it turned out as more informative to explain the individual calculus from the viewpoint of the culturally, socially and legally embedded commons (see Peters 1987). The construction of local knowledge related to resource use is embedded in culturally informed practices that draw upon local traditions. At the same time it is transforming them. The leaders of Toro, the former members of *OPANT*, the elders of Bolapapu and the members of the "youth group" are social actors who draw upon the same cultural tradition. Their "agency" – being closely related to the existent social structure – may well have the capacity to transform that structure in the long run (as happened with the election of a village head of non-aristocratic descent). Toro's success is not based on a local tradition, which is stronger or more authentic than the one of Bolapapu. By drawing upon the "right" property and environmental discourses and by constructing a "tribal situation" rather than a "village situation", Toro leaders were able to gain the necessary attention from significant outsiders. As the discussion showed, Toro can hardly function as a "model" to be copied one to one by other villages. Whatever direction the Kulawi property discourse will take, the robustness of any kind of resource use and property scheme adopted will depend on its capability to provide a minimum of "socio-economic security" for the peasant population.

References

Agrawal, A. and Yamada, N. (1997): How do Local Institutions Mediate Market and Population Pressures on Resources? Forest *panchayats* in Kumaon, India. *Development and Change*, Vol.28: 435-65.

Agrawal, A. (2000): Small is Beautiful? But is Large Better? Forest-Management Institutions in the Kumaon Himalaya, India. In: Gibson, C., McKean, M. and E. Ostrom (Eds): People and Forests. Communities, Institutions, and Governance: 57-87. London

Ahrne, G. (1994): Social Organizations. Interaction Inside, Outside and Between Organizations. London

Allen, J. (2003): Our Town: Foucault and Knowledge-based Policies in London. London.

Anthofer, J. et al. (2004): Adaptability Analysis and Risk Assesment of N fertilizer Application to Maize in the Napu Valley of Central Sulawesi, Indonesia. In: Fremerey, M., Gerold, G. and E. Guharja (Eds): Land Use, Nature Conservation and the Stability of Rainforest Margins in Southeast Asia: 409-26. Berlin.

Berkes, F. (2002): Cross-scale Institutional Linkages: Perspectives from Bottom Up. In: Ostrom, E. et al. (Eds): The Drama of the Commons: 293-317. Washington.

Bourdieu, P. (1977): Outline of a Theory of Practice. Cambridge.

Burns, P. (1989): “The Myth of Adat”. *Journal of Legal Pluralism* 28: 1-128.

Campbell, J. (2002): Differing Perspectives on Community Forestry in Indonesia. In: Colfer, J. P. and P. Resosudarmo (Eds): Which Way Forward? People, Forests, and Policymaking in Indonesia: 110-126. Singapore.

Evers, P. J. (1995): Preliminary Policy and Legal Questions about Recognizing Traditional Land Rights in Indonesia. *Ekonesia. A Journal of Indonesian Human Ecology*. No. 3: 1-23.

Fay, C. and M. Sirait (2002): Reforming the Reformists: Challenges to Government Forestry Reform in Post-Suharto Indonesia. In: Colfer, J. P. and P. Resosudarmo (Eds): Which Way Forward? People, Forests, and Policymaking in Indonesia. Singapore.

Foucault, M. (1971): The Order of Discourse. London.

Foucault, M. (1972): The Archeology of Knowledge. London.

Foucault, M. (1991): Gouvernmentality. In: Burchell, G., Gordon, G. and P. Miller (Eds): The Foucault Effect: Studies in Gouvernmentality. Chicago.

Foucault (1980): Power/Knowledge: Selected Interviews and Other Writings, 1972-1977. London.

Fremerey, M. (2005): Local Communities as Learning Organizations: The Case of the Village of

Toro, Sulawesi. In: A. Neef (Ed): Participatory Approaches for Sustainable Land Use in Southeast Asia: 253-2 Bangkok.

Gibson, C., Ostrom, E. and M. McKean (2000): Forests, People, and Governance: Some Initial Lessons. In: Gibson, C., McKean, M. and E. Ostrom (Eds): People and Forests. Communities, Institutions, and Governance: 227-242. London.

Hefner, R. (1990): The Political Economy of Mountain Java. Berkeley.

Huwono, M. (2005): Forest Resource Management: Battlefield of Local Knowledges among Local Leaders. A Case Study of Local Leadership in Kulawi Community of Bolapapu, Central Sulawesi. Unpublished MSc Thesis, Graduate School, Bogor.

Li, M. T. (1995): Images of Community. Discourse and Strategy in Property Relations. *Development and Change* Vol. 27: 501-527.

Li, M. T. (2000): Locating Indigenous Environmental Knowledge in Indonesia. In: Ellen, R. et al (Eds): Indigenous Environmental Knowledge and its Transformations. Critical Anthropological Perspectives: 121-147. Amsterdam.

Li, M. T. (2001): Masyarakat Adat, Difference, and the Limits of Recognition in Indonesia's Forest Zone. *Modern Asian Studies* 35: 645-676.

Long, A. (1992): Goods, Knowledge and Neer. In: Long, N. and Long, A. (Ed): Battlefields of Knowledge. The Interlocking of Theory and Practice in Social Research and Development: 147-170. London and New York.

Long, N. (1992): From Paradigm Lost to Paradigm Regained. In: Long, N. and Long, A. (Eds): Battlefields of Knowledge. The Interlocking of Theory and Practice in Social Research and Development: 16-46. London and New York.

Malonowski, B. (1926). Crime and Punishment in Savage Society. London.

Martens, M., Zeller, M. and R. Birner (2004): Does Technical Progress in Agriculture Have a Forest Saving Effect? Evidence from Central Sulawesi. In: Gerold, G., Fremerey, M. and E. Guhardja (Eds): Land Use, Nature Conservation and the Stability of Rain Forest Margins in SE-Asia. Berlin

McCay, B. (2002): Emergence of Institutions for the Commons: Contexts, Situations and Events. In:

Ostrom, E. et al. (Eds): The Drama of the Commons: 361-402. Washington.

McPherson, M. S. (1983): Want Formation, Morality, and some "Interpretive" Aspects of Economic Inquiry. In: Haan, N. et al (Eds): Social Science as Moral Inquiry. Columbia.

Meinzen-Dick, R. S. and R. Pradhan (2001): Implications of Legal Pluralism for Natural Resource Management. *IDS Bulletin* Vol. 32 No. 4: 10-17.

Mosse, D. (2001): "People's Knowledge", Participation and Patronage: Operations and Representations in Rural Development. In: Cooke, B. and U. Kothari (Eds): Participation: The New Tyranny? 16-35. London.

Nuijten, M. (1992): Local Organization as Organizing Practices. In: Long, N. and A. Long, (Eds): Battlefields of Knowledge. The Interlocking of Theory and Practice in Social Research and Development: 189-210. London and New York.

Ortner, S. (1984): Theory in Anthropology Since the Sixties. *Comparative Studies in Society and History* 26: 126-166.

Ortner, S. (1989): High Religion: A Cultural and Political History of Sherpa Buddhism. New Jersey.

Peters, P. E. (1987): Embedded Systems and Rooted Models: The Grazing Lands of Botswana and the Commons Debate. In: McCay, R. and J. Acheson (Eds): The Question of the Commons: 171-194. Tucson.

Posey, D. A. (2000): Ethnobiology and Ethnoecology in the Context of National Laws and International Agreements Affecting Indigenous and Local Knowledge, Traditional Resources and Intellectual Property Rights. In: Ellen, R. et al (Eds): Indigenous Environmental Knowledge and its Transformations. Critical Anthropological Perspectives: 35-54. Amsterdam.

Rodman, M.C. (1992): Empowering Place: Multi-locality and Multivocality. *American Anthropologist* 94, 1992: 640-52.

Sangaji, A. (2001): Penghancuran Masyarakat Adat dalam Pengelolaan Sumber Daya Alam di Sulawesi Tengah (The Destruction of Customary Communities in Resource Management in Central Sulawesi). Palu.

Shackelton, S. et al. (2002): Devolution and Community-based Natural Resource Management: Creating Space for Local People to Participate and Benefit? *Natural Resource Perspectives* No. 76: 1-8.

Sunito, S. (2005): Continuity and Discontinuity of Local Institutions in Community-based Natural Resource Management. PH thesis, Kassel and Bogor.

Tsing, A. (1993): In the Realm of the Diamond Queen. Marginality in an Out-of-the-Way Place. Princeton

Tsing, A. (1999): Becoming a Tribal Elder and Other Green Development Fantasics. In: Murray Li, T. (Ed.): Transforming the Indonesian Uplands. Halifax.

Weber, M. (1922): Wirtschaft und Gesellschaft. Berlin.

Weedon, C. (1987): Postfeminisms: Feminism, Cultural Theory and Cultural Forms. London.

Zerner, C. (1994): Through a Green Lens: The Construction of Customary Environmental Law and Community in Indonesia's Malukku Islands. *Law and Society Review* Vol. 28: No. 5: 1079-1122.

Chapter V

From "Wild West" to "Learning Organization": Processes of Institutional Change in Watumaeta, Central Sulawesi, Indonesia

Günter Burkard

"People know what they do; they frequently know why they do what they do; but what they don't know is what what they do does" (Michel Foucault, quoted in Dreyfus and Rabinow 1982: 187).

Abstract

This paper documents the processes of organizational change in the village of Watumaeta, located in the Napu valley, 103 km to the South of Central Sulawesi's provincial capital, Palu. Having a long established image of being a "worst case" in regard to leadership practices, social disintegration and deforestation in the area, within the last years Watumaeta has made substantial improvements in regard to community organization and the management of natural resources. The Watumaeta case shows that – far from being constant facilitating factors – one and the same "design principles" may have contradictory effects under different circumstances. The changes observed cannot be explained by conservationist motivations or equity concerns, but by the threat posed by socio-economic insecurity and mutual vulnerability. In contrast to Neo-institutionalist approaches based on "methodological individualism", this case study focuses on the cultural, socio-economic and institutional "embeddedness" of the common pool resource forest margin.

The Question of the Commons: State of the Art and Major Challenges[1]

Much of the theoretical reasoning on the commons is governed by a well-known dilemma: are resource users primarily motivated by self-interest or are they also motivated by a concern for resource maintenance and co-operation as a whole? The so-called "rational actor model" which assumes strict self-interest as the primary motivator of action, has been the dominant conception among the early resource economists and informs much of Garret Hardin's thesis of the "tragedy of the commons" (Hardin 1968). In short, Hardin's viewpoint is that the individual actor is forced to make his decisions within the framework of a system that "compels him to increase his herd without limit – in a world that is limited" (ibid: 1244). Inherent in Hardin's conception is the belief that those who restrict themselves in the use of a commons resource will inevitably lose out in comparison to those who are not willing to do so. There are two far-reaching implications in Hardin's analysis: the first is that only *coercive mean*s of control may be able to sustain common resources. The second implication is linked to the importance of *unitary ownership*, suggesting that only the state is capable of designing effective rules to prevent the impending "tragedy" (see the respective critics of Dietz et al 2003 and Gibson 2001).

However, an increasing number of field stúdies challenged Hardin's assumptions in showing that users in different parts of the world have developed a long-term interest in their resources. Confronted with often-complex institutions and rules of access in the field, several authors emphasized that Hardin had obviously confused *common property regimes* with *open access* situations (e.g. Gibson 2000). Fur-

[1] There exists some confusion in the literature in regard to the term "commons". The term "common property" relates to a resource characterized by joint ownership. A "common pool resource" on the other hand is a resource from which it is difficult to exclude potential users. Thus, "common property" relates to a certain management arrangement, whereas the term "common pool resource" is linked to the characteristics of a resource. Further confusion is added by the fact that both conceptions are usually abbreviated as "CPR". In this article, the term "commons" and CPR are used in the sense of "common pool resource". The term "common property" is only used, when its joint ownership is addressed.

ther, the effects of unitary ownership in form of government property were often disastrous. Due to a lack of enforcement and monitoring, former common property resources degraded often to de facto open access resources.

Until the present, the most influential theoretical orientation in the investigation of the commons is the Neo-institutionalist school of economics. Highly devoted to the assumptions of the rational action model, neo-institutionalist scientists are concerned with "incentive structures" in order to change the individual costs and benefits so that more co-operative action becomes possible (Bromley 1992, North 1990).

In contrast to these broader theoretical approaches, case studies of sustainable CPR-management focused primarily on the "design principles" of commons institutions. Albeit these studies were very useful in identifying facilitating conditions in regard to resource system characteristics, user group aspects and institutional arrangements, they tended to ignore the contextual factors, which facilitate a sustainable performance of commons institutions over time (Agrawal 2002: 45). Case studies often put more attention to the characteristics of functioning institutions than to the wider contextual situations within which they function (laws, market incentives, state policies and other "context variables" can only be ignored when they remain constant). Whereas the internal validity of these case studies is usually very high, their external validity was often limited. As Strathern (87: 5) points out, our concepts must be "dually constructed" which means that they must advance our comparative insight at the same time they should help to enhance our sensitivity to local realities. In his analysis of three prominent case studies, Agrawal (2002: 64) lists at least 36 conditions that "count" in successful CPR-management. Following Agrawal, there are several shortcomings of the design-principle approach. First, there is no agreed upon method of analysis and categorization of design principles. Second, several variables which are presented as facilitating group characteristics in different case studies such as "social capital", "group identity", "itinerancy" and "reciprocity" represent often not only "indicators" of a single underlying construct (e.g. "social cohesion"), but may also produce similar outcomes along the same causal chains. Therefore, some researchers have argued that analysis of CPR-management must shift from "correlation" to "causation" (e.g. Stern et al. 2003: 449). This is important because the effects of a certain variable may well depend on the condition of another variable or even on inter-actional effects between variables (see Agrawal 2002: 65).

Third, facilitating factors are often portrayed as if they would be relevant for *all* commons and institutions. Thus, despite the large number of case studies in the field, CPR-research has in most cases remained ahistorical. Fourth, in establishing facilitating principles and conditions, authors sometimes tend to adopt a simplistic view of reality. For example, Olson (1965) and Baland and Plateau (1999) believe that successful collective performance is better achieved if groups are small. Such a view however is of little value in a situation where the preservation of a large forest reserve needs an adequate group size in order to raise the necessary funds for monitoring and enforcement (Agrawal 2000). The same holds true for the "simple rules"-design principle adopted by Ostrom (1986: 611). Whereas simple rules may work out better than complex rules under conditions of large and heterogonous communities, they may not do so in case of small forest using groups who are familiar with their environment and who are in possession of the relevant knowledge to understand complex rules of use and access (see Stern et al. 2003: 458). Fifth, the value of positive or negative correlations of sustainable resource management with several "key variables" (e.g. poverty, demography, commercialization, heterogeneity) remains limited if such correlations do not consider the wider social, economic and cultural environment. Thus, does "poverty" lead to a higher dependence on CPR's and resource degradation (Jodha 1986, cited in Agrawal 2002) or does increasing wealth foster higher investments and uncontrolled use of the commons? Whereas some authors stress that commercialization destroys the social fabric of local communities, others point out that only commercialization can generate the necessary surplus for investment in resource regeneration (Dolšak and Ostrom 2003: 19).

However, the insight is gaining ground that such conceptions tend to link environmental degradation in a rather simplistic manner with mono-causal variables. This finding is supported by an increasing number of macro- as well as micro-level studies. Thus, in their overall discussion of the main causes of deforestation in developing countries, Allen and Barnes (1985) could find no linear relationship between resource degradation and population pressure. Similarly, Varughese (2000) sees no clear connection between demographic changes and deforestation in his comparison of 18 hill communities in Nepal. Contradicting examples where "heterogeneity" induces or impedes cooperation are also to be found in the literature (Olson 1965, Dietz et al. 2002). In regard to the complexities of heterogeneity, Baland and Plateau (1999) suggested a threefold distinction between heterogeneity of *endowments*, heterogeneity of *identity* and heterogeneity of *interests*. However, as Agrawal (2002: 60) reminds us, in real life situations these categories are not at all mutually exclusive. More often than not, natural resources are managed by communities that are divided along multiple lines of difference (ethnicity, class, gender, descent), which may have quite contradictory effects.

Whereas design principle studies tend to analyze static institutions on a single layer, recent approaches have stressed the need to look into nested sets of institutions in an attempt to find out how institutions are vertically and horizontally inter-linked (e.g. Berkes 2002). As Ostrom 1990) points out, it is very important that local communities obtain a right to devise their own rules and sanctions. Of the same importance however is the formal recognition of locally developed rules by government institutions because it puts local leaders into a better position to enforce them. Horizontal integration between local communities is crucial as well in order to avoid that rule breakers shift their activities to neighboring settlements. Berkes (ibid: 316) has emphasized that if natural resource management becomes too decentralized, the necessary feedback between user groups of adjacent areas may be lost. Similarly, McCarthy (1999, cited in Agrawal 2002: 64) suggests that in the case of larger resources, authority relations are better organized in a nested fashion. Just as analysis of context variables must shift from correlation to causation, analysis of institutions must shift from static institutional features to institutional change.

In contrast to the "methodological individualism" of the Neo-institutionalist paradigm, McCay (2002) has argued for the adoption of a "situated choice"-view, which sees preferences as being framed by larger commitments and socio-cultural phenomena. In her view, choices are always embedded into situations and contexts, which structure the preferences of actors. The Neo-institutionalist tendency to reduce specific local situations to "free-floating" individuals fits well with another trend in social and economic science: the use of experimental methods in CPR-research. Experiments however cannot simulate an important design principle: the relation between the actors involved! A social network with its overlapping and multiplex relations (including aspects of "social capital") cannot be simulated in experimental situations. However, if we assume different "situated rational choices" under different circumstances, it follows that decisions cannot be understood as a part of predictable processes (McCay 2002: 363). This makes it difficult to define cause and effect in local CPR-management. In general, research on CPR-management assumes implicitly that people believe that their actions have real effects on common resources. This however is far from being granted. Smith (1990, cited in McCay 2002) has emphasized that New England fishermen do not believe that their actions have direct effects on resources. Neither can they function as *causes of*, nor as *solutions to* problems of the commons. In their world view, the *natural processes* will always produce cycles of abundance and decrease of fish. Thus, there was little interest in the development of institutional devices and design principles in order to change the condition of the commons. Similarly, the genesis of institutions for the commons is often less informed by an awareness of resource degradation but by conflicting claims among resource users or by an attempt to reassume control over natural resources. There is no doubt that group characteristics such as "reciprocity", "trust" and "shared identity" all work in the direction of co-operative solutions. As pointed out by McCay (ibid: 384) however, there is a fatal lack of knowledge about what creates, sustains and reproduces these conditions. Further, a one-sided concern with "reciprocity" or related social constructs obscures the fact that it is often less a lack of community identity as such, but a struggle over the legitimate rights and claims which defines local situations.

Despite the meaningful progresses that have been made in our understanding of CPR-management within the last two decades, there remain major challenges for research of the commons. (1) There is a serious lack of information about the question how organizations and institutions for the commons learn and adapt. McCay (2002: 375) has argued for a "muddling-through"-approach, characterized by small steps with low initial costs in order to give communities the necessary room for maneuver and experiment. According to her, such steps should not be guided by larger values and goals in the first place. (2) Among other important questions, Dietz et al (2002: 14) point out to the need to look into initial situations and how they affect the emergence, performance and sustainability of institutions. Further they highlight the need "to understand how spatial and temporal heterogeneity in resource endowment creates opportunities for some to benefit at the expense of others". (3) Stern et al (2003: 457) point out that "success" of the commons may be judged differently by researchers and resource users. Whereas the former may judge success primarily in terms of sustainability of resources, the latter may judge commons institutions by e.g. their capacity to provide income, to maintain community relations or to elevate poverty. (4) As will be pointed out below, a major challenge is not only to reconcile different interests related to the use of natural resources, but how to deal with heterogeneity in terms of different priorities and conditions of security. (5) An important question is how an increasing integration into markets may affect the management of CPR-resources. This is of special relevance in situations where new market actors (e.g. migrants) obtain access to a common pool resource and create a situation of "access pluralism". (6) Last but not least, there is the concern of many researchers in regard to the establishment of sustainable linkages among institutions (Berkes 2002, Dietz et al 2002, McCay 2002, Stern et al 2003). Of course, the following case study cannot give answers to all the questions raised and it may suffer from the same shortcomings in regard to external validity as any other case study. However, it is hoped that the "Watumaeta Case" may shed some light on the importance of the social, cultural and legal political embeddedness of commons institutions in showing how certain design principles may have contradictory effects over time, depending on the wider social and institutional environment in which they operate.

Watumaeta: A Forest Margin Community in Transition

The village of Watumaeta (ca. 400 households) is located in the Napu valley, about 100 km south of the provincial capital of Palu. After the island of Sulawesi experienced an unprecedented "cacao boom" in the late nineties, the fertile forest border zones of the Napu valley became one of the major "cacao frontiers" in the region. Since then, competition for agricultural land among local and migrant small holders has resulted in a large-scale conversion of secondary forests, often involving encroachment into the protected zones of the National Park (Burkard 2002b). In regard to our understanding of the interrelationship between issues of social organization on the one hand, and processes of ecological stabilization and destabilization on the other, Watumaeta is of primary concern. Not only are displacement, competition over resources and forest encroachment most pronounced in this village; the availability of more "in-depth" information allows us also to look in more detail into the micro-processes, relations, power constellations and social practices through which the observed differentiation has been created. While it is by no means suggested that Watumaeta is representative of all communities in Central Sulawesi, the problems of this village reflect to some extend those that may beset other communities in the future. Reflecting more or less the situation of the "tragedy of the commons" described by Hardin (1968) when our research started in March 2001, the development Watumaeta has undergone within the last six years shows that resource degradation is not inevitable, but depends on the responsiveness of the population to outer incentives of institution building and rule creation. Whereas this is linked to the overall "design principles" prevailing within the community, such as legitimacy of leadership, ethnic heterogeneity and the willingness of the actors to recognize the coexistence of different security interests in the community, the Watumaeta case highlights that the emergence of such "design principles" in turn depend on the wider socio-cultural environment in which regional socio-political events play a decisive role.

The foundation of Watumaeta dates back to 1930 when several families from the Napu communities of Wuasa and Sedoa as well as from the Kulawi valley on the western side of the National Park moved to the Watumaeta plain. The inhabitants of Wuasa, Sedoa and Kulawi speak three different languages and are regionally perceived as different ethnic groups or categories (*suku*). There are no data about the number of spontaneous settlers and migrants resettled by the Dutch, but there is no doubt that Watumaeta was an ethnically mixed and spatially dispersed community from the beginning. Thus, farmers asserted private rights of the first clearer to the new land they opened, but did not claim a collective right to ancestral territories (*hak ulayat*) as it is often the case in other parts of outer Indonesia (see Evers 1995). Consequently, traditional rules on forest conversion were nearly absent, and no well-defined mechanisms to regulate resource use among community members could be found. Within the cyclical agricultural system (*shifting cultivation*) forested areas were embedded in the local tenure system and did not form a separate legal category. Preferential rights of a first clearer lapsed in case a plot/secondary forest was left idle for a certain time which was defined by the emergence of certain indicator trees (*mayopo*).

From the perspective of the settlement, the surrounding lands were cross-cut by spheres of control of different "swidden groups" (*robo*). The economic core unit however was not the "swidden group", but the individual household. Families worked their plots on their own responsibility and transferred ownership and use rights to their offspring. The high degree of independence individual families enjoyed is also reflected in the fact that they could join and leave the *robo* freely (Sunito 2004). Families lived from and consumed the products of their own fields, but they lived not from the products of a common resource regulated by deliberate rules of access to products. In spite of the fact that ownership rights were not very elaborate under the given circumstances, the dominant property type was one of "individual ownership" rather than "common ownership". These individual ownership rights emerged from an "open access-system" rather than a "common property regime". Individual plots were managed by strict reciprocal co-operation between households, but not by some sort of communal labor. Given the prevalence of a principle of free access to the forest, no institutional demand for the creation of "grass-roots organizations" in resource management existed. Thus, the role of the "customary community council" (*Lembaga Adat*) in resource management was not a paramount one under traditional conditions. Unlike secondary forests, primary forests were usually opened by the spiritually- and physically-robust community members (*tadulako*) who granted their plots to others on a long-term borrowing basis. The arrangement was often encompassing several generations. The distinction between *borrowing* and *owning* a plot was as blurred as was the notion of *inheritance*. In an open access regime, inheritance is not crucial. Thus, it was handled in different ways by different families.

Most families did not regulate it in a formal oral or written manner. Instead one generation transferred the fixed plots to the next, family members redistributed the "family pool" seasonally, according to the needs of the individual households.

The first outside intervention in local land tenure was the land registration in the seventies that was conducted within the scope of the implementation of the Basic Agrarian Law (BAL). Whereas according to the BAL (UU 5/1960, Art. 67) only land under cultivation can be registered, it was common practice to register forest patches marked by coffee trees and fallow plots of different ages as "reserve land" (Sunito 2004). With the establishment of the *Lore Kalamanta Wildlife Reserve* in 1981, Watumaeta villagers were *de jure* excluded from parts of the surrounding forests. Secondary forests located within the village territory however were still sufficient so that neither the land registration nor the establishment of the wildlife reserve involved significant disputes over land. The same holds true for in-migration and market penetration, which did not seriously affect the social fabric of the community for several decades. Whereas the first Toraja migrants settled in Watumaeta in 1945, the first Buginese from South Sulawesi entered Watumaeta in 1961. Since then, the number of both ethnic groups has risen steadily, but it was not before the mid-nineties that a combination of large scale in-migration, erratic land transactions and the closing of the forest margin as a major land provider caused an enduring reconfiguration of the agrarian structure. In difference to their predecessors, most of the Sudanese and Buginese migrants that entered Watumaeta since 1995 have already been successful export crop farmers before their arrival. It should be noted that most of them migrated not from their homelands, but from the village of *Kebun Kopi* some 150 km to the North of the Napu valley. Ironically, the immigration stream coincided more or less with the integration of the *Lore Kalamanta Wildlife Reserve* (and thus a significant part of Watumaeta's forest margin) into the newly established *Lore Lindu National Park* (Taman Nasional Lore Lindu) in 1995. In contrast to the popular view, this administrative change did not imply a change for the worse in regard to the use rights granted to local people by the law. More important was an increase in public attention as well as a tighter supervision of activities in the forest through the establishment of the local National Park branch office in nearby Wuasa.

Lured by low land prices, a second wave of (predominantly Buginese) in-migrants came during the economic crisis (*krismon*) in 1997/98, when the US dollar tied returns to cacao increased seven fold (Ruf and Yoddang 1999: 248). As can be observed all over Central Sulawesi, there is a tendency of the Buginese in Watumaeta to establish themselves as a large and economically powerful group in the Diaspora. Our village census conducted in July 2001 showed that at this point of time the Buginese made up already 47.8% of all Watumaeta villagers. In contrast to the various "mixed cropping strategies" and the delicate "subsistence-cash crop balance" adopted by the local population, the Buginese economy is dominated by the export sector with a preference for cacao production. The most seriously felt impacts of in-migration in Watumaeta were thus of demographic, not socio-economic nature. Supported by the high cacao prices during the economic crisis and the availability of cheap land, Buginese and Sundanese migrants became not only the wealthiest peasants in Watumaeta, but also the biggest landowners.

There are four major peculiarities linked to the process of displacement in Watumaeta. First, as pointed out by Li (2001: 90), displacement from resources occurs not via the appropriation of forest by commercial concessionaires. Instead it occurs piecemeal, as a result of individual negotiations and transactions. The second feature of the processes observed is the small portion of permanently cultivated plots in these transactions. Thus, more than 90% of all plots transferred to migrants by locals (households and village head) have been secondary forests and fallow land of different ages. In only one case a wet rice plot (*sawah*) was alienated. In contrast to the Palolo valley, there has not occurred a significant conversion of wet rice fields into perennial estates. Third, the lion's share of the plots transferred concerned not land inside the prohibited forest area, but consisted of forest and grassland located within the village boundaries. Watumaeta migrants do not fit the image of the land hungry, spontaneous settler who encroaches randomly into the forest. Being outsiders, migrants are taking pains in obeying the law and are reluctant to open a plot without consent of the village administration. As Potter (1996, cited in Li 1999) observed in Kalimantan, rather than spreading into the forest interior, Watumaeta migrants "tend to stay close to the road itself as their point of access to markets" (ibid: 29). The most striking fact however is the small amount of plots sold to outsiders by individual local households in comparison to those which shifted hands via the village head. Our survey on the acquisition of 75 plots operated by migrants (see table 1.) supports rumors in the village which accused the village head of transferring large tracts of fallow land claimed by local families to outsiders.

Thus, only 13.3% (n = 10) of the sample plots transferred to migrants have been sold by individual households with seven out of 10 transactions taking place from 1997 to 1999. The concentration of private land sales within this time span is indicative of the overall situation non-tree crop farmers faced during the economic crisis (see Sunderlin 2000). In contrast to the price increase of cacao, returns to the locally dominant non-rice annuals (*palawija*) diminished significantly at the same time input prices for fertilizer rose steadily since the end of 1998 due to the exhaustion of imported stocks (Ruf and Yoddang 1999: 250). Thus, the increase of private land sales during the crisis cannot be understood in isolation from the wider economic circumstances.

In our database, 37.3% (n = 28) of all plots owned by migrants have been obtained from the village head. This is 43.8% of the whole land area transferred. In terms of share of plots this figure is almost identical with the portion of plots provided by fellow migrants (n = 29). In regard to the total land area involved however plots provided by fellow migrants amount to only 36 ha in comparison to 55 ha which were obtained from the head. Further, 9 out of 36 ha were not transferred on a permanent basis, but on the basis of borrowing (*pinjam garap*). In addition, one must keep in mind that much of the land obtained from other migrants shifted hands via the village head before. For example, the first Buginese migrant obtained almost 10 ha from the head, which he sold afterwards to his companions. Table 1. shows from which providers and in which manners migrants acquired their land *prior to* the economic crisis (phase 1), *during* the economic crisis 1997 – 1999 (phase 2) and *after* the crisis (phase 3). Before the crisis, the common form of land access for migrants was either purchase or a "grant" from the village head, which usually included an administration fee. Whereas the head's practice of selling land against cash continued, no grants of plots occurred during the crisis and afterwards. Instead the opening of a new frontier area in 1998 was accompanied by a new agreement, called *bagi tanah*, literally "land division". Within this arrangement, a migrant obtains a tract of secondary forest without immediate compensation against the obligation to convert it into a cacao plantation, which will later be divided between land provider and receiver on an agreed division scheme. Such exclusive land-sharing arrangements exist only with newcomers; locals are definitively excluded from this kind of access. Thus, the way migrants obtained land changed, but it is obvious that the village head exerted a strong continuity in his role as major land broker. Whereas land acquisition from other migrants (either by purchase or inheritance) is rising between 1997-99 and then diminishing again, since 2000 migrants borrowed their land more and more from other migrants. In the same time, the role of absentee landlords living in the provincial capital of Palu increased considerably, who rented out 20% of all plots worked by migrants. Absentee landlords own relatively large tracts of land between 4 and 10 ha each, which they cultivate with long-term tree crops such as candlenut, teak and agathis. New migrants – often coming from Palu – plant vegetables between these trees and, at the same time, they fulfil the task as stewards or guards of the absentee landowner's plots (in general, absentee residents purchased the land from the head). There is a clear differentiation process within the migrant community: whereas first generation migrants are powerful landowners, second generation migrants have to start as borrowers of the longer established migrants or the absentee landlords.

That migrants have to rely increasingly on land borrowed from their fellows has something to do with the fact that forest reserves outside the National Park are meanwhile exhausted. A supporting factor for the leaders' position in the delineation of land is his knowledge about non-registered land. Because the population data record is kept under lock by the village head, other village officials have no reliable information on the actual number of plots transferred and the number of migrants living in their community. During our fieldwork in 2002, neither the village officials, nor the local population seemed to realize in how far the Bugis and Sundanese migrants had already established themselves as a large and economically powerful group in Watumaeta. Watumaeta villagers are not only ignorant about the actual number of migrants living in their village, they are also ignorant about the real extend and the location of secondary forests which were delivered to migrants by the head.

time plot obtained	access type	land provider				total
		migrant household	*local household*	*village head*	*absentee landlord*	
PHASE I *before 1997*	purchase	1	2	4	0	7
	inheritance	1	0	0	0	1
	grant	0	0	5	0	5
total		2	2	9	0	**13**
PHASE II *1997-1999*	purchase	6	7	4	0	17
	inheritance	7	0	0	0	7
	bagi tanah	0	0	2	0	2
	pinj.garap	0	0	0	3	3
	grant	1	0	0	1	2
total		14	7	6	4	**31**
PHASE III *since 2000*	purchase	5	1	6	0	12
	inheritance	1	0	0	0	1
	bagi tanah	0	0	7	0	7
	pinj. garap	7	0	0	4	11
total		13	1	13	4	**31**
		29	**10**	**28**	**8**	**75**

Table 1. *Land access and land providers*

In defending his practices, the head emphasized the disappearance of *shistomiasis* after migrants have planted the concerned areas with cacao. Deriving his legitimacy from a specific blend of traditional features such as descend from the settlement founders on the one hand and bureaucratic power on the other, the customary and official bases of the leader's power are highly intertwined. Being perceived as the legal authority by the newcomers, migrants have been told that the land they are offered is unused land, without an owner and therefore under the jurisdiction of the village leadership. In this regard, he is even supported by state law. Art. 27 of the Agrarian Law (UU 5/1960) states clearly that land must be cultivated actively and can revert to the state if left idle. In some cases migrants have been informed that local people have delegated the disposal of their land to the head because they cannot read and write and are not used to dealing with outsiders. Backed by the "national law" in his dealings with migrants, the reference to "customary law" served as a legitimating force in dealing with critics from within the community. According to customary law, an heir cannot sell inherited land without the consent of his co-heirs. In case of divergent opinions, it is the youngest daughter who has the final say in this matter. Given the fact that the wife of the head is the daughter of a famous *robo*-leader and land clearer, he could argue that his wife has already given her consent and that the transaction is thus in line with customary law (see Abdulkadir Sunito 2004).

What does this mean in terms of security? It is not only that the head was not fulfilling his task in protecting his fellow villagers from outside threats, but also it was he himself who endangered the security of the villagers by erratic land transactions. Despite the fact that on average migrants own significantly larger holdings than locals (3,2 ha against 1,7 ha per household), the major threat is less the difference in individual land holdings, but the closing down of the access to locally defined reserve land which has been converted to cacao plantations by migrants. This process in turn is not an outcome of land transactions between individual households, but the outcome of the relationship between migrants and powerful people inside (village head) and outside (town based landlords) the community. The villagers of Watumaeta have not been able to organize and channel in-migration and market incentives within the village in a constructive manner. It can be argued that such erratic land transfers by individual leaders are only possible where both, collective land rights as well as village cohesion are only weakly developed. Being confronted with this kind of situations, it has become popular to blame the Indonesian government under the New Order Regime (1965-1998), which deprived forest dependent communities of their self-governing capability and sanctioning patterns by imposing a uniform, bureaucratic structure in the countryside and by concentrating the full management authority over forest resources in the hands of the state. This is certainly true, but as far as the Napu valley is concerned, a great deal of disintegration existed already before Indonesian independence. According to local history, the first serious village friction happened already in 1945 (Togea 2004: 16). Originally composed of dispersed groups, which have either been resettled by the Dutch or which migrated spontaneously from the highland interior to the valleys, the inhabitants of the Napu valley lacked a meaningful identification with their villages from the beginning. Thus, "village guardian spirits" which usually function as a community integrating cohesive in rural areas of Eastern Indonesian (Scholz 1962, Mischung 1984) are totally absent. As pointed out above, in the customary understanding, forest is no independent legal category, but is embedded in the land tenure system. Whether the utilization of a tract of forest was restricted was exclusively based on the fact whether it belonged to someone. There were no further restrictions imposed on the individual by the community, nor did outsiders need the community's consent to make use of virgin land within its territory. The "customary council" (*lembaga adat*) did not play an important role in land use and resource allocation in the past. Until the present, the central conflict resolving institution remains the village administration represented by the village head (*kepala desa*). Under the given circumstances, there have been neither clearly defined community rights, nor local institutions, which could exclude in-migrants and counterbalance the leader's erratic land transactions and interpretations of "law".[2]

Within the last decade, the main point of controversy was less the establishment of the National Park as such, but the fixing of its borders. Boundary placing was a unilateral action taken without participation of the local communities on the premise that the border is a temporary demarcation and that its concrete course will be negotiated afterwards. Thus, from the beginning, the establishment of the park

---

[2] There is a provincial regulation from 1993 (SK 529.2/8158) which denies explicitly the existence of community controlled land in Central Sulawesi. Though heavily opposed by local NGO's and part of the scientific community in Palu, at least as far as the Napu valley is concerned we see no reason to question the premise of the regulation, albeit we know that our standpoint is not very popular.

led to a high unpredictability of the forest resource. It is an established fact that as long as access to a common pool resource is not effectively restricted and, at the same time the future legal situation is not predictable, people will not only try to extract as much as possible from it, but they will also try to obtain benefits within a short time period. This is especially to be expected in the Watumaeta situation where land claims are to a high degree linked to processes of displacement and issues of "intergenerational access" (see Burkard 2002a). This trend has further been intensified by the vague promise of the Park management that lands can still be returned to the villages. With no clearly defined rights to the forest resource, "legal unpredictability" preceded "ecological unpredictability". In our interviews Watumaeta villagers were rather doubtful whether their fellows would comply with any rules restricting resource use in the future. However, if a high number of "free riders" is perceived as being given, the predictability of a "healthy forest" must be judged as rather low and people will not feel much incentive to support its preservation (see Gibson et al. 2000: 213). In contrast to the rather limited role in resource management provided by local custom, the "customary council" *(Lembaga Adat)* emerged to become a sort of "voice of the people" and involved itself actively in the struggle over resources in claiming so-called "ancestral lands" (*tanah adat*) called *powanuanga* located inside the National Park. The council legitimates its claims with various proofs of former settlements (planted bamboo and coffee stands, stone mortars, graves, relicts of housings etc.) that have been "found" inside the protected area. In line with local practice, the kind of resource management envisioned by the council was not one of a village controlled common property regime of the *powanuanga*, but a division of the *powanuanga* plain among landless local households with each household receiving a fixed share of 2 ha. Under these circumstances it is not astonishing that a significant part of the local population encroached into the Park area planting markers (*gliricida sepium*) in order to secure their share in advance. On the lands to be divided, however, it was common sense that the beneficiaries should not be allowed to alienate their plots in perpetuity. It was further agreed that infractions of this rule must be punished severely and that the sanctioning capacity should be vested in the customary council.

The "local approach" to the problem was thus quite different from the one adopted by the migrants. In contrast to locals, migrants try to obtain certificates immediately after ownership rights have been established. Once a plot becomes recognized as a migrant's private property (*hak milik*) by certification, the plot is irreversibly removed from the local family pools. Security of tenure is especially important in the case of cacao because of the long-term character of the investments for perennial cultivation. If a migrant cannot be sure whether his newly acquired plot will be reclaimed afterwards, certification becomes crucial. The fact that it is the migrants who suffered most from unsolved land conflicts may partly explain their preferential change to buy land from other migrants rather than from locals. Migrants do not only enjoy a higher degree of socio-economic security due to their better access to markets and safety of tenure by certification; living predominantly in nuclear households characterized by small household sizes makes them also less vulnerable against ecological and market adversities than the relatively large extended households of the local groups. On average, migrant households in Watumaeta consist of 4,6 persons in comparison to 7,2 persons among the local households.[3] Given the higher productivity of their perennial plots and the smaller number of mouths to be fed, migrants operating 1-2 ha of cacao groves are still well endowed if only a part of the product can be harvested or sold (Burkard 2002a: 17). Given these imbalances, economic conflicts within the village centered on striving for security, are unavoidable. Thus, since the late nineties, relationships between locals and migrants became increasingly strained. Migrants often claimed that locals are lazy and incapable to till the land in a productive manner whereas local people accused the immigrants of seizing (*rampas*) their lands and forests by making collusive arrangements (*kolusi*) with the village elite. In contrast to the Protestant Napu people, Buginese and Sundanese migrants are Muslims, so that competition for resources tended to mix up with religious resentments and the wider reification of ethno-religious identity in Post-Suharto Indonesia. The ensuing tensions were thus first of all about the "political economy" of being a Protestant/local or a Muslim/migrant (see Aragon 2001: 47). Besides, locals became more and more aware of the demographic

[3] Refugees from the regency's capital Poso not included (see next chapter).

changes affecting their village. A serious issue was a significant increase of steeling of cacao plants and fruits after local people realized the economic value of the crop. When I revisited Watumaeta in September 2003 after I left the village one year before, one of the first changes one realized was the (relatively) large number of cacao trees planted in the formerly agriculturally neglected home lots. The farmers had transplanted parts of their cacao seedlings from the fields to their home gardens in order to safeguard a better supervision against theft. Such "shifts of infractions" are not uncommon in situations where the imposition of sanctions and the monitoring of encroachment are disapproved or perceived as unfair by the related population (see Kopelmann 2002). Developments within the last years however indicate that neither resource degradation, nor a continuous deterioration of social relationships is inevitable. The emergence of a co-evolutionary equilibrium between developments in the sphere of social organization on the one hand and natural resource use on the other will depend to a large extend on the willingness of the related population to respond to outer incentives of institution building and rule creation and to recognize the coexistence of different interests within the community. This, in turn, is intrinsically linked to the ability of the institutional environment to safeguard security.

Mutual Vulnerability: Changing Relationships between Locals and Migrants

Administratively, Watumaeta is located within Poso regency (*Kabupaten Poso*). The regency's capital Poso was shattered by communal violence since December 1998.[4] The conflict started with youth quarrels between a Christian and Muslim neighborhood in Poso town. However, the conflict escalated rapidly as it mixed up with regional political events as well as religious and socio-economic structures.[5] It

[4] The Poso crisis attracted relative little international attention as it was thrust into the background by more drastic violent events in other parts of Indonesia, especially the Moluccas.

[5] For more detailed information on the chronology of the crisis and its backgrounds see Aragon (2001).

was not before May 2000 however, that the conflict seized the rural areas of the regency, after Christian henchmen attacked a Javanese transmigration site south of Poso. Later followed by further attacks on Muslim settlements and houses, this event marked the beginning of a new phase in the crisis, which – in contradiction to former peaks of violence – hit more seriously the Muslim than the Christian community. One estimate is that up to 800 people – mostly Muslim – have been killed between May and August 2000 (see Aragon 2001: 66ff). In the aftermath of the attacks, Christian and Muslim refugee streams from Poso town and the rural areas around Lake Poso dispersed in all directions. Whereas Christians tended to hide themselves in the Protestant villages in the highlands (or went to the Christian Minahasa region in North Sulawesi), the majority of Muslim refugees fled to the provincial capital of Palu. As Aragon (ibid: 59) reminds us, establishing a local chronology of events always entails problems of accurate sequencing. The observations here of the Poso crisis in general, also hold true for the micro-processes that shaped the course of events in the areas affected indirectly. Several stationary field-stays over the period of almost six years and the establishment of an excellent rapport with the villagers and the community representatives however, should allow a reliable reconstruction of the incidents and processes which reshaped the social fabric of Watumaeta.

During and after the peak of the crisis, an estimated 2000 refugees had to be sheltered in the Napu valley, the main burden of which had to be carried by Watumaeta and the neighboring village of Wuasa. Whereas only a relative small number of Muslim refugees had to be provided for by the migrant community, the majority of Christian refugees had to be managed by the local population. Given the rather limited support from the provincial government, most refugees had to live with (*numpang*) Christian and Muslim households. This transformed many local extended households as well as several migrant nuclear households into "multiple family" units. These arrangements for the refugees often tended to overstrain the economic capacity of the host families. The more skirmishes that were reported from home and the more it became obvious that their sojourn in Watumaeta may probably last for years, the more the refugees felt themselves forced to resume farming activities in Watumaeta (Lieb 2003). Thus, albeit in indirect manners, the Poso crisis once again reinforced the pressure on the forest margin. It is not clear how extensively the village head was involved in the allocation of forest areas inside and outside the National Park that were cleared by refugee families. The legitimacy of the leader and his relations with the local population, however, eroded further after he allowed the refugees to use the Protestant "church land" (*lahan jemaat*) for the cultivation of seasonal crops (*palawija*) and temporary housing without permission from the Protestant church leaders (see Abdulkadir Sunito 2004). In mid-2003, after some refugees had certified portions of the church land as private property (*hak milik*), some returning refugees even sold their plots to their fellows.

Neither on the community level when faced with Muslim immigrants, nor on the level of religious congregation when faced with Christian refugees, had Watumaeta villagers been able to organize collective resistance against the appropriation of their lands by outsiders. Similarly, Muslim and Christian residents did not unite against a new common competitor for resources. Given the ethnic-religious tensions that dominated Christian-Muslim relationships in the regency, both, migrants and locals became increasingly aware of their *mutual vulnerability*.[6] This vulnerability, however, did not primarily originate from the "accumulated resentments" (Aragon) between locals and migrants within the community itself. Instead, the major fear was that "religious provocateurs" (*provokator agama*) from both sides might hide among the refugees in order to export the Poso conflict to Napu. Keeping in mind that a major facilitating factor for the outbreak of communal violence in Poso town was the "ghettoization" of ethnic-religious alignments (Aragon 2001: 74), the more thoughtful community representatives began to realize that the religious territorialization of Watumaeta's neighborhoods, once intentionally created by the leader in order to avoid clashes between customary and religious practices (e.g. the joint drinking of the local sago palm liquid *saguer* versus the migrants disgust for alcohol), could now become a facilitating factor for the genesis and spread of communal violence. Migrants

---

6 Following Singleton and Taylor (1992; cited in McCay 2002), in this paper *mutual vulnerability* is understood in terms of a "context variable" of the social and political environment, not as "the extend to which people are subject to peer pressure because they rely on good opinion and friendship as well as co-operation of others" (McCay 2002: 385).

were well aware that it was not their religious adherence *per se* which endangered their security in the Diaspora, but the potential of religion (*agama*) to interact with socio-economic structures and social envy (*kecemburuan sosial*). At the beginning of our field research in March 2001, about seven months after outbreak of the most violent bloodshed in Poso, the migrant community – headed by a far-sighted leader – had already made substantial offers to the local population in order to reduce its security fears and to adjust its own security interests to those of the locals.

Those contents of the offers which serve the interests of the migrants are primarily linked to issues of *tenure security*, whereas those which serve the interests of the locals are broader in scope and relate to issues of *displacement* in general. As pointed out above, the agrarian system of the migrants, especially the Bugis is dominated by the export sector and the cultivation of cacao. Given their preference for cacao monocultures, the migrants want the owners of neighboring plots to use their land compatible with their own in order to minimize crop failures and to optimize productivity. One important issue of the "deal" crucial for the migrants was that the preliminary cultivation of unused land should be aimed at fallow plots surrounded by cultivated land. Further, fallow plots within a cultivated area must be cultivated in a manner compatible to the established plots in a given cultivation area (*hamparan*). These parts of the agreement aim primarily at the maintenance of *security of production,* which must be distinguished from *security of land tenure.* Because in most cases the village leader himself sold "conflict land" of unclear status, migrants are not only more frequently involved in land conflicts than locals, but they also suffer most from the fact that their problems have not been brought to an end. In general, there are four major types of conflicts: (1) conflicts about the exact borders and actual size of the acquired plots; (2) the plot or a part of it is claimed by third parties; (3) problems related to location and legal status, e.g. a plot is located within the prohibited area of the National Park; and (4) plots have been sold several times. Migrants suggested several clear *formal procedures* for land transactions to be included in the village rules. First, village authorities must approve every private land transaction. Second, the agreement on borders and ownership rights of all farmers owning adjacent plots is to be prescribed. Third, land transactions have to be witnessed by several community leaders (e.g. hamlet heads, neighborhood heads), with the number and formal position of witnesses being fixed.

These rules of land transfer, in the drafting of which the head of the migrants' hamlet (*kepala dusun*) played a crucial role, are similar to the rules for land share arrangements (*bagi tanah*). The prescribed frame for the conditions of share-arrangements has been constructed under his advice with a "model contract" being kept in his house. The various types of conflict, which occurred in the past, have been collected, scanned and worked into the final formula in order to anticipate possible eventualities in the future. Whereas these rules of contract serve primarily the security concerns of the migrants, their concerns are by no means exhausted with migrant's interests. It was pointed out that the migrants' concerns are intrinsically linked to the compatible use of land. Therefore, they suggested that the active cultivation of plots is to be prescribed in the village rules. This suggestion is in line with the still-valid 1960 Basic Agrarian Law (UU5/1960). Under article 10 (1), a person who owns agricultural land is "in principle obliged to cultivate or exploit it actively by himself", though other legislation has left room for sharecropping arrangements (UU2/1960). The notion that land must be utilized rather than neglected is emphasized in article 27, which states that a plot can revert to the state if left idle, in which case the rights of ownership are annulled (Harjono 1993: 67). Because most people who own fallow land (*lahan bero*) and secondary forests (*belukar*) are locals, migrants made the offer that any migrant purchasing land and not cultivating it within a period of three months after acquisition, will lose his residence rights and is liable to be expelled from the village. One of the major rationales behind this rule is to safeguard the priority of *cultivation* over *accumulation*.

A further suggestion was that migrant newcomers should not obtain full residence rights within the first three years and that only a semi-permanent residence card (*KTP sementara*) should be granted. This residence right can be withdrawn at any time in case of rule infraction or in case a newcomer disturbs public peace. Thus, troublemakers and "free riders" are liable to be expelled from the village in case they transgress any rules during that time. The second part of this rule is of the same relevance for both Muslim residents *and* Poso refugees, and serves the security needs of all Watumaeta residents. If land sales that do not fulfill the conditions stated above lead to conflict, both parties involved – local seller and migrant purchaser – will be punished in the same manner. The solemn sanctioning right is vested in the *Lembaga Adat* that consists exclusively of local people. Besides, the migrant community declared its willingness to restrict the number of newcomers.

The rationale behind all these suggestions is a threefold one: (1) Security is not possible on a one-sided basis; (2) therefore, local security interests should be considered in the rules as long as they do not impair the security of migrants; (3) Security is best achieved when people are willing to share risks, consequences and sanctions not only for others, but also for members of their own group; and, when (4) Enforcement of rules is vested in a local institution (in this case the *Lembaga Adat*). Thus, in 2001 the Watumaeta situation was a rather grotesque one: in the indigenous perception the village head did not fulfill his task to protect the local population from outside threats, whereas the outsiders themselves made substantial offers to reduce the security fears of local people.

Realizing that long-term security cannot be upheld under the existing ethno-religious and socio-economic conditions and that in order to close the "economic gap" one had to close the "knowledge gap" first, Buginese migrants started to share their experience and assisted local farmers in the establishment of cacao plantations. The migrants were well aware that their efforts might be hampered by the life-cycle of perennial crops. This problem is especially pronounced in the case of cacao, because during the first 3-4 years trees do not bear fruits and investment costs often exceed the benefits of the first harvests. Thus, in order to bridge the unproductive years, the inter-planting of fast-yielding cover crops became vital. Here, the compatibility of the expertise between Buginese and Sundanese migrants was crucial. Whereas the Bugis are specialized in perennial crops, several Sundanese families practice a seasonal crop rotation of carrots (*wortel*) and leek (*daun bawang*) on mounded vegetable beds (*bedengan*).[7] The

[7] The *bedengan*-system was first practiced by a Sundanese migrant who learned the cultivation technique in a horticultural enterprise in Kalimantan. The role of the migrants as agricultural innovators is comparable to the role of individual return migrants who introduce their acquired skills in their home villages (Burkard 1999).

vegetables are sold to a company in East Kalimantan via a well-established market network. Thus, the *bedengan*-vegetables offered an ideal cover crop for local families. Given the risks related to perennial cultivation and the need for fast cash among many local families, it is not surprising that it was first of all the *bedengan*-system which attracted the interest of local families.

Inevitably, the Watumaeta case raises questions about the role of socio-cultural heterogeneity and leadership in self-organization. It is often assumed that in regard to collective action, homogenous communities are more effective than heterogeneous ones. To compare Watumaeta with the resettlement community of Sintuwu in Palolo valley which accommodates also a significant portion of migrants (see Burkard this volume), it becomes obvious that the crucial "design principle" is not the existence of an ethnically accentuated socio-economic heterogeneity as such, but how this heterogeneity is structured. From a cognitive point of view, Sintuwu is characterized by a "blurred heterogeneity". Since most of the "local" Kaili residents are Muslims, there is no interaction of socio-economic structures with religion. Although leadership practices, processes of displacement and the degree of land alienation are comparable to Watumaeta, migrants and locals live in the same neighborhoods and have often intermarried. There is no clear established leadership that could represent different interest groups or certain categories of villagers so that people do not know whom to address in case of conflict. This situation differs with Watumaeta, where socially positioned groups are characterized by clear boundaries. Bugis and Sundanese are perceived as migrants (*pendatang*) by the locals and define themselves as such in relation to the Napu people and other regional groups. Migrants and locals live in clearly demarcated compounds or hamlets and no intermarriage occurs. The point is, that clearly identifiable groups can formulate and defend their interests more easily in regard to others because they do not have to consider multi-layered relations. Further, leadership in Watumaeta is well-established. Due to careful politics and clever strategies the local population meanwhile accepts the migrants' head. This is reflected in the fact that regular village meetings are also held in his house. Migrants clearly formulated their points for the crafting of village rules and, at the same time, they combined their concerns with meaningful offers to the local population. What was first perceived as a major threat to security, i.e. the ethno-religious compartmentalization of neighborhoods, had suddenly positive effects on people's capacity for self-organization. The lesson to be learned from the Watumaeta case is that the organizational capacity of heterogeneous groups can be quite high when the group boundaries are clearly defined and when legitimate leadership represents the groups. Further, competing groups must be willing to share risks and consequences for rule breakers among their own group. If these preconditions are given, marked boundaries between groups can be helpful in the rule making process. If, on the other hand, social cohesion is too strong, the process may well be impaired by long-lasting commitments between the actors involved. A great deal of the intra-local conflicts prevailing in Watumaeta could not be solved because people did not want to risk troubles with their relatives, especially in cases where members of the family of the village head were involved. Bennett's conclusion that the proximity of village heads "to the members of their constituency and its natural resources made it difficult for them to be as detached and autocratic as the heads of higher levels of government" (Bennett 2002: 60f) does not hold true in the case of Watumaeta where tight familial relations often hampered actions against power abuse.

A second innovation to redirect local-migrant and inter-religious relationships was the foundation of the so-called "religious forum" (*forum agama*). Indicative of the situation, the impulse to action came from the migrant community after the violent incidents in mid-2000. The major function of the forum is to maintain regular contacts between the religious leaders (priests, imams etc.) and to act as an arbiter in case of conflicts between Christians and Muslims. The forum is also involved in the investigation of suspected provocateurs. The forum does not manage conflicts over land in order to avoid interventions into the state's domain. Involvement in land issues is confined to situations in which the help of the forum is explicitly requested by the county administration (*kecamatan*). The major objective of the forum however, remains to establish regular inter-religious connections through the collective performance of collaborative work assignments (*kerja bakti*). As can be observed all over Central Sulawesi, there was a visible competition among Christians and Muslims concerning the construction of prayer houses. In late 2000, a new mosque in the Buginese hamlet and a new Protestant church were under construction. What started as a form of zealot competition one year before was now turned into a medium for co-operation by the religious forum. As a forum member once explained to me:

> *"If you come into a village and want to know if it is good organized, look at its mosque or church. If the prayer house is in decay, the village is not well ordered. How can we order the community, if we are not able to maintain our prayer houses".*

Thus, Watumaeta was the first village in the Lore Lindu region where Christians participated in the collective construction of a mosque and where Muslims were actively involved in the erection of a church. And – almost unbelievable for most Indonesian citizens – probably the first village in Indonesia where an imam preached in a church and where a Protestant priest preached in a mosque. Whereas these meetings were held only twice with a rather limited number of participants belonging to the other religion (only three Christians participated in the mosque meeting), they had an inestimable value for the building of trust and reciprocity among the religious leaders. The leader of the religious forum, a former Catholic who converted to Islam after his marriage, addressed the participants with the words:[8]

> *We are all creations of God in the same way and our life in this world is only temporarily. We should not look at each others religious background, tribal background... this life is only temporary... so let us colour this life with respect to each other... let us enjoy the richness of nature* ***(kekayaan alam)*** *together... We have only a limited time to make a living* ***(cari nafkah)****...*
> *I need rice to eat, you need rice to eat, and you drink sugar as I drink sugar, so let us enjoy this life....*

It is out of question that the *forum agama* has gained important merits in regard to religious tolerance and in strengthening leadership ties within the village. From an organizational point of view however, its merits exceed by far the scope of religion and social relations. As pointed out in section 2, the formation of village institutions for the

---

[8] The quotation is the recapitulation of his words given during an interview on February 3rd, 2005.

management of common pool resources cannot build up on existing "grass-roots"-organizations. In a certain sense the *forum agama* became the model for organization building as well as vertical and horizontal integration. After neighboring villages followed the Watumaeta example, the *forum agama* was first established on the county level (*kecamatan*), later the regency level (*kabupaten*) and within a time span of two years on the provincial level. Until the foundation of the *forum agama*, village institutions in Watumaeta have been "top-down" organizations created on behalf of the state within the framework of the governments discourse on "development", such as the farmer contact groups (*kelompok tani*), the neighborhood groups (*rukun tetangga*), the village credit co-operative (*koperasi unit desa*) or the family welfare organization (*pembinaan kesejahteraan keluarga*). Further, in integrating several villages horizontally before higher levels of administration have been addressed, it provided the core structures of horizontal linkages that should become essential for the successful management of the common pool resource "forest margin". As pointed out by McCay (2002: 370), the "existence of institutions that can be adapted for new purposes" involves lower transaction costs because people can build up on existing decision-making structures and because the actors involved already know the social and cultural expectations. In Napu the process of institutional change did not necessarily involve a redirecting of existing institutions created for other purposes as described by McCay (ibid: 370), but – as will be shown in the next section – the forum's success facilitated the process towards mutual regulation of land use.

The overall defining context of this self-organization has been the Poso crisis, especially the violent events in mid-2000 which resulted in an estimated number of 300-800 victims, mostly Muslim. Knowing that a conflict between migrants and locals over resources could easily turn into a religious conflict, these institutions have their roots in the political economy of Central Sulawesi as well as in conflicting claims, but are by no means motivated by concerns for forest preservation or sustainability. These organizational innovations started under the leadership of the long established village head who was to a great extend responsible for the transfer of local reserve lands and the ensuing tensions in the village. The most significant event within the last years was the election of a new village head in late 2002. After the winner

winner of the election, who happened to be the son of his predecessor, was announced, Watumaeta was shattered by turmoil when the house of the newly elected head was attacked several times by local residents. Due to a courageous engagement of the *forum agama* however the situation calmed down and the culprits were handed over to the local police station. Because the new head was actively involved in several *bagi tanah* land transfers in the past, locals feared that neither in terms of personal loyalty, nor in terms of his policies, the head would substantially differ from his father. This fear seemed understandable and was to a certain degree shared by the researcher. To my own surprise however, the institutional environment, socio-economic configurations as well as the wider circumstances under which the election took place, forestalled such a development. The new head owes his straight win over the other two candidates (all of them of prominent Napu descent) first of all to the demographic constellation. In contrast to his opponents, he was not chosen by his local fellows, but almost exclusively by the immigrant community, which, meanwhile, outnumbered the locals. Thus, while the (numerically less) local votes were divided among his opponents, he could easily unite the (numerically dominant) migrant votes. Given the fact that majority of the land operated by migrants was obtained from his father and that they could not know which kind of policies other candidates would apply in dealing with migrants, migrants felt their security most guaranteed if the administrative power remains in the hands of the ruling family.

At the beginning of his tenure, the leaders' legitimacy among his Napu fellows was extremely low. Against common assumptions however, in this case a serious lack of legitimacy was not to the community's disadvantage. Instead, in order to achieve (or "create") legitimacy, the newly elected head took pains to involve the villagers in the decision-making process. Whereas the migrants had "their" representation in the village head, the locals on the other hand were represented by the newly created BPD (*Badan Perwakilan Desa* = village representative body), which was installed shortly before the head was elected. The BPD forms the basic rural institution within the new area of regional autonomy (*otonomi daerah*) and was established as a substitute for the former village council called LKMD (*Lembaga Ketahanan Masyarakat Desa* = body of the village society). In contrast to the LKMD however, which mainly served as a forum to promote government programs, the functions of the BPD as elucidated in the *Law on Regional Autonomy* of 1999 (UU 22/1999, Art. 104) are as follows: (1) to collect, organize and give voice to the aspirations of the common people, (2) to maintain and foster local custom (*adat*), (3) to explain regional regulations to the people and (4) to plan and install village regulations, which means that the BPD is actively involved in legislation (*legislasi*). According to the law, the village head must be appointed by the BPD.[9]

Fortunately for Watumaeta, the BPD was installed *before* the appointment of the head, so that the head depended to a great extend on the good will of the (locally dominated) village council, which could exert a considerable degree of control in counterbalancing the leaders' power. It is important to note that this institutional configuration was neither the outcome of a deliberate decision, nor the outcome of a

---

[9] In theory all relevant groups in the village should be represented in the BPD according to sex, religion, age group and profession. Of course, this can seldom be achieved in reality. But what is to be stressed is that the BPD works close together with the village head, whereas on the other hand it has a relatively high degree of autonomy compared with the LKMD. In case of misbehavior of village officials the BPD is obliged to send a report to the regency head (*bupati*) with a copy sent to the head of the county (*camat*). The broad parameters that define the crafting of rules and the management practices of this institution are based on the rules of the Minister of Internal Affairs No. 3 from 1981 about village decisions and the National Law No. 22 on regional autonomy, which are laid down in the district rules from 2001. Since October 2004, the Law on Regional Autonomy UU 22/1999 has been substituted by Decentralization Law UU 32/2004. However, many of the paragraphs in UU 22/1999 which foster decentralization are still existent in the new law. However, important functions of the BPD (e.g. legislative functions) have been dropped totally or been limited in scope. Whereas co-operation between the village administration and the BPD is enhanced; control of the administration by the BPD is almost absent in the new version of the law (see UU 32/2004, Art. 202-215, especially Art. 209). While the state seems to maintain its "participatory approach" in allowing local variations in village organization and local regulations (UU 32/2004, Art. 211), the BPD (now called *Badan Permusyawaratan Desa* = village *consensual* body) is formed by the traditional principles of "discussion and consent" *(musyarawah dan mufakat)* which do not necessarily match the democratic principles of a common vote. Unfortunately, neither UU 22/1999, nor UU 32/2004 explicitly addresses issues of natural resource management. Because most villages in the region now claim "ancestral land" inside the National Park and the protection forest, this paragraph is of special relevance for the villages concerned.

conscious construction of "design principles", but the synergetic effect of an accidental inter-play of several decisions motivated by security concerns. As Sherry Ortner points out, "social change does not for the most part come about as an *intended* consequence of action. Change is largely a by-product, an *unintended* consequence of action, however rational action may have been" (Ortner 1989: 258). During several short-term field stays in Watumaeta from April until October 2003 the atmosphere in the village had definitively improved. Almost all interviewed community members attested the leader good will and honest commitment to the community as a whole. This does not mean that conflicts have been eliminated from community life, but different from the past, they were actively handled by the village administration. In Watumaeta conflicts have often either been fought out by the incendiary use of fire ("burning plots"). A draft for the new village rules from 2003 entailed clear regulations for the use of fire, prescribing the presence of all owners of adjacent plots during the land clearing.

As pointed out by Stern et al. (2003: 474), in their focus on "design principles", researchers of the commons paid more attention to the characteristics of institutions than to the contexts within which they function. We may add: and to the situations and events within which they are created. As this section illustrates, independent of contextual factors, the impact of one and the same "design principle" on the organizational capacity of local communities can be positive, negative or curvilinear even within one and the same community. In the autocratic system of village administration of the Orde Baru government (1965-1998), the village head was the solemn source of power in natural resource distribution. Under these circumstances, ethno-religious and socio-economic heterogeneity had definite negative effects on local self-organization and the management of natural resources. Under the more decentralized system of the present however, this heterogeneity leads to a better community management in case the related sub-groups are sufficiently represented and willing to share equal risks and consequences for their own group. Our own basic assumption that "legitimacy of leadership" is crucial for the organizational capacity of local communities has proved to be mistaken. Rather than being a fixed "design principle" valid under all circumstances, it turned out that under certain constellations of the institutional environment, even a massive lack of legitimacy (or the need to *create* and *maintain* legitimacy) may improve organizational performance. The Watumaeta case emphasizes also the limited value of theoretical approaches, which base their deductions on individual motives for co-operation. Thus, the motivation of the Buginese migrants to co-operate was not individualism, competition, co-operation, altruism or any other "individual trait". Their co-operation was exclusively based on subjectively perceived security concerns. Within the last years, the capacity for self-organization, which has its roots in the anticipation of communal violence, became itself a major "design principle" in the creation of institutions for conflict resolution and natural resource management. Whereas there was a strong source of motivation for common pool institutions from within, external agency was involved, as well as the will be documented in the following section.

What has been Learned? Negotiating the "Community Conservation Agreement"

Since the declaration of the Lore Lindu National Park (*Taman Nasional Lore Lindu*) in 1995, *The Nature Conservancy* (TNC) was actively involved in the management of the park. In principle, the management plan is based on the notion of zoning with different zones subject to certain restrictions, ranging from enclaves, limited use rights of old coffee stands, harvesting rights of non-timber forest products and zones for eco-tourism and research to totally prohibited areas. The adopted pattern mirrors, more or less, the zoning scheme prescribed in the *Law on Conservation of Natural Resources and the Ecosystem* (UU 5/1990 §32).[10] Though TNC rejects the idea of the sustainability of indigenous land use systems, which are perceived as insufficient for bio-diversity preservation, it has revised its earlier top-down resource policies in favor of more collaborative management schemes. The core institutional devices of the adopted strategy of *on-site protection* are the "Community Conservation Agreements / CCA" (*Kesepakatan Konservasi Masysarakat*), in which local communities enter into a negotiated arrangement with the Park Authority. In the 2nd half of 2001, TNC implemented the first CCA's in five pilot villages in Napu.[11] After the CCA's were socialized in the villages in July 2001, steering committees (*panitia*) had been established in all villages in September 2001.[12] The local masses, however, were still excluded from participation in the meetings in order "to avoid the misperception that decentralization is already achieved" (TNC 2002: 16). An "open-for all" public meeting however was held in December 2001, shortly before the agreements were signed in January 2002. Watumaeta was not part of the 1st five pilot villages, but received a CCA in August 2004. It was one of the first villages covered by a large facilitating effort to establish CCA's in all 60 villages bordering the park.

Starting in 2000, the customary council (*Lembaga Adat*) tried to regain control over so-called "ancestral lands" (*tanah adat*) in the forest with the promise that the area will be divided among landless local families with each household receiving 2 ha for cacao cultivation. This led to a race of encroachment with individual households marking their plots by clearing in order to secure their share in advance.[13] In line with TNC's conservationist orientation, the general content of the CCA is to avoid further encroachments and to reforest open gaps inside the park, which have already been opened by individual farm-

---

[10] TNC's involvement in the Park management precedes the establishment of the National Park Authority (*Balai Taman Nasional Lore Lindu*) in 1997. Besides operating more than 1300 nature sanctuaries in the USA, TNC made agreements for the co-management of nature reserves with several governments in the Southern hemisphere.

[11] This was not the first agreement between a village and the Park Authority, but the first one facilitated by TNC. Before TNC's new approach, at least three agreements had been successfully facilitated by an Indonesian NGO called the "Free Land Foundation" (*Yayasan Tanah Merdeka = YTM*). In contrast to the conservationist orientation of TNC, YTM focuses on the empowerment of local communities and supports customary land use patterns, which are believed to be intrinsically sustainable.

[12] The discrepancy in the official document (TNC 2002) between the legitimization that CCA is a demand from the villages and the "need" that the necessity of CCA must be explained to the villages in a "socialization effort"!

[13] It is important to note that the division of the land into equal, individually hold parcels as envisaged by the council is quite at odds with the notion of *tanah adat* and the collective right to ancestral territories (*hak ulayat*).

ers. Plots already opened within the National Park must be replanted according to a fixed reforestation plan, whereas all types of secondary forests will be preserved and all opened plots close to rivers and slopes have to be abandoned. These regulations are relatively easy to monitor and practicable in every village where the forest is already disturbed. From the beginning, it was doubtful, however, whether individuals would participate willingly without the certainty that benefits would accrue to their families. Meanwhile, after cacao prices fell significantly since 1998, a shift from cacao to timber, candle nut and other trees could be observed in all villages, so that the agreement followed in fact the existing economic trend which helped to support its implementation. The CCA's are granted for three years; subsequently, they are either prolonged, revised or rejected. The reforestation species (teak, resin, candle nut, agathis, eucalyptus etc.) are defined by the natural surrounding, which is divided into eleven reforestation sites of different extend, topography and vegetation. Besides reforestation regulations, the agreement entails clear rules and sanctions in regard to the collection of forest products and timber for private use, the clearing of agricultural land, watershed management and the tapping of resin (*damar*). Whereas the sanctioning capacity is vested within the *adat* council, sanctions can only be enacted in the presence of the village administration and the BPD. The border of the Park is demarcated with the economically valuable candlenut trees (*kemiri*).

In order to guarantee a better supervision and co-ordination of the CCA, TNC established a separate "Village Conservation Council" (*Lembaga Konservasi Desa* = *LKD*) at the village level. Since the declaration of regional and local autonomy in Indonesia, the villages in the research area often suffered from the lack of responsible regulation of village organizations prevalent also in state organizations. Leaders could exercise undue power by making their decisions outside of formal organizations because the authority of local institutions was often contested (see Burkard 2002b). Guided by the principle of co-management, the LKD is constituted by one member of all major village institutions (village administration, BPD, *Lembaga Adat*), one official park ranger as well as "other" personalities whose opinion is perceived as relevant (§ 23, CCA Watumaeta 2004). Paragraph 21 of the agreement summarizes the functions of the LKD as follows:

- To socialize (*mensosialisasikan*) the conservation agreement to the related society,
- To carry out participatory planning (*perencanaan partisipatif*) with the Park-Authority,
- To supervise the implementation (*mengawasi pelaksanaan*) of the agreement,
- To evaluate (*mengevaluasi pelaksanaan*) the conservation agreement,
- To report the evaluation results (*melaporkan hasil evaluasi*) to the village head,
- To provide an umbrella for communication (*mewadahi komunikasi*) between (local) society and the Park Authority.

In regard to the practical realization of the agreement, the last point is probably the most important one. As McCay points out: "the existence of a political, social, and physical space for learning from and arguing with one another is one important design principle that should not be taken for granted. In many nations and at many times, political repression makes it nearly impossible to find and use places for talking and arguing about the commons" (McCay 2002: 383). Whereas the LKD is concerned with the implementation of the CCA on the village level, TNC envisaged that conflicts between villages (several sites in the forest are simultaneously claimed by different communities) or problems, which cannot be solved by the LKD alone, are to be handled by a corresponding "County Conservation Council" (*Lembaga Konservasi Kecamatan* = *LKK*). However, the LKK never did materialize and was substituted by the establishment of the so-called "Buffer Zone Forum" (*Forum Wilayah Penyangga* = *FWP*), facilitated by the ADB-sponsored "Central Sulawesi Integrated Area Development and Conservation Project" (*CSIADCP*).[14] In contrast to other facilitating NGOs

---

[14] Here, the implementation of the CCA reveals something of the "conservation business" in Central Sulawesi. Although the watershed management is included in the CCA, the CSIADCP project facilitated the signing of a separate watershed management agreement called *Kesepakatan Konservasi Masyarakat Adat* in the villages. Despite its reference to *adat,* the agreement means the realization of government regulation PP 31 / 2001, § 27 and fulfils major obligations of the conservation

and institutions in the research region which tend to treat villages as islands unto themselves, TNC does not assume that the local (village) level is most important. Instead, the crucial role of vertical and horizontal "cross-scale inter-linkages" (Berkes 2002), "intermediate" levels of organization such as the *kecamatan*-counties and the "nesting" of organizations located at various levels has been taken into account from the beginning of the facilitating process. Unfortunately, TNC's far-sightedness to transcend institutional boundaries in establishing the LKD was not reflected in the perception of the forest resource as an integrated ecosystem. Given its strong concern with the National Park as a legal entity and the integrity of its borders, the CCA is only valid for the forested landscape encapsulated by the park.[15]

It is important to note that the CCA is neither "devolution", nor "community-based resource management" in the real sense. The points listed in §21 of the agreement are concerned with implementation, not decision-making and represent a transfer in the locus of action. But they do not entail a transfer of power and authority. Whereas a certain degree of freedom is granted to the villages to develop their own sanctioning systems, the overall objective of the CCA remains strictly defined by the state: rules must aim at the conservation of the forest cover of the Park and sanctions must be enforced for those who transgress the rules.

The most important aspect of the CCA, however, is neither the agreement as an institutional entity, nor its suitability as an organizational device to safeguard the stability of the forest margin, but the processes of self-organization and community discourses it set in mo-

---

law UU 5/1990. After TNC established a co-ordinating organization at the county level (LKK), CSIADCP established a similar organization in form of the FWP at the same level. In order to avoid competition between the two organizations, TNC was forced to abandon the LKK as a supervising institution with the FWP being the exclusive organization at the county level. CSIADCP in turn abandoned its plans to install a watershed management organization on the village level.

[15] TNC conceded to the researcher that in several villages an intensified shift of activity could be found in adjacent protection forests (*hutan lindung*) and limited production forests (*hutan produksi terbatas*) after the agreements had been signed. There is no overall resource management plan for the whole forest resource as an ecosystem. Whereas the management of the National Park remains largely under control of the central government, protection and production forests are under control of the regency authorities.

tion. In April 2004 a three-day workshop was held in the village of Wuasa, Napu's administrative and economic center. The workshop was organized and prepared in a common effort by TNC, Care International and the Park Authority. Besides Watumaeta, the CCA's of four other villages were on the agenda. Fortunately, three of the four villages were actively involved in the *forum agama* and faced similar – though less intense – problems in regard to deforestation and socio-economic imbalances. This factor, combined with a shared cultural background, seems to have helped the leaders to engage in "meaningful communication" (McCay 2002: 384) in advance. Even before the facilitation meeting took place, the five villages made an account of the existing rules and sanctions. During the meetings they found that certain sanctions in cash for same infraction in the forest differed 1:10 between the villages. The need to harmonize sanctions between villages was given highest priority. Every village made a draft to be continuously discussed and adapted. Fortunately, the local NGO *Yayasan Pekurehua* facilitated the alliance-making process.[16] Founded in the late nineties, the major goal of the NGO is to maintain and sustain the *adat* and identity of the Napu people. According to the NGO, the prohibition against selling family land (*budel*) was one of the major customary principles in the past, which tied the Napu people to their surrounding resources. The active members of the NGO are exclusively young people (below 35), so that their concern with the customary ban on land sales clearly reflects the interests of the young and disinherited members of Napu society.

One of the most facilitating factors, however, was the fact that the Buginese and Sundanese migrants of Watumaeta declared that they would neither operate nor monitor any land covered by the CCA-agreement. Instead, the area should be under total control of the local population. The offer included even the willingness of the migrants to abandon plots they have bought within the related area in the past. As an ongoing process, leaders of the local community have been engaged in a debate about the migrant's rights to settle. Whereas some members of the customary council asserted the right of local people to exclude outsiders from settling within their territory, the village head

[16] *Pekurehua* denotes the ethnic group that is believed to be indigenous to the Napu valley.

and some other opinion leaders pointed out that migrants are Indonesian citizens and, as such, are free to settle everywhere in the archipelago. However, they should respect local custom. The question in how far Buginese and Sundanese migrants can also participate in the agreement was solved with the migrants' self-determined abstinence from the area, which provided the village leaders some room for maneuver.[17] More importantly, these one-sided concessions opened the way for the *Lembaga Adat* to withdraw its plans to divide the *powanuanga*-plain among poor local families. So far, this was one of the most serious obstacles in the negotiation process. A well-known feature of Indonesia's decentralization efforts is the existence of two competing discourses which are generally referred to as the *indigenous rights* and the *people's economy* discourses (Campbell 2002), usually perceived to be located on the upper stratum, urbanized section of Indonesian society. Conservationist NGOs, regional adat-leaders and social scientists, heavily advocate the focus on indigenous rights, whereas the discourse on *people's economy* is rather supported by the bureaucratic elite and social movements. However, the issues which were raised during the facilitating process showed that both discourses are more "grass-roots-discourses" as it is generally believed and produce struggles within local communities themselves. In contrast to the competing, major CCA-facilitating NGO in Central Sulawesi, the "Free Land Foundation" (*Yayasan Tanah Merdeka*, abbr. *YTM*), TNC did not make use of indigenous terms for forest zoning in the agreement. In Indonesia, the representation and mapping of traditional zones of land use has become a well-honed tool to prove the inherent sustainability of customary land use practices in order to regain control over natural resources (see Zerner 1994). There are important reasons for TNC to refrain from indigenous zoning schemes:

- Indigenous zoning schemes are usually associated with claims to common "ancestral" or "customary" lands (*tanah adat*), the existence of which is officially denied by a provincial decree from

[17] As Aragon (2001: 78) points out: "The 1999 Regional Autonomy Laws asserted that the rights, origins, and customary traditions of villages and regions are to be respected, without saying exactly how that is to be done in multi-ethnic areas experiencing just the opposite lessons during the Suharto period".

1993 (SK 529.2/8158/1993). The governor recognizes only so-called *swapraja*-lands. The term *swapraja* however, denotes the former domains of the Napu (and Kulawi) kings, not common property regimes of individual communities. As an official partner of the government, TNC can hardly conduct its facilitating efforts in opposition to the government's discourse of development and conservation.

- TNC does not recognize the sustainability of indigenous resource management practices, which are believed to be insufficient to sustain ecological bio-diversity.
- The various (indigenous) categories of forest (in Napu language: *holua*, *lopo*, *pandulu*) are meanwhile mixed up due to a shortening of the cultivation cycle and the fact that people have cut vast tracts of primary forest as well. The common pattern of distribution is often for patches of secondary forest to be scattered throughout areas of primary forest or cultivated area. It is almost impossible to develop resource use rights in regard to a given type of forest. With no physical correlation between larger areas and vegetation type existing, deliberate rules of resource use which have proofed as successful means in resource preservation in other settings (temporal restrictions of harvest, multiple species management, resource succession and rotation) are difficult to establish. In short: there is no evidence to establish an overlap between **legal boundary** and **resource boundary**.

Indeed, the conception of customary lands (*tanah adat*) does not match the requirements of the CCA. The agreement aims first at biodiversity conservation and the reforestation of opened plots. The acknowledgment of this erratically created sphere as "customary area" would indirectly invite others to make use of their common customary territory. Claims based on community membership would be unavoidable, thus fueling the deforestation process the CCA is designed to redirect. Further, within the traditional tenure system of the Napu people, "forest" is no independent legal category, but is perceived as being embedded in the local tenure system. Regulations do not apply in terms of vegetation (forest cover which can be converted against forest cover that needs to be preserved), but in terms of ownership. Restrictions in the utilization of forest are first of all based on the fact whether a forest patch belongs to someone. It cannot be taken for granted that all community members would understand their rights and obligations in regard to a (newly created) common resource.

During the last facilitating efforts in April 2004, there was a significant difference in opinion between TNC and the Park Authority in regard to the layout of the CCA. TNC tried to avoid claims to customary territories for the reasons cited above. On the other hand the National Park official, in pointing out the successful empowerment of the community of Toro which had based its claims to ancestral lands inside the Park with a vigorous revitalization of customary law (see Shohibuddin, this volume), encouraged the villagers (1) to put local terms for man-man and man-environment relationships into the draft, (2) to use local zoning terms (albeit they do not fit the situation there) and (3) to point out more clearly that it is community lands which are debated in order to raise the possibility that Watumaeta can be officially recognized as customary law community (*masyarakat hukum adat*). However, the participants did not really understand this.[18] The village leaders stopped his speech and pointed out that it is sanctions and regulations (*sanksi dan aturan*) that they want to discuss, and urged the Park Management to provide for an information and extension meeting on the new Forestry Law and its implementing regulations in the villages.

Given the high number of immigrants in his village and a serious lack of relevant knowledge among the present population, the head of Watumaeta made clear that the use of indigenous zoning schemes is not favorable in his village. Other leaders pointed out that they have enough problems with the socialization of the existing drafts and do not want to confuse the people further with additional concepts and constructs. Here, the policies of the village leaders is in line with the scientific finding that in heterogeneous communities with weak community ties, simple rules often work out better than complex rules (see

[18] There was already a consensus established that opened plots will be reforested and that secondary forests will not be touched in the future (irrespective of their age), which is reasonable because of the scattering of opened plots and secondary forests of different ages. One wonders why the Park official came up with the issue of zoning at this point of time.

Stern et al. 2003). The reliability and clarity of the agreements are important, and the way they can be covered by a single monitoring scheme in the five villages, not the number of local terms and constructs they entail. In the draft for the Watumaeta agreement, the head was not sure about the use of the term *tanah adat* in the agreement. Village leaders hoped for the Governor's recognition of the agreement. The Governor however, rejects the existence of *tanah adat* in his province. In this regard, the head's reasoning makes a lot of sense. Given a high number of free riders and the enduring land conflicts in the village, the head needs the backing of higher authorities in the implementation of the agreement. Despite the common trend of "community empowerment" in Central Sulawesi, it remains a fact that locally developed rules are easier to be enforced internally if the government also recognizes them. Understandably, the head does not want to take the full responsibility for what happens inside his village boundaries alone. The large size of the forest resource and the resulting negative externalities involved require that institutions and responsibilities are devised at multiple levels, including the provincial government. As pointed out by Berry (1988, cited in Li 1995: 501), struggles over resource control always entail conflict over meanings. Whereas the *Lembaga Adat* legitimized its claims to the so-called *powanuanga*-plain as an old settlement of the ancestors with the existence of stone mortars, bamboo stands, remnants of houses and stone tombs in the forest, the village head pointed out to the possibility that *powanuanga* was not an old settlement, but a place of retreat for criminals. Interestingly, they used the same "proofs" to underpin their thesis, such as the spatial distance between the two tombs and housings, which indicate a dispersion of a limited number of people, symptomatic of criminals who hide themselves in the forest.[19] In respect of the *Lembaga Adat* however, the part of the agreement devoted to sanctions is called "sanctions for the transgression of management rules related to the *customary territory* of Watumaeta".[20] Thus, some aspects of common lands, such as the prohibitions to sell, rent out or

---

[19] The head of the local branch office of the Park Management also adopted this point of view.

[20] Sanksi terhadap pelanggaran ketentuan pengelolaan wilayah adat desa Watumaeta (CCA Watumaeta 2004: 29).

mortgage land have been integrated into the sanctioning paragraphs. The necessity to forestall further land transactions is shared by the *Lembaga Adat* and the village head.

Influenced by the Toro success story, the Park official did not realize that in Napu it is *community security*, not the recognition as *customary community* that is at stake. The village of Toro has successfully reclaimed ancestral lands inside the Park. Constituting an enclave inside the Park and less affected by market penetration (cacao) and immigration streams, the stability of the forest margin in Toro is relatively easy to maintain in comparison to Watumaeta. Copying the Toro agreement and transferring it to Napu however would not only be a deviation from the necessary site-specific adaptation, it would also lead to higher conflict potentials with the state and undermine the legitimacy of the agreement in the eyes of the Napu people who are not interested in connecting their claims with customary re-vitalization. The Napu-discourse on customary land use is more than a dilemma between "legalists" and "those who would empower", but reveals something of the representations of "self" and "other" in natural resource management. Informed by the national discourse on indigenous rights and customary communities as well as by his own experiences in Toro, the Park official overestimated the locals' concern with customary re-vitalization and indigenous land use schemes. However, he underestimated the leaders' concern with law, legal predictability and governmental support. Ingrained with the stereotype of the *local community*, he did not realize how far the local leaders had already built their regional network and learned about the need to make horizontal alliances between villages and to safeguard vertical administrative integration. He further overlooked the fact that the orientations of local communities towards institutions for local resource management are anything but homogenous. Whereas the customary councils of Watumaeta and some other villages wanted to claim "customary lands" in the forest (though without the use of forest zonation), village heads were rather reluctant to do so. Overshadowed by the Poso crisis, village heads feared that a re-vitalization/formulation of customary law that is not in line with the governor's policies may lead to a withdrawal of the provincial police and military support in case of conflict. Village leaders thus found themselves in a paradoxical situation: if the villages would be more empowered in managing their surrounding resources autonomously, simultaneously, their leaders would be less empowered to safeguard overall security.

So far, however, a major problem remains unresolved. This is the issue of internal resource distribution, which is not addressed in the agreement. Who should be responsible to guarantee a minimum of socio-economic security for all community members? Though the *Lembaga Adat* had become a sort of voice for local interests in reclaiming ancestral lands, it must be stressed that equity *per se* is not a goal of *adat*. If plots already cleared will be planted with valuable trees, the question arises as to who has the right to plant and use these trees. What categories of persons should be given priority to use these plots: the poor, the young, and the disinherited? Confronted with this question, almost all Watumaeta leaders answered that the one who cleared the plot must be given the preferential right to use it. The leaders' viewpoint is in line with the *de facto* customary law which is based on the rights of the first clearer. This, however, would mean that in the end, illegal encroachment would be legalized by the agreement and that others who did not invade had no benefits over those who entered the forest. It is too early to conclude and to predict in whose interests the process will work. However, the way the process went so far and how Watumaeta leaders managed it deserves full respect, remembering the organizational situation some years ago. Based on several field stays in 2000/2001, we described Watumaeta as one of the "worst cases" in the Lore Lindu area in regard to leadership practices, social disintegration and resource degradation (Burkard 2002a: 37). Though in another paper we made the point that resource degradation in Watumaeta is not inevitable (Burkard 2002b: 10), we could not imagine a development like the one documented in this paper. Stern et al. (2003) have raised the important question of what "success" means to different people. So far, research on common pool resources focused almost exclusively on questions of sustainability. For social and natural scientists "success" was more or less identical with the "sustainability" of the investigated resource. This however is not the only, and often not the most important question for the populations who use these resources. As Stern et al. (ibid: 457) point out, institutions may be judged by how well they provide jobs, how they contribute to maintaining good relations within the community, how far they provide access to resources from outside, for instance. In regard to the Watumaeta agreement, it is too early to judge its effects on the stability of the forest margin. If, however, we broaden our attention beyond the scope of "forest sustainability", the major positive effects of the CCA are definitely that it provided a forum for discussion, village-internal negotiation, self-organization and – last but not least – the formulation of the necessary community discourses which are needed to develop into a genuine "learning organization". These capacities themselves are indispensable "design principles" for the development of sustainable resource use patterns in the future.

Conclusion

Being linked to the Central Sulawesi "cacao boom", it is tempting to explain forest conversion in the Lore Lindu area primarily in terms of a "capital-driven encroachment". Similarly, the fact that migrants for cacao production now operate vast areas of land formerly used by local families for a mix of food and cash-crop cultivation, suggests a displacement scenario where local people have become marginalized by economically powerful in-migrants. The preceding pages however reveal that encroachment and displacement are linked to a complex set of factors and cannot be related to a single phenomenon. Keeping in mind that in-migration, market penetration, and the technological change they imply do not affect individuals and communities directly, but are "mediated" by local institutions (Agrawal and Yamada 1997), it shows that the responsiveness of local communities towards outer incentives is largely shaped by the socio-cultural, organizational and legal environment, in which local leadership plays a crucial role.

The "Watumaeta case" reveals the limitations of the "methodological individualism" (McCay 2002) adopted by researchers of various scientific backgrounds. The postulation of individual motives for co-operation is misleading in explaining the behaviour of the actors involved. The migrants' interest in co-operation is not informed by variables such as individualism, profit-maximization, competition, co-operation and altruism, but is first of all motivated by a common strive for security. It is not some "individual traits", but the incapability of the institutional environment to safeguard security and the ensuing situation of "mutual vulnerability" which changed their behaviour in the direction of more co-operative action. As pointed out by McCay and Acheson, "common dilemmas must be explained in terms of the dynamics of conflict and competition between different social groups located in history and social systems rather than between the rational economizing individual unspecified and the group also unspecified" (McCay and Acheson 1987: 22). Our case study showed further that one and the same "design principle" might have contradictory effects under different circumstances. Under an autocratic and centralized government, where the village head represents the solemn de facto source of power in resource distribution, immigration and the concomitant socio-economic and cultural *heterogeneity* definitively had negative effects on the social cohesion in the village and the management of CPR-resources. Under the situation of political decentralization, where the power of the head is controlled by the BPD, socio-cultural and economic *heterogeneity* was advantageous for the community. Of course, the question of how this heterogeneity is structured is crucial. In Watumaeta, the existence of clearly defined groups with clear boundaries enabled locals and migrants to formulate their aspirations more precisely vis-à-vis each other and to engage in meaningful deals and negotiations. Accordingly, studies on the management of common pool resources must shift from the analysis of "design principles" to the analysis of the contexts within which these design principles function in one or the other direction. Similarly, analysis must shift from the investigation of "indigenous" or "local" institutions to the contexts within which self-organization can happen (see McCay 2002).

The case of the "forum agama" shed some light on the importance of "win-win experiences" in the process of community building in formerly socially fragmented communities. It further helped to facilitate the building of horizontal alliances between individual villages that could organize important steps (e.g. the concordance of rules and sanctions). Such alliances are not only important for future common monitoring, but they primarily enhance the bargaining power of the communities involved. The negotiation process revealed also that local communities and their leaders may be reluctant to formulate their resource policies in opposition to state directives, instead of "empowerment for its own sake" they may rather seek for recognition and support from higher government institutions because rules recognized by the government may be easier to enforce internally. Of course, the factors that were important for the emergence of self-organization are not necessarily the same that are important for the sustainability of the institutions created (e.g. the LKD) and the sustainable management of the forest margin. However, much has been achieved on the way from "wild west" to "learning organization".

References

Abdul-Kadir, M. (2004): Orang Kampung and Pendatang: An Analysis of Demographic Structure and Migration in Two Forest Margin Villages. In: Fremerey, M., Gerold, G. and E. Guharja (Eds): Land Use, Nature Conservation and the Stability of Rainforest Margins in Southeast Asia. Berlin.

Agrawal, A. (2002): Common Resources and Institutional Sustainability. In: Ostrom, E. et al. (Eds): The Drama of the Commons: 41-86. Washington

Agrawal, A. and Yamada, N. (1997): How Do Local Institutions Mediate Market and Population Pressures on Resources? Forest *panchayats* in Kumaon, India. *Development and Change*, Vol.28: 435-65.

Aragon, L. V. (2001): Communal Violence in Poso, Central Sulawesi: Where People Eat Fish and Fish Eat People. *Indonesia* 72: 45-79.

Allen, J. C. and D. F. Barnes (1985): The Causes of Deforestation in Developing Countries. *Annals of the Association of American Geographers* 75(2): 163-84.

Berkes, F. (2002): Cross-scale Institutional Linkages: Perspectives from Bottom Up. In: Ostrom, E. et al. (Eds): The Drama of the Commons: 293-317. Washington.

Burkard, G. (1999): Bauern, Kader und Migranten. Ländliche Modernisierungsprozesse am Beispiel einer Dorfgemeinschaft in der Sonderregion Yogyakarta/Indonesien. Pfaffenweiler

Burkard, G. (2002a): Stability or Sustainability? Dimensions of Socio-economic Security in a Rain Forest Margin. Storma Discussion Papers No. 7. Bogor

Burkard, G. (2002b): Natural Resource Management in Central Sulawesi. Past Experience and Future Prospects. Storma Discussion Papers No. 8. Bogor.

Bromley, D. W. (1992): Making the Commons Weak: Theory, Practice, and Policy. San Francisco

Campbell, J. (2002): Differing Perspectives on Community Forestry in Indonesia. In: Colfer, J. P. and P. Resosudarmo (Eds): Which Way Forward? People, Forests, and Policy Making in Indonesia: 110-126. Singapore.

Colfer, C. J. P. and I. A. P. Resosudarmo (2002): Introduction. In: Colfer, J. P. and P. Resosudarmo (Eds): Which Way Forward? People, Forests, and Policy Making in Indonesia. Singapore.

Dietz, T. et al. (2002): Introduction. In: Ostrom, E. et al. (Eds): The Drama of the Commons: 3-37. Washington

Dolšak, N. and E. Ostrom (2003): The Challenges of the Commons. In: Dolšak, N. and E. Ostrom (Eds): The Commons in the New Millennium. London.

Dreyfus, H. L. and P. Rabinow (1982): Michel Foucault: Beyond Structuralism and Hermeneutics. Chicago

Evers, P. J. (1995): Preliminary Policy and Legal Questions about Recognizing Traditional Land Rights in Indonesia. *Ekonesia. A Journal of Indonesian Human Ecology*. No. 3: 1-23.

Eijkemans, C. H. (1995): Profitability or Security. Decision-Making on Land Use among Toba Batak Peasants in North Sumatra, Indonesia. Nijmegen

Gibson, C., Ostrom, E. and McKean, M. (2000): Forests, People, and Governance: Some Initial Lessons. In: Gibson, C., McKean, M. and E. Ostrom (Eds): People and Forests. Communities, Institutions, and Governance: 227-242. London

Gibson, C. (2001): Forest Resources: Institutions for Local Governance in Guatemala. In: Burger, J. et al. (Eds): Protecting the Commons: A Framework for Resource Management in the Americas. Washington

Hardin, G: (1968): The Tragedy of the Commons. *Science* 162: 1243-1248.

Hardjono, J. (1993): Land, Labor and Livelihood in a West Java village. Yogyakarta.

Kopelmann, S., Weber, J. M. and D. M. Messick (2002): Factors Influencing Cooperation in Common Dilemmas: A Review on Experimental Psychological Research. In: Ostrom, E. et al. (Eds): The Drama of the Commons: 113-156. Washington

Li, M. T. (1995): Images of Community: Discourse and Strategy in Property Relations. *Development and Change*, Vol. 27: 501-527.

Li, M. T. (2000): Locating Indigenous Environmental Knowledge in Indonesia. In: Ellen, R. et al (Eds): Indigenous Environmental Knowledge and its Transformations. Critical Anthropological perspectives: 121-147. Amsterdam.

Li, M. T. (2001): Agrarian Differentiation and the Limits of Natural Resource Management in Upland Southeast Asia. *IDS Bulletin* Vol. 32 No. 4: 88-93.

Lieb, R. (2003): The Adaptation of Internally Displaced Persons in a Rain Forest Margin area. The Case of Kecamatan Lore Utara in Central Sulawesi, Indonesia. Diploma thesis Faculty of Organic Agricultural Sciences, University of Kassel.

McCay, B. (2002): Emergence of Institutions for the Commons: Contexts, Situations and Events. In: Ostrom, E. et al. (Eds): The Drama of the Commons: 361-402. Washington.

McCay, B. and J. A. Acheson (1987): The Question of the Commons. Tucson.

Mischung, R. (1984): Religion und Wirklichkeitsvorstellungen in einem Karen-Dorf Nordwest-Thailands. Wiesbaden.

North, D. (1990): Institutions, Institutional Change and Economic Performance. Cambridge

Olson, M. (1965): The Logic of Collective Action: Public Goods and the Theory of Groups. Cambridge

Ortner, S. (1984): Theory in Anthropology Since the Sixties. *Comparative Studies in Society and History* 26: 126-166.

Ostrom, E. (1986): Issues of Definition and Theory: Some Conclusions and Hypotheses. In: Proceedings of the Conference on Common Property Resource Management: 599-614.Washington

Ostrom, E. (1990): Governing the Commons: The Evolution of Institutions for Collective Action. New York

Ruf, F. and Yoddang (1999): The Sulawesi Cocoa Boom and its Crisis. *Plantations, Recherche, Development*, July-August: 248-53.

Sahlins, M. (1981): Historical Metaphors and Mythical Realities: Structure in the Early History of the Sandwich Islands Kingdom. Michigan.

Scholz, F. (1962): Der Herr des Bodens in Ostindonesien. Köln

Sitorus, F. (2002): Land, Ethnicity and the Competing Power. Agrarian Dynamics in Forest Margin Communities in Central Celebes, Indonesia. Storma Discussion Papers No. 5. Bogor

Stern, P. C. et al. (2002): Knowledge and Questions After 15 Years of Research. In: Ostrom, E. et al. (Eds): The Drama of the Commons: 445-490. Washington.

Strathern, M. (1987): Dealing with inequality: Analyzing Gender Relations in Melanesia and Beyond. Cambridge

Varughese, G. (2000): Population and Forest Dynamics in the Hills of Nepal: Institutional Remedies by Rural Communities. In: Gibson, C., McKean, M. and E. Ostrom (Eds): People and Forests. Communities, Institutions, and Governance: 193-227. London

Togea (2004): Sejarah Desa Watumaeta (History of the village of Watumaeta). In: Kesepakatan Konservasi Masyarakat Desa Watumaeta (Community Conservation Agreement of Watumaeta). TNC, Palu

Zerner, C. (1994): Through a Green Lens: The Construction of Customary Environmental Law and Community in Indonesia's Malukku Islands. *Law and Society Review* Vol. 28: No. 5: 1079-1122.

Chapter VI

Aristocrats and Democrats – Leadership and Resource Use in the Villages around Lore Lindu

Reinald Döbel and Momy Hunowu

Abstract

This chapter, drawing on our own interviews and observations, as well as those of other researchers, describes differing leadership styles in the Lore Lindu region and their consequences for resource use, and therefore the sustainability of the conservation efforts of the Lore Lindu National Park Authority. We argue that some of the ideals of older leaders fit well with the new democratic environment and with ideals of participatory democracy promoted worldwide, while others contradict them. On the positive side, old and modern ideals both emphasise mutual respect and listening in the local con text, expressed through *musyawarah mufakat*. On the negative side, local myths sometimes accorded ruling families a quasi-supernatural origin, fostering autocratic attitudes. Dutch colonial government as well as the Suharto regime both relied on and further enhanced the autocratic aspects of local rule. Which side of tradition present local leaders follow, affects the success of reconciling local development needs and conservation efforts: it is difficult to muster local support for voluntary conservation agreements when imposing leaders give an example of resource misuse. Both autocratic leaders and democratically minded opposing groups may refer to tradition to justify their course of action. Examples on both sides show that the tradition of listening and respect, which conforms to ideals in other areas of the world and the discourse of participatory development, is more likely to produce sustainable solutions.

Introduction: Language as an Entry Point to Development Research

The present chapter's focus on leadership and use of resources is the result of the fruitful association of the research project LAGSUS (Language, Gender and Sustainability)[1] to STORMA – fruitful in terms of complementing theoretical perspectives on the tension between development and conservation, and fruitful in terms of the discussions between researchers of different backgrounds.

LAGSUS understood language as a resource for development in its own right (see Bearth 2008). For its sociology component – represented by the German author of this chapter – language also provided an entry point to questions which are often not easy to approach even though they are burning issues for all concerned: questions about leadership and justice, and how these contribute to or hinder development. The narrower the natural resource basis, the more pertinent these questions become: when there are fewer resources to use, distribute and manage, any decision about access to and use of these resources touches people's livelihoods more directly – unless there are alternative sources of income from the "modern" economic sector. The comparative perspective of LAGSUS, which was also active in resource-poor areas of communal reserves in Namibia, the Baruli lands south of lake Kyoga in Uganda, and the Toura lands in Ivory coast, provided a background for seeing similarities in people's hopes and conceptions across continents: Similarities in the hopes most people in both African and Indonesian research sites place in a "development" which is synonymous with the expansion of a modern economy and modern institutions of education, health care, and transport. Such similarities make the global enthusiasm for the recognition of local knowledge as a basis for *sustainable* development look more like an idealistic attempt of well-intentioned groups from industrialized countries to reconcile the belief in "development" with the need to manage natural and human resources without undermining the long-term basis for human economic activity (see also Sunito, this volume). The contrast between this idealism and the emerging awareness of the conflict

[1] The four-year LAGSUS project was funded by the Volkswagen Foundation whose support is gratefully acknowledged.

between the present need to use natural resources for development, and the need to preserve natural resources for the needs of future generations also became apparent in the local discourses in the villages surrounding the newly created Lore Lindu National Park on which this chapter will focus.

The association also had practical consequences. The need to share research time between different LAGSUS research sites resulted in the unusual constellation that the main foreign researcher had only four periods of field research of about a month each at his disposal to get a grasp of the complex situation of the forest margin areas on the edge of the Lore Lindu National Park. He spoke none of the several local languages and his command of *Bahasa Indonesia* only achieved an acceptable level through the relationships with his co-author and their other collaborators.[2] Without growing sympathy and trust neither the research perspectives nor the methods imported by the foreign researcher could have achieved anywhere near the results presented in this chapter. Conceptually, we focused on issues of the relationship between local institutions, leadership, resource use, language, conservation and development in all the interviews in six of the sixty villages surrounding the Lore Lindu National Park. During one of the field research periods these interviews were based on the additional stimulus of videotaped dramas, developed and performed locally (with a lot of improvisation) by school children, about the violation of village conservation agreements.[3] And we tried to learn about local networks of actors by asking knowledgeable villagers to draw maps of the villages, in order to locate the houses of members of farmers' groups and of village functionaries whom we tried to interview individually, in addition to the group interviews following the showing of the videotaped performances of school children.

While only two members of the research team participated in the writing of this chapter, we believe that all would agree that some of the most important information came about not through the planned schedules, but through chance meetings – and the peculiar timing of

---

[2] Because they were so important to the success of the team, their names should be mentioned here: Bani Susanto, Uus Supriyadi, and Marten Lancia

[3] This was the responsibility of Uus Supriyadi who, as a theatre expert, had been seconded to the team from a state agency for cultural affairs in Palu.

one of the periods of field research, which lasted from shortly before Christmas 2004 into the first two weeks of the new year 2005. On the one hand, research activities could not follow a very strict schedule because people were busy with preparations and celebrations. On the other hand, we were invited to stay as a team in the houses of several *Kepala Desa* with two of whom we also went to church on Christmas Day and New Year's Eve, and we were invited to attend the church service and village celebration organised by a third. This provided us with unique opportunities for relaxed conversations and casual observations, which complemented our discussions of the results with other long-term researchers. We believe it is useful to mention these circumstances because they show the value of accepting unplanned events during field research as opportunities for insights, even if they do not fit the more controlled model of gathering information often taught to students of the social sciences.

Lore Lindu: From a Heroic Past to a Sustainable Future?

Other researchers too were struck by how certain local leaders regard the preservation of their history as an important task. Being a leader and being a guardian of oral tradition belongs together in two respects: first, guarding unrecorded local history contributes to a sense of identity (Sohibuddin, this volume, Sunito 2005), and second, from that history derives the legitimacy of the "traditional" order and their leaders – as expressed in orally transmitted *adat* regulations. Hence, guarding history implies guarding their own position. As in other parts of Indonesia, "traditional" leaders in Central Sulawesi were deliberately sidelined and the local forms of *adat* discouraged during the autocratic regime of the Suharto years (1965-98).

As a result of this situation, the revival of *adat* seemed to express the spirit of democratization in Indonesia after the fall of Suharto: local customs which had been suppressed in the name of a strong central state building effort were treated with a new respect as one of the sources for the creation of "unity in diversity", which had become and remained an empty slogan in the eyes of many people. In the eyes of Indonesian NGOs as well as in the eyes of an international movement for indigenous rights, this *adat* revival could also be used to reinstate their rights over the forests to local people who have made their homes there for as long as they can remember – a right which the Indonesian government's forest laws repealed "with a stroke of the pen" in the early 1980s. The result: virtually all forests outside of Java were declared to be "state land" (Mappatoba 2004: 46).[4] The laudable intentions and understandable enthusiasm behind the support for the revival of *adat*, however, ignored the fact that not all of the traditional orders held together by *adat* regulations were based on values which deserve this support from the point of view espoused by most present NGOs, which is to consider human rights and democracy as essential elements of all development. In the case of Central Sulawesi, some present practices can be understood better in the light of oral traditions which show that adherence to *adat* may mean support for the privileges of a certain social group, privileges which can hardly be defended as just.

This is a conclusion suggested by the oral traditions we encountered in the course of our field research, particularly as they relate to *adat*. The oral historical traditions concern the origin of the indigenous societies having lived in the areas surrounding the Lore Lindu National Park, particularly the role of founding figures who had first established settlements in then virgin forest. One story has princess *Bunga Manila* follow her beloved water buffalo *Tolelembunga* over the

---

4 Later we will come back to one of the seemingly overlooked aspects of this legislation: depriving many indigenous people of the right to manage what they saw as *their* lands which they felt perfectly capable of managing with the use of their various *adat* laws. Indonesian and International NGOs and other organizations began to see, in the revitalization of these *adat* laws, one way of reinstalling these indigenous people into their role of "stewards" of the forests. In the process, they overlooked the detail that stewardship of forests was only one role of *adat*: the management of forest resources was only one aspect of the overall regulations of social life. Other aspects concerned the regulation of marriage relationships and, most importantly, of the role different social groups played in the respective societies, which were not all egalitarian "close-knit" communities, as the now famous international image portrays them. Where these societies were organised in a hierarchical manner, the *adat* laws also guarded the privileges of certain groups. Hence, the revitalization of these laws implied a revitalization of these privileges. This is the aspect overlooked most in the enthusiasm created by the climate of democratization following the fall of Suharto.

mountain and establish a new village in the Napu valley. A few years later she meets and then marries the warrior *Sadunia*, born from a bamboo stem, thus founding a line of *magau* from whom the leading families in this village still claim descent (Tokare 1990, Tokare 1999, and our own data from Sedoa).

In this as in other stories and oral histories collected by TNC (The Nature Conservancy) in the course of their participatory negotiations for voluntary village conservation agreements, one hears of heroic men and women. Princesses do play a role, as in the story above, but are overshadowed by the exploits of the male warriors. The picture emerging from oral history as well as the stories about the functions of *adat* in present society shows rather independent local societies, each led by an aristocratic figure, and often at war with each other.[5] Extant stone statues called "megaliths" as well as the stories themselves point to a long history of settlement. When the founding figure of the regional ruling families, *Manuru*, gave the current names to the important locations of the present *Kecamatan* Lore Utara and Lore Selatan, he obtained them by asking people already living there. The foundation story also attributes the origin of the aristocratic ruling families to a marriage of *Manuru* with a princess from heaven found in a bamboo (Tokare 1990: 8 f.).[6] Not surprisingly, some of their descendants went back to heaven according to some stories (Tokare 1990: 10). It is reported that before his death his sons were instructed which lands they would rule – with some of these lands having to be conquered first (Tokare 1999: 6). Such stories not only mention heroic defenses against invading armies from places as far away as a Mongolia ruled by *Khu Bhi Laihkan* (TNC-LLFO 2001: *Sejarah Desa Wuasa*), but also trace the descent of the families of rulers back to *Manuru*, thus attributing a common origin to the network of important families in the whole area between Palu and Poso (Tokare 1999). Many, if not most present-day *Kepala Desa* and other functionaries belong to that network, as noted by Sunito (2005: 53 ff.).

---

5 The life-long fight between *Mpebiaro* and *Latandu* is reported differently in the oral traditions collected by TNC and earlier collectors. See Tokare (1990) vs. *Sejarah Desa Wuasa* in TNC-LLFO (2002)

6 Sunito (2005: 190), quoting Mattulada, sees in the figure of Manuru a clear sign of the Bugis kingdom of Luwu.

Settlements formed around particularly strong leaders and these were free to engage in warfare against each other. The fortifications of these settlement and their location on elevated places (Sunito, this volume; and Sunito 2005: 81) show that warfare was sufficiently widespread. The high esteem for male bravery in war is shown in the oral accounts of the war with the Dutch[7] and the remark that in those days a young man would be ashamed to return from war without part of the skin of a slain enemy.[8] The spoils of war included prisoners, who played an important role in the economy of the *bangsawan* (aristocratic class) or *maradika/tuana*[9] as *hamba* or *budak* (slaves). They provided services as domestic workers, such as cooks, as labourers, and also as hunters. While the heroic stories do not mention slaves at all, the collected oral traditions clearly show their place in society: slaves had to be included in the bridewealth for women of the *maradika* class (Tokare 1990, quoted after Sunito 2005: 68). Memories of how slaves were treated also still exist: some of the descendants of former slaves remember that a slave could be sold for a sack of salt, if the owner was short of that commodity. How strictly the segregation between *maradika/tuana* and *budak* was enforced is illustrated by the story of a slave who was observed eating from the plate of a *maradika/tuana* – which was strictly prohibited. Without being told the reason, he was removed immediately from his duties in the kitchen, given other light tasks and fed well. The night before an oncoming feast he was locked into a house, still without any knowledge of what would happen. It was only when he heard the songs of sacrifice being sung outside that it dawned on him that he would be the sacrifice. He still did not know why. Only when he was taken out of confinement the following morning was he informed about his mistake and the death penalty to be meted out to him that very day.[10]

Death penalties for misbehaviour also happened elsewhere: while a member of a *Lembaga Adat* told a group of visiting students[11] with a

---

[7] One of these accounts we listened to in the interview with Virginia Nello on 26 December 2004, another is reported in Hunowu (2005: 36 ff.)

[8] Interview with Edy Lumalo, 26 December 2004

[9] The term *maradika* was used mainly in the Kulawi area, while the term *tuana* is mentioned by Sunito (2005) for the Napu valley

[10] We are very grateful to Laksmi Savitri for mentioning this story.

[11] One of the authors happened to be present during this visit in September 2003

curious kind of approval that in former times certain offences carried the death penalty, a more critical member of the village elite one day remembered how death penalties were administered: by binding a slave to a post and hacking him to death with a *parang* (slashing knife). The villagers were obliged to take turns in slashing the wrongdoer. Psychologically, this probably ensured that all took responsibility for this death – as presumably all had been involved in arriving at the decision for the death penalty in the preceding meetings establishing the culpability of the wrongdoer in a *musyawarah mufakat*.

We do not know to what extent pre-colonial life was characterised by such violence. The recorded historical stories are silent about such events,[12] and mention instead the abilities of strong leaders not only to be brave and outspoken when necessary, but to show patience and justice to their followers.[13] Such qualities are still evoked today to describe "good leadership" (*Kepemimpinan yang baik* – see below) and show that ideals have as much historic continuity as the network of relationships of aristocratic families.

Similar qualities of initiative, generosity, and patience were required in the past for the more mundane tasks of organising the agricultural activities in the forest. The organization of *robo* working groups – which has been described in some detail by Sunito (2005: 72 ff.) – required their leaders to be able to organise the work of cutting, clearing and planting, but also to keep their followers satisfied by providing food, organising celebrations, and assisting members when needed: they were free to chose which *robo* they wanted to belong to and could always "vote with their feet" if they were unhappy. The ability to establish relationships of trust both with the members of the group and with outsiders to whom he represented the group was among the important personal qualities necessary to establish oneself as a *robo* leader – in the eyes of Sunito's informants (Sunito 2005: 77). Together with wet rice cultivation and privately owned gardens, the *robo* form of shifting cultivation in the surrounding forests secured

---

[12] The *Sejarah Desa Wuasa* does not mention the human sacrifice which had been deemed necessary for the foundation of this village, as reported by Sunito (2005: 52)

[13] *Tindarura juga memiliki kepemimpinan yang kuat, berwibawa, jujur, adil, sabar dan sangat bijaksana, walaupun sangat berani bertindak keras dan tegas.* (Tindarura also was a strong leader who had authority and was honest, just, patient, and very resourceful, although he also dared to act firmly and harshly. Tokare 1999:6f.)

the food supply of the pre-colonial societies in the *Pekurehua* (Napu) part of pre-colonial Central Sulawesi. While other forms of shifting cultivation were important elsewhere, the *robo* organization seems to have been peculiar to this part. Nevertheless, work groups – called *mapalus* – were and are a common feature of social life here as in other parts of the region, and the qualities cited for present-day leaders conform to what has been reported by Sunito. At the same time, it is important to keep in mind that the *robo* form of organization clearly worked to the advantage of *robo* leaders: On average, the system allowed those leaders to cultivate about twice as much land as others managed to cultivate – 1.5 ha as compared to the 0.5 to 0.75 ha of other people – because the work done by the members was worth more than the services provided by the *robo* leader in terms of food, festivities, and other obligations. We do not know whether critical minds of those days perceived the injustice of the system as clearly as one of Sunito's present-day respondents, but find an indication of a more widespread dissatisfaction with the systen in the information that it was gladly abandoned in the 1970s (Sunito 2005: 79).

Obviously, a *robo* leader needed sufficient resources and it was therefore mainly the owners of large herds of water buffaloes who could undertake such a task. As in other parts of Sulawesi and beyond, water buffaloes played an important part in the *adat* regulations in the form of payments for fines and also in the form of bridewealth. It may be that the peculiar conditions of the Napu area with its unusually vast grazing lands provided the conditions for a higher number of water buffaloes than elsewhere, but it is difficult to imagine that the role they played in securing the position of the elite can not be taken as exemplary for the whole of pre-colonial Central Sulawesi which, because of geographic isolation, remained largely untouched by the more centralized kingdoms in the South and by Dutch colonialism. We can therefore assume that the water buffalo's role in "maintaining the social structure" (Sunito 2005: 71 f.) was similar in other areas, such as the Kulawi valley on the other side of the mountain range which now forms the Lore Lindu National Park. Water buffaloes were much more than work animals, to which function they were reduced in other rice growing societies. *Tolelembunga*, the mythical *kerbau* whose choice of place led to the founding of *Sedoa* is vested with magical powers in the story recorded by Tokare (1999). Water buffaloes were the only acceptable payment for brides, and the difference between the price for a slave wife (two water buffaloes), a commoner wife (4-6 water buffaloes) and an aristocratic wife (8-12 water buffaloes and a slave) is an obvious expression of pre-colonial stratification (Tokare 1990, cited after Sunito 2005: 68). Water buffaloes were also essential for certain rituals, particularly in connection with death. Furthermore, fines for transgressions of *adat* regulations had to be paid in water buffaloes. The system thus ensured that ownership of water buffaloes was concentrated among aristocratic families, whom others had to approach for help to meet the stipulations of *adat.* This, in turn, gave rise to widespread relationships of bondage between aristocrats and commoners.[14]

The superior position of the nobles was thus ensured materially through a system of unequal exchange concerning labour practices in the *robo*, bond relationships and bride prices,[15] and also through the spoils of war, for which there seem to have been frequent enough opportunities. The position also found expression and support from the *adat* system, which prescribed behaviours expressive of respect for the nobles and contained regulations determining the payment of bride prices and the sanctions for transgressions. To the extent that the *magau* depended on the support of other important members of the nobility, who might suggest replacement of a *magau* they found unfit for ruling,[16] their authority came at the price of conforming to certain standards themselves, which had nothing to do with the bravery and strength required for warfare.

Considering the elaborate rules of respect surrounding the nobles, this aspect must have remained rather invisible to the commoners, some of whom still remember that bowing down deeply before members of the nobility, or waiting for them to start eating, were "natural" behaviours.[17] The respect for the special position of the nobility was also visible in the elaborate taboos surrounding the death rituals for

---

[14] Sunito (2005: 71) quotes a Dutch investigation of 1903 which reported that 90% of the population were in a state of bondage.

[15] We find Sunito's (2005) detailed description of the reproduction of the social system of *Pekurehua* (the Napu valley) particularly enlightening.

[16] This aspect of the *Lembaga Adat* was stressed in two interviews in Toro in September 2005

[17] Interview in Watumaeta, Dec. 25, 2004

nobles: fines were demanded of anyone who performed certain mundane tasks in the vicinity of a house where a noble had passed away (Sunito 2005: 68) or who failed to dismount his horse when passing the arcs erected to mark ongoing mourning rituals,[18] fines which were specified in *adat* regulations.

That these regulations were indeed an instrument of rule and not, as a present romantic misconception presents it, simply an expression of a "traditional way of life" became clear when a descendant of a *magau* stated that she could not be a member of the *lembaga adat* because the *adat* was made by the *bangsawan* (nobles) and then guarded by the commoners sitting in the *lembaga adat.*[19] With the exception of a few rules for the protection of river banks and sources of drinking water, *adat* was not concerned with what today is called natural resources management. While the guardian spirits of the forest – of particular places and of the forest as a whole – were honoured, this happened in a spirit of deference, rather than in the spirit of "guardianship" which characterizes some of the modern conservation efforts. *Adat* regulated how people lived together, how they related to each other, and how property was transmitted through inheritance. The one set of regulations that has been committed to writing shows an effort to revive the spirit of adherence to *adat.* Of the fourteen chapters, four are concerned with organizational matters, two are concerned with marriages, illicit relationships, and marriage payments, one each with division of property (inheritance), mourning regulations, theft, public commotion, mischief, slander, "other matters" – and one with transport on motor bikes. *Tanah Adat* (*Adat* land) does not figure in these regulations which evoke "togetherness" (*watu tambaga momepanimpu*) as the essence of a harmonious life guided by *adat*, which "breathes the philosophy" of

- Do not envy (Mome Hingki = Jangan ada denki),
- Do not hurt each other (Mome Hubi = Jangan saling menyakiti),
- Do not hate (Mome Podaa = Jangan ada benci)

(Majelis Adat Kecamatan Kulawi, April 2003: 2).

---

[18] Interview in Watumaeta, Dec. 25, 2004

[19] Interview with Virginia Nello in Watumaeta, Dec. 25, 2004

Thus, the written code does not bear out the claim we encountered surprisingly often: that one of the functions of *adat* had always been to ensure a responsible use of natural resources, and that therefore the revival of the traditionally strong adherence to *adat* in danger of being lost in the modern world would benefit modern conservation efforts. The claim was well received by NGOs and the National Park Authority for whom the various *Lembaga Adat* in the village became "natural" cooperation partners in the efforts to establish the voluntary village conservation agreements. This may have been facilitated by the international reputation of indigenous peoples as responsible and self-conscious "stewards" of their natural environment, who should be reinstated in that role, instead of being deprived of their use-rights in the name of conservation.[20]

Such advocacy for local communities or "traditional people" is sometimes also linked to efforts to protect what is believed to be a more contented life. From that angle, *adat* can be viewed as preserving this contentedness, which adds to the attraction of supporting *adat* in Central Sulawesi. It should be noted, however, that many of the songs sung during the *dero* round dances, are songs about tears and longings for a lover who has left or is impossible to reach, tears caused by the impossibility of being together. These dances are performed at all festivities, and men and women may freely join the big circle at any place they choose, thus conforming to the image of "togetherness". For a man and woman to really come together, however, meant either becoming involved in an illicit affair, which carried a heavy *adat* fine if discovered, or paying the bride prices stipulated by *adat*. Listening to DVD recordings of such *dero* songs afforded us with the opportunity of hearing one of the more knowledgeable local men admit that these *adat* rules have indeed made it very difficult for

---

[20] These arguments form part of an emerging new paradigm of nature conservation and park management. In 2002, IUCN (the *International Union for the Conservation of Nature*) published a special issue of its journal *parks* on *Local Communities and Protecte Areas*. The introduction states: "That local communities can and must play a critical role in protected areas is not new; what is new, is the way that, in practice, their role is being accepted, encouraged and, indeed, embraced in very different parts of the world. Community involvement is central to an emerging new paradigm for protected areas (PAs) worldwide." See also Feyerabend et al. (2004).

men and women to come together, and that this has indeed been a major cause for the sufferings depicted in the *dero* songs.

Aristocracy and Colonialism

With the exception of Palu, Central Sulawesi, particularly the mountainous area which has now become the Lore Lindu National Park, had remained untouched by Dutch colonial efforts until the beginning of the 20th century, when the Dutch shifted to a policy of direct intervention even in the outer islands under the title of an "ethical policy" aimed at the "emancipation of the native population". In practice, this meant that the Dutch colonial government would attempt to get local power holders to sign a *Korte Verklaring* (short declaration) with which they would acknowledge Dutch supremacy while remaining responsible for their area's administration[21] (Weber et al. 2002: 10). The greater political and economic participation of indigenous populations – the "natives" as they were called in colonial documents, under the "ethical policy" in practice meant support for increased agricultural production, such as for irrigation in rice cultivation, but also including enforced cultivation of crops such as coconut, sugar, and coffee (Weber et al. 2002: 12 f.).

While the intention was to avoid violence as much as possible, resistance against the Dutch overlordship both in *Pekurehua* (Napu) and in Kulawi was determined, and strong enough to require military intervention, locally known as the Peore war in Napu and the Bulu Momi war[22] in Kulawi. The first Dutch attempt to enter the Kulawi valley was blocked by the army gathered by *magau* Touwalangi (To woalangi) at mount Momi in 1905, but he surrendered after the Dutch army, composed mainly of Minhasa soldiers from North Sulawesi, had gained access via another route in 1906 (Hunowu 2004: 36 f.; Weber et al. 2002: 15). On the other side of the mountain, in Napu,

[21] "With the signing of the *Korte Verklaring* the created *zelfbesturende landschappen* agreed not to sign contracts with any other powers and to obey all instructions of the Dutch queen and her officials (Spit 1991: supplement VII)" (Weber et al. 2002: 10).

[22] *Peore* is the name of the location in the Napu valley, while *bulu* is the local word for "mountain" and *momi* means "sweet" in the Kulawi language. In both instances the name of the war refers to the location where the decisive encounter took place.

several attempts to get *magau* Ama to sign a *Korte Verklaring* failed to overcome his determination not to accept foreign rule, and when he finally appeared for negotiations at Peore in 1907, a cultural misunderstanding of the European custom of a welcoming handshake, which *magau* Ama took as an aggressive act because of its vivacity, erupted into severe fighting which was ended only by the waving of a white flag by Ibu Polite, the sister-in-law of Ama (Tokare 1990:43 f.).[23] Later she married a commoner whom the Dutch allowed to rule as *magau* Kabo[24] under their jurisdiction.

In the comparatively short period between 1905 and 1942, Dutch rule left important marks on the populations and landscapes of Central Sulawesi: the extension of irrigated rice farming, the obligation to grow coconut trees, the introduction of maize intercropping, the prohibition of shifting cultivation and the enforced resettlement of hill people in the valleys,[25] the improvement of the road infrastructure, the introduction of education, and the Christianization of the region, mainly through officers of the Salvation Army, who in 1931 in the Kulawi valley, entertained 50 officers in 71 villages and opened some 18 schools (Weber et al. 2002: 25). With respect to the stated goals of the "ethical policy", however, these interventions seem to have been a failure: there was no general improvement of the economic conditions of the population, (Weber et al.: 28) and education, apart from that offered by the mission schools, was restricted to descendants of the local elites, for whom it was seen as a preparation for administrative service (Weber et al.: 24).

Despite Christianization on the one hand, and the active spread of Islam through Bugis raja who cooperated with the Dutch on the other,

---

[23] We heard this story also from a fourth generation descendant of Ibu Polite. In that account the number of soldiers was greater, the unequal fight between Dutch rifles and local swords more vivid, and Ibu Polite's waving of the flag more heroic, including a touch of magic: how she had managed to run through flying bullets unharmed (Interview with Virginia Nello, 26 December 2004).

[24] Interview with Virginia Nello, 26 December 2004. It is interesting that another third generation descendant of *magau* Kabo, Sam Kabo, had been assistant Governor in Palu in 1968, according to the headmaster of the school in Watumaeta, interviewed 25 December 2004.

[25] Sometimes this had disastrous effects on the health of the resettled people, to the extent of forcing the Dutch to allow them to return to the hills and to abandon the policy altogether (Weber et al. 2002: 22 f.).

adat also continued to exercise its influence under the leadership of those local nobles who had agreed to pledge allegiance to the Dutch queen. The only real change in the local power structure, according to colonial records, was the abolishment of the position of *tadulako* – i.e. warriors. Having noted the close links between powerful families across the region ourselves, we therefore agree with Sunito (2005: 190) that

> *"From the time the Dutch got hold of the Napu valley in the beginning of 20th century, the political power in the valley and later of the village Wuasa were circulated only amongst a small number of families. The descendants of the nobles were also obviously represented in other important position in the community and beyond, such as head of sub-districts, head of school and teachers, church leader, and later in the new village institutions such as the village representative body and in the village and sub-district conservation bodies."*

We only need to add that the same applies to the Kulawi valley, on the other side of what is now the Lore Lindu National Park, where even the settlement pattern in villages still clearly reflects divisions between *bangsawan/maradika*, commoners, and slaves (Hunowu 2004: 29, 47).

Local Government under the Suharto Regime

After a brief spell of Japanese rule at the end of World War II, Indonesia declared independence in 1945, but had to fight by diplomatic as well as military means for the recognition of the former colonial master, the Netherlands, in 1949. Nevertheless, the first Indonesian Constitution and the first law about local government dates from 1945.[26] It is remarkable to find a considerable continuity in the emphasis of Indonesian law on local autonomy, from the early times of Independence through the Suharto times, as seen in the creation of three levels

[26] The United Nations Economic and Social Commission for Asia and the Pacific hosts a country paper on Indonesia within the framework of the Comparative Study on Local Government in Asia and the Pacific, (online at: http://www.unescap.org/huset/lgstudy/country/indonesia/indonesia1.html) from which this information is drawn.

of autonomous local government in the Local Government Acts 22 (1948) and 1 (1957), with Act 18 (1965), and Act 5 (1974) again stressing local autonomy and the "decentralization and deconcentration principle".[27] This is remarkable because the general image for the years of authoritarian rule under Suharto is one of strong, centralized government machinery controlled by and serving the center in Jakarta. Locally, stories abound about the appointment of officers who originated from different areas, and sometimes did not speak the local language, at the district (*Kabupaten*) and sub-district (*Kecamatan*) levels.

It is generally acknowledged that these years brought a uniform system of administration and services all over Indonesia, offering general education and health care even to remote villages, and achieving the spread of the national language *Bahasa Indonesia* over the whole nation. In Central Sulawesi as elsewhere, this uniform system of administration officially had no place for an *adat* council (*Lembaga Adat*): all laws concerning finances, land ownership, and forest management were issued from Jakarta, and supposed to be applied uniformly all over Indonesia. However, members of locally powerful families managed to maintain a certain continuity even under this system,[28] and according to present members of various *Lembaga Adat* in different villages, these councils never stopped functioning, albeit in an unofficial manner: they continued to regulate matters of marriage and inheritance, and administer sanctions for transgressions. They also continued to deal with land matters, where the newly introduced land laws were not sufficiently known or trusted for people to register their land for secure titles (Sunito 2005).

Decentralization of Local Government and the Effects on Resource Use

The era of *reformasi* after the Indonesian "Revolution" of 1998 saw a renewed emphasis on regional and local autonomy, which was reflected in Law No. 22 (1999).[29] The most important change was to

---

[27] http://www.unescap.org/huset/lgstudy/country/indonesia/indonesia1.html

[28] Sunito (2005: 53 ff.) gives some details for Napu

[29] For details on this law and its locally never mentioned companion Law No. 25 (on finances) see CBNA (2000)

make the *Kepala Desa* accountable to the elected local council (BPD – *Badan Perwakilan Desa*), who were even given the mandate to ultimately ask for his withdrawal if his administration was found wanting.[30] NGOs welcomed the new law: for the first time in Indonesian history it provided for a truly democratic government at the local level, with a structure encouraging responsible leadership.[31] In 2004, however, Law No. 32 replaced the fully elected *Badan Perwakilan Desa* with the less powerful *Badan Permusyawaratan Desa*, which NGOs and members of these councils saw as a partial return to more authoritarian practices.

Members of the village governments, and their advisors from the *Lembaga Adat* – saw the change as a correction of mistakes in the earlier law: rather than *overseeing* the work of the *Kepala Desa*, the revised BPD was expected to work *together* with him and the village government, particularly in devising plans for local development. One aspect of this cooperation concerned the raising of income, which was encouraged by the new legislation of Law No. 22 to an extent that a SMERU study team saw a danger of a return to an economically stifling "expensive" environment (SMERU 2001). We also found evidence of this trend in at least one of the villages, where the village government and the *Lembaga Adat* had agreed on fees for issuing local licences for holding *pesta* (feasts), building houses, cutting trees for personal use, etc. Another important aspect of the legal changes was the provision of space for the creation and inclusion of local institutions into the village administration:[32] this made it legal to integrate the *Lembaga Adat*, as well as newly created surveillance teams for the Voluntary Village Conservation Agreements (LKM, see below) into the formal structure. It appears that the revival of *adat* and the recent rise to prominence of the *Lembaga Adat* in the villages was enhanced by this change. Looking back at the history of the *Lembaga Adat*, it is

---

30 While there had already been a similar committee – the LKMD (*Lembaga Kethahan Masyarakat Desa* – Village Defense Council; sometimes shortened to *LMD* = *Lembaga Masyarakat Desa – Village Council*) during the Suharto era, its members had been appointed and the *Kepala Desa* headed the Council. In Bolapapu, specifically, at one time both the *BPD* and the *LKMD* existed side by side.

31 Interview with the Chief Officer of CARE in Palu, 21 September 2005.

32 Interview with Ir. Helmy, then working for BTNLL (*Balai Taman Nasional Lore Lindu* – Lore Lindu National Park Authority), 14 January 2005.

remarkable that a modern *Kepala Desa*, who has a reputation for good relations with NGOs, stressed that he saw in the *Lembaga Adat,* itself a democratic institution: following the classic model of *musywarah mufakat.* They reached decisions through public deliberation until they achieved a consensus.

At the regional level, the decentralization and devolution of financial responsibilities also extended to the management of forest resources and the issuing of licences for forest use. While this provided welcome opportunities for the generation of income for provincial governments, it also opened new doors for corrupt practices[33] (in addition to consolidating old ones) and it presented some people with the opportunity to establish strong and uncontrolled positions of power, thus reversing the intention of the decentralization move. Combined with a decreasing presence of state power in the form of military presence this seems to have contributed to illegal logging and other encroachments on the National Parks in Indonesia – and thus created the conditions in which the BTNLL took recourse to cooperation with NGOs to protect the Lore Lindu National Park, created in 1997, against encroachments. This situation of a newly created National Park threatened by encroachments of a population which had a strong tradition of using forest resources freely for hunting, collection of forest produce as well as for agricultural activities in a shifting cultivation system, also provided the rationale for the research project "Stability of Tropical Rainforest Margins".

Voluntary Village Conservation Agreements and Local Leadership

Other researchers have investigated various reasons for the threat to the stability of the rainforest margins and the National Park, such as immigration from Southern Sulawesi and from other parts of Indonesia, the rapid spread of cocoa cultivation because of its success on the world market, land hunger leading to illegal settlements within the park area, and the population increase against a background of poverty in general. This chapter focuses on the link between local leadership and the sustainability of resource management, which is especially

[33] This assessment is the result of many conversations with other researchers, members of NGOs, and other people encountered in the course of the research.

pronounced in the case of the *voluntary* conservation agreements (*KKM* = *Kesepakatan Konservasi Masyarakat* – People's conservation agreements) for the implementation of which the National Park Authority cooperated with ADB-funded CSIADCP (Central Sulawesi Integrated Area Development and Conservation Project), CARE, TNC (The Nature Conservancy), and YTM (*Yayasan Tanah Merdeka* – Foundation for free earth).

The different approaches to the village agreements have been described in detail by Mappatoba, who comes to the conclusion that the "intensive processes of deliberation on self-government" initiated by the attempt to reach voluntary conservation agreements, "can be considered as 'schools of democracy' in themselves" (Mappatoba 2004: 178). The legal room for these negotiations had been created, as mentioned before, by Laws No. 22 (1999) and No. 25 (1999) on local governance, in conjunction with Forestry Law No. 41 (1999) which for the first time contained the concept of "customary forest" (Mappatoba 2004: 177). In our own interviews, it was this concept which constituted the justification for members of *Lembaga Adat* as well as *Kepala Desa* to demand renegotiations of the boundaries of the park as promised during the first setting of the park's boundaries, so that they might be able to manage their *tanah adat* as in the past. At the same time, they consistently declared that *adat* had had a conservation function even in the past: that they were perfectly capable of managing the forest resources sustainably, based on their "traditional knowledge" (*kearifan lokal*). As examples, they consistently referred to the protection of the river banks through the prohibition of planting activities within fifty meters, a prohibition aimed at preventing erosion and flooding.[34] Another example of "customary management" was provided by the zoning which resulted from the practice of shifting cultivation: the different names for primary forest, a (permanent) forest garden, an area for dryland farming, and finally secondary forest[35] were presented as names for deliberately planned forms of use, complete with sanctions for non-compliance. In reality, however, all

[34] It is interesting to note that floods and landslides had caused the Dutch to start replanting forests and to resettle people in the valleys (Weber et al. 2002: 20).

[35] For Kulawi, these names are *Wana ngkiki* (primary forest), *wana*, *pangale* (secondary forest), *pahawa pongko, oma* (forest gardens), and *balingkea* (dry land farming) (Hunowu 2004: 50).

plete with sanctions for non-compliance. In reality, however, all forest areas, possibly with the exception of protected steep areas on hilltops and around sources of rivers, had in the past probably been simply common property, access to which could be had for the asking.[36] "Customary areas" did exist, however: they referred to the area belonging to a certain village. Thus, the village of Sedoa protested the (illegal) occupation of the Dongi-dongi area because they considered it as their *tanah adat.*[37] Another case concerned the struggle over the boundary between the territories of the villages of *Toro* and *Katu* – in an area deep within the Lore Lindu National Park; a struggle which could only be settled through outside mediation.

When we arrived, a number of village agreements had been successfully negotiated between the National Park Authority and the respective village administrations, and special enforcement teams (*LKD – Lembaga Konservasi Desa* = Village Conservation Committee) under the umbrella of the respective *Lembaga Adat* had been set up. The question then was: to what extent would these institutions be successful in ensuring compliance to the regulations stated in the *KKM,*[38] regulations which included certain use rights as well as a number of prohibitions?

Our sample consisted of six villages: four in the Napu valley, and two in the Kulawi valley. Using an open-ended discussion format following the showing of the video of a short theatre play about the violation of a regulation, performed by local school children, and based on their own script, we sought to engage as many members of the bodies concerned, and as many ordinary villagers as possible. In a follow-up visit designed to elicit more information on the relationship between development and conservation, and on issues of leadership, we focused on members of farmer groups in one village each in the Napu and Kulawi valleys, viz. *Kaduwaa* and *Toro*, respectively. Observations from these case studies are presented in the section below.

---

36 We are grateful to Günter Burkard (personal communicaton) for his views on this point, and find confirmation in Sunito's description of how *robo* leaders gained access to the areas they wanted to clear for cultivation (Sunito 2005: 64 f.)

37 The special case of the occupation of Dongi-Dongi and the opposite roles played by various NGOs in the events is described in Adiwibowo (this volume).

38 The difference between *KKM* and another similarly named project, *KKMA: Kesepakatan Konservasi Masyarakat Adat* is described by Burkard (this volume).

Before that, we would like to develop a model for the relationship between development, leadership, and sustainability.

Our starting point is the observation that the poorest segments of the population seem to be consistently overlooked in negotiations concerning the distribution of development assistance as well as in the negotiations for the village agreements. We found this in individual interviews with poorer individuals who complained about "forgotten" promises for inclusion in housing assistance, about being called to contribute to community work (*gotong royong*) – but being forgotten when it came to the distribution of assistance, and about never having received any assistance at all. In other words: it is the poorest group that leads international and national initiatives to design and implement special programs to "combat poverty", and it is precisely the same group who seems to be most overlooked at the local level. This observation was confirmed in an interview with a senior CARE[39] employee, and also in interviews with several village leaders. From CARE we learnt that their approach to participation is limited to accepting whoever is introduced to them as representing a certain group of inhabitants at the *dusun* (= "hamlet" – subsection of a village) level in the villages where they work. Therefore, the participatory village development plans which resulted from CARE's efforts are in reality not much different from the planning documents resulting from government procedures, which also include detailed guidelines for participation.[40] A second confirmation came from a *kepala dusun* who reported that the government assistance designed as a compensation for the abolishment of the subsidy on oil, and publicly declared a special assistance to poorest sections of the population, had not reached a number of people in the village who were definitely in need of it, while other people who were in a reasonably comfortable position, had received it.[41]

---

[39] In September 2006 we learnt that CARE had given up its offices in Palu altogether.

[40] See the relevant section at http://www.unescap.org/huset/lgstudy/country/indonesia/indonesia1.html.

[41] One of the reasons for this marginalization may be that people were afraid to be labelled "poor" in the Suharto era, because this carried with it a reproach and a consequent *social* marginalization by fellow villagers. The creation of many new assistance programs in the *reformasi* (reformation) era, however, made it profitable to be

A very similar process seems to have been at work in the negotiations leading to the voluntary conservation agreements. A statistical analysis of participation in the process, performed by Mappatoba, yielded the result that irrespective of the NGO involved, the people who were more likely to participate in the negotiations were those who were better connected, who possessed more social capital, or who were older and more educated – i.e. those who had more human capital (Mappatoba 2004: 179). We believe that this was reflected in our group discussions, where, after a while of polite confirmation of the official image projected by the NGOs involved, of the village agreements as a positive development, the importance of which our dialogue partners understood, particularly as they themselves benefited from it, someone would get up and start to tell that there was also another side to it: that the agreements were hard especially on the poorer people in the village, because their *only* source of income – apart from piecework for others – was often the collection of forest produce, particularly rattan. This even happened in *Betue*, the one village which was supposed to be a model for a successful alternative, as it boasted of a group of farmers growing vanilla, which at the time fetched quite a good price.[42] This group had been started by the village secretary whose initiative was supported by the *Kepala Desa.* They, and other officials, had praised the growing of vanilla as a successful alternative to collecting rattan in the forest.

In another village, where the schoolchildren had made a moving representation of the plight of the *burung allo* (hornbills), the opposing faction freely admitted that hornbills had become scarce indeed – but would it not also be appropriate to think of the plight of the humans? In the same discussion, whether by the big project CSIADCP

---

officially categorized as "poor" – and therefore attractive to many who are not actually poor, but close enough to the village administration to be registered as "poor".

42 The price subsequently deteriorated, and sources in the internet refer to an oversupply of vanilla on the world market, which has hit traditional producers hard. The extension officer of another village knew of a case of an individual grower who had spent a considerable amount of money for establishing a vanilla garden. But when the crop was ready for marketing, the price had dropped so low that he could not even recover his investment. He destroyed the garden, but after some years tried again – and failed in a similar manner. Whether his subsequent suicide was connected to these failures was left pending in the air …

or the NGO CARE, all outside development efforts were squarely dismissed as failures. We take this to mean that the feeling of being disadvantaged, which also led to the massive illegal settlement in Dongi-dongi described in detail by Adiwibowo (this volume), is widespread and has real grievances as its cause that did not start with the establishment of the Lore Lindu National Park, but which were aggravated by it. The first steps in defining the boundaries were not taken with sufficient care, and the promises to review those boundaries were never honoured. Even the "soft" participatory approach by the NGOs, in which they took great care to involve as many people as possible and to reach a real consensus concerning the outcome of the written agreement – never touched the issue of the boundaries: the only negotiable items were the details of what might still be permissible within the boundaries, if properly regulated through the *KKM*.

In this situation, the statement of one of the *kepala desa* whose villagers were in a particularly difficult situation because they had sold most of their land to immigrants, made perfect sense: "We need some development to come here: we need first of all a continuous supply of electricity, and second, some enterprise which can give us employment – or at least ideas and the skills for starting something ourselves. Otherwise our people will not be able to stop needing to depend on forest resources." Incidentally, he was also the one who expressed most clearly his hope that the report resulting from our research would be more successful in eliciting support from higher levels of bureaucracy than his own pleas had been so far. In this sense he embodied some of the qualities of leadership which most people saw as characteristic of a "good" leader: clarity, vision, and initiative. In (to us) surprising unanimity, good leaders were expected to be well-equipped with *SDM* (*sumberdaya manusia* = literally "human resources", referring to education and intelligence), to be patient, to be able to master their feelings, to be better able to manage the relationships between others, and to be honest and just (*jujur dan adil*) in their actions towards others. With variations in wording and emphasis, virtually all of those we asked agreed on this short list. Two aspects we found particularly remarkable: the emphasis on openness and fairness[43] which

[43] By this we do not mean "fair-play" in the sense of playing to the rules of the game, but rather adhering to a sense of justice shared between the leader and his followers.

was expressed by Christians as well as by Muslims, and by more traditionally-minded leaders as well as by more modern-minded people.

One former *kepala desa* who belonged to a family tracing their descent to the mythical *Manuru* and who seemed to us to have incorporated these values in his career of several re-elections, succinctly summed up the necessary qualities: A successful leader

1) Has no need of being feared by the people;

2) Understands the wishes of the people; and

3) Holds deliberations until consensus is achieved.[44]

We find this unanimity in the characterization of desirable leadership qualities the more remarkable as they echo statements encountered in the course of the LAGSUS research both in Uganda and in Namibia. In Namibia, one of the respected and successful leaders of a farmer's association agreed with his father – who held the post of a headman – that a good leader has no need for fear.[45]

It is interesting to note that modern managers are re-discovering the same "virtues", as expressed in the title of a book advising on the "rebuilding of trust, performance, and commitment in the American Workplace": *No Fear Management*. The authors contrast their recommended style of *performance-driven leadership* with a fear-based *command and control* type of leadership, which they describe as "mismanagement": it decreases rather than enhances organizational performance, because it obliterates trust, demeans people, tells people what to do, fails to listen, and values organizational growth and survival above the well-being of all others in the environment. This, Chambers and Craft (1998: 12f.) say, "is a management style whose hour has passed because it produces outcomes that are counterproduc-

---

[44] Interview with B. Tamambali in Sedoa, Dec. 28, 2004:

1) Tidak perlu ditakuti oleh masyarakat

2) Memahami keinginan masyarakat

3) Musyawarah untuk mufakat

[45] The same leader responded with agreement to another remark of the Indonesian former *kepala desa*: "A good leader is in front, in the middle, and at the back at the same time."

tive to organizational success and survivability" because it leads to increased lies, to hiding mistakes, to hesitation to take initiative and risk, to absenteeism, to blaming others, to hostile interpersonal relationships and more (Chambers and Craft 1998: 68ff.).[46] Good leaders, on the contrary, are truthful (Chambers and Craft, 1998: 88), are attuned to other people's feelings (Chambers and Craft, 1998: 89), communicate clearly (Chambers and Craft, 1998: 113ff.), and exercise a participative style of decision making (Chambers and Craft, 1998: Chapter 7). They also create "community" by being consistent and fair (Chambers and Craft, 1998: 130f.), allow "rituals" for group identification, celebrate positive outcomes, give opportunities for success, and give recognition for achievements (Chambers and Craft, 1998: 136f.).

In the context of implementing voluntary conservation agreements the three qualities described by the former village head gain additional gravity: it is beyond doubt that restrictions in the use of natural resources are necessary for the sake of conservation. A leader who does not embody these qualities, however, a leader who does not "lead by example", will not be able to gain the support and commitment necessary for a long-term adherence to the rules now negotiated. We saw this on two occasions: in one of the villages the *kepala desa* himself had established a maize field on a steep hill near his house and inside the park area with the help of young male relatives. Pointing to this field, a young man casually remarked that in his village the leader did not give a good example of what was needed. In another village the discontent with the handling of grievances had grown to such an extent that a former *kepala desa*, and member of the local *Lembaga Adat*, remarked that he himself would soon stop admonishing people to adhere to the rules of the conservation agreement, and instead, simply allow them to use the forest inside the park boundaries at will, including the establishment of new cocoa plantations.

Before turning to some specific observations emerging from our research, we would like to point out that poverty as a result of unequal ownership of and access to resources aggravates the situation: as noted in the case of the organization of the *robo* working groups and as im-

46 These are also some of the twelve "traits" for identifying the presence of fear in an organization.

plied in the young man's remark about the *kepala desa's* maize field,[47] the incongruence of more well-off people asking for further sacrifices from those who are already disadvantaged engenders a feeling of injustice that is often not expressed and hence, probably more widespread than visible at the surface. It is possible that even the perpetrators themselves share this feeling: this might explain why they find it difficult to express the openness seen as essential for a "good" leader. During our research we could not quite suppress the suspicion that there is a link between the fact that it was only in two villages that we failed to obtain a bigger group discussion, which needed the invitation from the *Kepala Desa*, after showing our video of the school children's performance, and our impression that it was precisely in these two villages that the leaders themselves had put their own interests over "the wishes of the people".

Encounters with Present-Day Leadership in Villages around Lore Lindu

Autocratic Leadership Changing to Democratic Leadership?

While aristocratic descent cannot be invoked any more as a legitimation for office, and families successful in the modern economy have appeared as rivals for influence on public life,[48] observers should resist the temptation to see an unambiguous trend here, as a brief overview of the cases encountered during the field research reveals. There are cases of change, like in *Bolapapu*, where the new *Kepala Desa* could not rely on well-established networks of old aristocratic relationships. This may have contributed to the more open climate of debate in that village, which surfaced in a struggle over the meaning of both *adat* and conservation, a struggle between a faction composed of well-established elder village members and a group of younger challengers. In other villages, such as *Sedoa* and *Watumaeta,* the office remained within the circle of well-established old families, with oppo-

[47] As a successful trader, the *Kepala Desa* of that village is clearly way ahead of other inhabitants: not only did he own both a car and a truck, he was also close to completing the building of the biggest (and only two-storey) house in the village.

[48] Sunito (2005: 55f.)

site changes in the style of leadership. The shift from father to son in *Watumaeta* also brought a decidedly democratic turn, while an opposite shift to more autocratic leadership accompanied the election of a leader from a different family in *Sedoa*. The most pronounced democratic orientation of our sample was exhibited by the two *Kepala Desa* of *Betue* and *Kaduwaa*, but the leader of *Kaduwaa* belonged to an old and well-established "important" family. And there is the case of *Toro*, which we will describe separately and in more detail because the more extensive data we gathered during repeated visits allowed us to find a more complex picture beneath the surface of a "model marriage" between tradition and democracy for which that village became famous: a picture of an aristocratic and autocratic revival, in the cloak of a language of democracy.

It should be mentioned that of the six villages in our sample, only three can trace their roots back to pre-colonial times: the two villages on the Western side of the National Park – *Bolapapu* and *Toro* – and *Sedoa* on the Eastern side. Of the other villages on the Eastern side, *Kaduwaa* was founded at the beginning of the 20th Century,[49] while *Watumaeta* is of rather recent origin, having been started with the cutting down of primary forest in 1929 (TNC 2004). *Betue* was established only in the 1950s to allow refugees from Rampi to settle down permanently. There, no reference to aristocratic legitimation or privilege, or to their history before the resettlement in general, appeared in our interviews. In the old villages of *Sedoa* and *Toro*, oral history and myth remained important reference points, and the names of the important families in *Watumaeta* and *Kaduwaa* can be found in a list of dominant families in *Wuasa*, the old centre of the *Napu* valley, provided by Sunito (2005: 55). *Bolapapu*, the old centre on the Western side, is also the centre of an *adat* revival movement (see the quote from the written version for a regional *adat* above).

The following section presents some specific observations from the five villages where we did not have the opportunity for repeated visits, as in *Toro*. The style of presentation is connected to the research process which was guided by a perspective focusing on patterns of allegance to voluntary agreements, but had to make the best of a limited

[49] According to the *Sejarah Desa Kaduwaa* (TNC 2002a) the settlement was started in 1911, four years after the Piore war in 1907.

time and varying opportunities for observations and interviews. This seemingly haphazard process still led to a simple and clear general hypothesis about the connection between leadership style and resource use and, hence, conservation: only a truly democratic leadership style in a non-formal sense of mutal understanding and respect, and characterized by an absence of fear of the leader, can build the trust and commitment necessary to ensure sustainable adherence to conservation rules and sustainable resource use. Autocratic leadership based on fear of punishment, on the contrary, destroys not only trust and respect, but also adherence to conservation rules, which are then experienced as a more or less arbitrary imposition from the leaders. If this is coupled, in addition, with what people themselves perceive as unfair prerogatives for certain individuals or groups, they will seek out opportunities to break the rules in an attempt to share in the benefit of exploiting natural resources.

Sedoa was the only village – apart from *Toro* – where oral history was intervowen with mythical stories, and still played a role as a point of reference for present-day discourse and politics.We only had to show some interest in the history of the village, to be treated with a long story about princess *Bunga Manila* from the kingdom of Sigi who followed her beloved water buffalo *Tolelembunga* to the place where *Sedoa* is now located. Tokare (1999) and TNC (2002b) recorded a more complete version this story, the importance of which we realized when it was offered a second time over the discussion of the video, when the topic turned to the strength of *adat*: The *Kepala Desa* and the *Ketua Adat* seemed to present this history and its presence in the public consciousness as one of the reasons for the strong adherence to *adat*. They were adamant that the villagers were much more afraid of violating *adat* regulations than they were of breaking the law – a claim we heard in other places as well – and they were equally adamant that none of the villagers would illegally cut down a tree, as the school children had depicted in their story. In their territory, only outsiders would commit this offence. If caught, the logs would be confiscated and used for the benefit of the community. The two leaders were worried that the video might be presented as a research result and, as such, misrepresent the practice of the village: The offender would not, as depicted by the children, be turned over to the police, if found unable to pay the fine stipulated by the regulations, but would be reprimanded and asked not to repeat the offense. The political aspect of the enthusiasm for the preservation of the historical account became clear when we asked a younger man who had studied village relationships more closely after marrying into the village, whether identifiable groupings existed in *Sedoa*. He immediately responded with the names of five important families – which turned out to be the very ones which the story had depicted as descended from the marriage of *Bunga Manila* with *Sadunia*, and which comprised the families of both the present and the previous *Kepala Desa*.

Such mythical descent carries with it a strong attachment to the place, and an implicit obligation to defend the customary territory delineated with various place names in the story. The villagers had indeed done so even before the establishment of a *KKM*: They confronted the illegal settlers of Dongi-dongi and asked them to respect their boundaries, as the village head and the *Ketua Adat* proudly related to us. With the *KKM* established, they had organised regular forest patrols, also depicted as rather successful. We wondered, however, about the discrepancy between the assurance of successful control measures and the maize field on the slope just behind the *Kepala Desa's* house – a slope which was situated inside the park's boundaries. This discrepancy had also been noticed by other people in the village, as a young man's remark showed: "Our *Kepala Desa* himself is not giving a good example."

We had always been impressed by the openly forceful behaviour of this particular *Kepala Desa* who had just recently been elected, and presented himself as a benefactor of the village by inviting a priest to hold a Christmas Service, after the actual Christmas days. During the sermon the priest reminded the villagers how faithfully he and all of his extended family served the village. In the evening the whole village was invited to a *dero* dancing for which he had paid the band.[50] During this event he became somewhat drunk and fell over a chair – much to the amusement of the villagers who had observed it, but ob-

---

[50] His considerable wealth had its source in his successful trading activities and allowed him to finance this village feast out of his own pocket, giving him additional power and influence in the village. The rise of a new class – i.e lacking aristocratic descent – of influential families mainly through commercial activities has been documented by Hunowu (2004) and Sunito (2005: 55f.).

viously were afraid of him: whenever he turned round to look, their laughter froze. We left Sedoa with the impression of a danger: the danger that the social capital available for conservation, which had been built by the previous *Kepala Desa*[51] through a leadership style aimed at winning the support of people, would be damaged by the autocratic style of leadership of the present *Kepala Desa*. It had been this previous village head who had emphasized the importance of avoiding fear but understanding the people's concern in the interview cited above.

In the neighbouring village of *Watumaeta*, where our field research period over Christmas and New Year 2004/2005 began, the change of village leadership was in the opposite direction, from an autocratic leader to a democratically oriented one. At first, we did not understand this change for what it was, however: one of the implications of the democratic orientation of the present *Kepala Desa*, who had immediately offered to host us, was that he remained busy discussing village affairs with various visitors even during these festive days. His father, the previous village head, took the opportunity to tell us about his own term of office and impressed us with his friendly eloquence about his reasons for encouraging immigrants from other parts of Indonesia to settle in his village: according to him, particularly the immigrants from Java possessed superior working habits and agricultural techniques. They worked harder and introduced the method of the *bedeng*, a cultivation method using a raised bed for a variety of crops growing concurrently and in succession and thus making optimal use of a limited surface. To his chagrin, the local population had not taken up these superior methods. Only later did we understand that his readiness to volunteer this and other stories was probably also an attempt to present his own role in the process of immigration in a more positive light than in stories we might get from other people. As we learnt from another researcher, there had been irregularities in land dealings that had made him increasingly unpopular and almost brought him to court, quite contrary to his own claims towards us. Interestingly, he had used a peculiar interpretation of *adat* rules to justify some of the

[51] His commitment to conservation issues had led the assistant director of BTNLL to recommend him to us.

land sales to immigrants (for more detail see Burkard: "Wild West", this volume).

The main result of his policy was that the Javanese and Bugis immigrants, true to their reputation as hard workers and astute negotiators, established themselves as successfully in *Wautameta* as in other villages. Also, as in other villages, their success allowed them to start buying up the land of older local inhabitants, with the result that the vast majority of them were left landless. The chairman of the village council – the *BPD* – alerted us to the fact that most of the dwellings belonging to the "indigenous" population were shared between two or even three families. Lacking both land and opportunities for wage labour, their remaining hope was a change in the boundaries of the Lore Lindu National Park which would legalize some of the illegal clearings we saw on our way to a "monument" which we were shown as proof of the rightfulness of the village's claim to a larger "traditional" territory than presently acknowledged by the Park Authority. Seeing that the stone mortar and stone-bound terraces were located on terrain reckoned to belong to the valley, i.e. not yet "in the hills", we could understand the insistence of the members of the village council and *Lembaga Adat* that the setting of the boundary of the National Park had deprived them of the use of areas they had traditionally considered as their rightful property. They felt doubly cheated as they had been promised that the valley bottom would remain theirs to use in the initial negotiations before the establishment of the park. Against this background, the vehement claim of the members of the *Lembaga Adat*, who remained for discussion after showing the video became comprehensible: that they could *promise* to contain future transgressions of park boundaries – if only the Park Authorities would agree to a renegotiation of these boundaries. As we heard similar stories in other villages, particularly in *Bolapapu*, the "voluntariness" of the conservation agreements negotiated in a participatory way by TNC and CARE is shown to have been limited: a re-negotiation of the boundaries of the park was never part of the NGO's mandate and hence not part of these participatory processes.

This situation also explains the dismay expressed by one of the members of the *LKD* (*Lembaga Konservasi Desa* = Village Conservation Committee) about performing his duties of enforcing the regulations of the village conservation agreement: "I know how important it is to conserve the forest. On the other hand, I am a villager myself and I have the same needs." As in other villages we visited later, this job is difficult for another reason: despite general knowledge of the prohibitions, ownership of chainsaws is quite widespread. The idea to have just one or two chainsaws in a central location where people would be able borrow them for the uses permitted by the *KKM* (i.e. for cutting trees for personal house building and for coffins), forwarded by the chairman of the *BPD*, seemed very far from finding support. It is probably no coincidence that the violation portrayed in the school children's theatre play was enthusiastically cutting trees with a chainsaw. As far as the proclaimed unanimity in *understanding* the importance of conservation issues is concerned, a young man volunteered his assessment of the situation: "It is mainly the more well-off people such as the merchants and the government officers who support the protection of the forest. It is the poorer people who do not see alternatives and who cut trees for sale in the forest. I would probably do the same if I needed the money."

With respect to the difficulties of conforming to the standards of good leadership mentioned above, we found it telling that the *kepala desa,* whose aspiration to being a good leader was clear to us, was dismayed at the impossibility of even delivering official grants as expected: he had, for example, been forced to sign for having received 25 kg of *raskin* (*beras miskin* = "rice for the poor", distributed at no cost to needy households) when all he actually got were 10 kg. No similar difficulties were reported to us in *Betue*, which is a village inhabited mainly by people of *Rampi* origin who fled the violent upheavals in their home area in the 1950s. As refugees they were glad to be given a place where they could finally settle down after many months of travelling. The impression gained during our stay in that village was one of greater than average coherence. It was also one of hope – hope for success with vanilla farming, which the village secretary had brought to the village on his own initiative and which had attracted the interest of a sufficient number of people, including younger ones.[52] The village had also organised – and won – a football tournament against teams from other villages on occasion of the New

[52] In this respect he was true to the role he was expected to play as the most prominent member of the most prominent family in the village.

Year celebrations for 2005. As in other villages, however, the group discussion revealed that the hope was not shared by all, and that there remained a poorer segment of the population who felt hampered by the *KKM* in activities needed to support their livelihoods. Here, we also heard a local assessment of what is presented elsewhere as a more general theory about the effect of modernization: that the level of expectation has been raised to a point where the simple life which the *Kepala Desa* still remembered from his youth, has become impossible, because people are forced to work harder and to seek income more actively. As the *Kepala Desa* remarked: *"ekonomi kerja kita"* – "the economy works us".[53] The time with him also provided us with further insight into the balance between an occasional outward show of strength and a generally welcoming and receptive attitude required for a successful leader. On several occasions he emphasized how the management of people required paying attention to their feelings at all times, but requiring firmness (*kekerasan*) occasionally. It was one of his explicit principles, however, to use only subtle (*secara halus*) forms of force. As an example he mentioned a certain man whom he had noticed because of his repeated absence when called to participate in communal works (*gotong royong*). Instead of reprimanding him, he waited until the occasion arose to teach him a lesson. The opportunity came with the distribution of free seedlings of cocoa trees. The *Kepala Desa* simply did not inform him. When the man came to his house to ask why he had not been informed like the others, he was asked back: "What do you yourself think the reason is?" According to the *Kepala Desa*, the man realized his earlier mistake and promised to change. To make the lesson stick, the man received some free seeds for annual crops, but was still refused the cocoa seedlings.

We believe that this story illustrates a problem many leaders are confronted with – not only in Central Sulawesi, but even beyond Indonesia. And we believe, that the principles of leadership which the *Kepala Desa* of Betue summed up in three short maxims might have been mentioned in similar phrases by leaders in many different places:

---

[53] Alternatively, it might be translated as "the economy is our work" – implying that "we" have become responsible for the functioning of the economy. "In addition to being responsible for our own maintenance," one might add.

1) Never use force except in subtle ways;
2) If there are people who do not get involved in village affairs, they should be approached individually and asked for the reasons for their behaviour; and
3) Help the people in the village with their needs, even if you are not well-off yourself.[54]

We observed a similar orientation – though not expressed in such clear maxims – in *Kaduwaa*, where we learnt in detail about the several stages of negotiating back and forth between the *dusun* (subsections) of the village and the office of the *Kepala Desa* to make sure all opinions were considered in the final *KKM*. There was agreement between him and all those present in our group discussions that this had indeed been a very democratic process. His "democratic instincts" were also shown by the care he took in consulting the members of the *Lembaga Adat* as much as the members of the *BPD*. A telling detail was his concern about the feelings of one of the poorest members of the community, whom we had interviewed as one of the only three people who did not belong to any farmers' group: when we mentioned that we had forgotten to make a picture of that man, his immediate response was that the man would surely very much appreciate receiving a photograph of himself together with his musical instrument.[55]

Such details as well as his general manner indicated his leadership qualities, which had earned him unopposed re-election, and the loyalty

[54] We were inclined to believe his statements about leadership and about being contented with remaining a planter, as we saw him leave for work in his plantations, which he showed to us with obvious pride, and because of the fact that his was one of the few houses which still had a thatched roof – which is cheaper but considered somewhat outdated by most.

[55] We had interviewed him, because his household was one of the only three which did not belong to any farmers' group. He was an immigrant from South Sulawesi who had left his home area of Makassar because there he did not have any access to land, and he did not see how he could support a family there. In Kaduwaa he worked as a sharecropper, feeling secure enough in his tenure to compare his situation favourably with his former status in Makassar. Despite his frustration about not finding his active participation community work rewarded: he complained that he was never invited to participate in sharing the benefits of state or NGO programs brought to the village through the medium of the farmers' groups.

of the village's women's organization. The women claimed to have no ambition to enter the male domain of village politics directly. In this respect they wanted to be seen as different from the women of *Toro*, who had earned some regional fame for moving in this direction.. At the same time, they prided themselves on being much better able than men to organize themselves for the achievement of tasks, because, as they said, they more easily acknowledged gaps in their knowledge and asked for information or the help of others. With respect to current global debates and our own focus on leadership attitudes, we found it telling that they would sum up the difference between their attitude and that of the men in the sentence: "We don't have such big egos (*ego tinggi*) as the men." We are inclined to believe that similar statements can be heard from many women in very different corners of the planet.

More time would have been needed to dig deeper into the origin of the *Kepala Desa's* undoubted leadership qualities. It would certainly have been interesting to establish to what extent his attitudes had their roots in life experiences in modern contexts outside of the village,[56] or in his own socialization, precisely because he did belong to one of the regionally important families listed in Tokare (1990), and also because the discussion about leadership – in which various chairpersons of village groups participated – traced the concept of leadership back to a tradition of warriorship (*taduluko*) in former times. In the traditional concept evoked, both descent – *uwerena* ("from the peak" in the *Pekurehua* language) or *totoiuwiane* ("from the source" in *Pekurehua*), and personal qualities played a role. Among the latter, the quality of *wahena* caused some discussion, as the group remained somewhat uncertain about the extent to which this word referred to "charisma" as a personal quality or to "blood" as representing a mixture between descent and a personal quality. More concretely, a traditional leader had been supposed to be industrious (*madeta*) and trustworthy (*rapionalai* = *dapat dipercayai* in the Indonesian language). The latter is particularly interesting, as this quality of trustworthiness also surfaced as essential in *Toro*, where at least some of the leaders of farmer groups whom we interviewed, saw justice in the execution of a leader's functions as essential for building trust. One of our older re-

[56] We understood that his wife left a successful TV career to join him.

spondents in that village even mentioned three different words for and corresponding to concepts of "trust": *pepuea* for trust in the "creator", *kaparahaia* for trust among "ordinary" people, and *kamagana* for trust in a *magau* (i.e. aristocratic leader) – but specifically a *magau* who was honest (*jujur*), just (*adil*) and who "could be trusted" *(dapat dipercayai*). And only in that case would his word decide any disagreement people might bring to him for final arbitration. That such a view was not a peculiarity of the village of Toro is shown by the fact that one of our respondents in *Kaduwaa* simply stated that according to traditional *Pekurehua* ideals a "good" leader "had to be just" (*harus adil*), which would fit well with the traditional concept of a leader as the "father of the village" (*uma boya* in *Pekurehua*) mentioned in the group interview with the *Kepala Desa.*

In none of our interviews were the qualities just mentioned clearly linked to the "systematic" conception of "good leadership" as being "in front, in the middle, and at the back" at the same time, which two of the leaders in our sample referred to. This is hardly surprising if one takes into account that its Javanese form had been popularized and sloganized – i.e. emptied of real content – during Suharto's presidency. Yet the chairman of the *BPD* in Kaduwaa who was also a member of the *Lembaga Adat*, claimed that it had been an indigenous *Pekurehua* concept, which had "come down from our forefathers" and had been "much loved by the people". To prove his point, he gave us the equivalent of the Javanese terms in *Bahasa Pekurehua:*[57]

---

[57] We owe this version of the table to the assistance of Javanese colleagues who pointed out the full meanings and correct spelling of the corrupted Javanese terms, and the fact that the Javanese terms for "Pushing from behind" and "Educating in the middle" had been reversed in their order during the discussion in the village. The full meaning of the Javanese terms is as follows:

- Leading in front: tutwuri handayani
- Educating in the middle: ing madya mangun (*membangun* = develop) karso (*karsa = motivation, initiative*) – in the middle developing initiative – which had been shortened to *ingmadio*.
- Pushing from behind: ing ngarso sing tulodo (*teladan/tauladan* = example) – at the back (behind) you give good example – which had been corrupted to *ingarso suntulodo*.

English	**Javanese**	**Indonesian**	***Bahasa Pekurehua***
Give an example/ lead in front	Tutwuri handayani	Di depan memberi contoh	Yoru mewali panggin
Educate in the middle	Ing madya mangun karso	Di tengah membimbing	Itanga menini
Push from behind	Ing ngarso sing tulodo	Di belakang mendorong	Iboko mohum-banga

Table 1. *Leadership qualities and their expressions in indonesian languages*

While this is no conclusive evidence for the local *origin* of the concept, we do have a curious additional indication for the *appeal* of the concept – even in a completely unrelated culture: when this concept was quoted to a rural Herero chairman of a Farmers' Association in Namibia, he exclaimed: "But you are taking the words from my mouth: I just wanted to explain to you our concept of a good leader in the same words. We say that a leader has to be be in front to receive the first blow, in the middle to encourage the others, and at the back to receive the last blow to his heels." This man also exhibited some of the other qualities mentioned above in his own leadership style – and, most importantly in our view, confirmed that a "good" leader has no need of people fearing him. We certainly do not want to romanticize traditional concepts. Rather, we believe that the existence of such congruences indicates the existence of constellations of cognitive and emotional patterns which are valid *across* cultures and contexts: They might serve as an empirical warning not to succumb to a fashionable cultural relativism. In our view, this is supported by the recommendations for performance-based management in the book *No Fear Management* quoted above, which point in a similar direction.

We do not want to elaborate this point, but rather return to Kaduwaa, where the *Kepala Desa* took up his role as a "leader" also in the settlement of a conflict over the transgression of the voluntary village conservation agreement by inhabitants of a newly established neighbouring village. The inhabitants of *Dodolo,* who had not wanted to live under the restrictions placed on the inhabitants of *Katu*, the territory of which was completely within the boundaries of the national park, had felt too restricted by a swamp on one side and the limited area for cultivation on the other, and therefore taken to expand their land by clearing forest – which the Kaduwaa people considered *their* "customary land". The very active chairman of the *LKD* had asked for some material support from the NGO TNC, but insisted that they should resolve the conflict on their own: "What is the worth of our agreement and the *LKD*, if we cannot manage this on our own?" The settlement involved several meetings between villagers from both villages and an investigation of alleged cases of the sale of some of the cleared land to outsiders. It also involved meetings with the NGO which the Kaduwaa people saw as having instigated the transgressions in the name of "people's rights". In one of these latter meetings the *Kepala Desa*, with noticeable emotional engagement, acknowledged the importance of "people's rights" but insisted that they, the people of Kaduwaa should be accorded the same rights.[58] As a result of these meetings, a binding agreement – in fact, involving a revision to make it easier for the neighbouring village to accept the agreement, was signed by both sides in 2003 and seems to have been kept until the time of our last visit in 2006.

To us, *Kaduwaa* seemed to show what is possible when decentralization, the support from an NGO and the National Park Authority, and a more than formal democratic orientation of the village leadership work together. We refrain from calling it an example of "synergy" as this fashionable word still lacks a precise enough definition. The relative harmony of Kaduwaa contrasts starkly with the open debate between different factions characterizing the regional center of *Bolapapu* on the opposite, Western side of the Lore Lindu National Park. As mentioned before, *Bolapapu* had also been a center of resistance against the Dutch colonization, and had remained an administrative center after being defeated. The continuity of social divisions inherited from pre-colonial times is still observable today: not only are the areas of habitation clearly segregated according to the former classes of nobles, free people, and slaves, but all of the last ten village heads – and probably further back – were members of *maradika* families. The office-holder during our research was the first one who did not come

---

[58] We have a copy of the audio tape containing the relevant statement.

from a noble family.[59] Because of the separation of areas of habitation, the descendants of former slaves recently managed to establish their own village – after long years of struggle. Their *kepala desa*, however, is a descendant of a *maradika* family. The Indonesian researcher who knew the details corrected the German researcher's idealistic assumption that this might have been a clever strategy of image management on the part of the segregationists: the committee responsible for organising the election had consisted of old inhabitants from influential families and had come up with only this one candidate who had to be persuaded to return to the village from the place where he had moved.

Several groups have been active here with respect to *adat* revitalization and with respect to conservation. It is interesting to note that in both cases the competing factions consisted of older people from well-established families on one side, and of younger people with a less traditional outlook on the other. While the group of elders not only wanted to document and preserve existing *adat* regulations in writing, but also to make them binding for the whole sub-district, it was interestingly the group of younger men who argued that this plan went against the spirit of *adat*: first, *adat* belongs to an *oral* tradition and loses some of its essential character by the very act of writing it down, and second, any *adat* sentence is the result of local deliberations and therefore has to take local specifics into account – not only concerning the circumstances, but also concerning what constitutes a viable case and what are the appropriate sanctions (Hunowu 2004). As it was, the old elite won the day with a written code detailing regulations especially concerning marriage, and also specifying the sanctions for transgressions of these rules. Contrary to the claims we heard in other villages, the use of natural resources did not figure in these regulations (see above). And against the protestations of the younger group, an important part was given to specifying modern monetary equivalents to old valuables such as a specific kind of cloth, golden dishes/trays, water buffaloes, and cows.[60]

---

[59] His father was actually Chinese and was married to a Kulawi lady. His election was partly a result of the younger generation's fight for a different interpretation of *adat*. For details of this struggle see Hunowu (2004)

[60] The very fact that such valuables are mentioned points to *adat's* function of ensuring aristocratic privileges mentioned earlier: neither did all commoners possess water buffaloes, nor did they have easy access to those golden *dulang* (trays) or cloth

The struggle between a younger generation and the "village elders" (Hunowu 2004) can also be observed in different conceptions about the protection of natural resources through the *KKM.* Both groups agreed that the inhabitants could not be asked to sacrifice their livelihood needs to the protection of "wilderness" resources. But where the traditionally-oriented elder generation favoured an approach using detailed written rules, the younger generation was more in favour of a "hands-on" approach of judiciously using forest resources as needed (Hunowu 2004: 96). Through the intervention of the regional project CSIADCP, the younger generation has found a precarious institutional basis: it is mainly the members of this group who form the membership of a new organization for coordinating conservation efforts at the district level. The *Forum Wilayah Penyangga* (FWP – Buffer Zone Forum) was created and funded by CSIADCP, but suffered from a lack of operating funds after the termination of the project. Because they were determined to continue what they believed to be important work, they had started to seek external funding, and asked the researchers for assistance in this endeavour. More than the members of the older established elite they were aware that successful conservation required alternative income, and they therefore cooperated with NGOs in Palu who ran small programmes.[61] It was members of this group who were among the more vocal and critical in the group discussion following the showing of the video: while the "dignitaries" – including the newly elected *Kepala Desa* – praised CSIADCP for the assistance provided,[62] the chairman of the FWP called the whole project, and other efforts, such as those by CARE, an outright failure, saying that they had not "touched the people" (*tidak menyentuh masyarakat*).

(*mbesa*). Thus it was clear that any commoner committing an offense had to turn to an aristocratic sponsor for being able to pay the fines – possibly having to put himself in debt for repaying

[61] Among them are *Bantaya* which supports the production of traditional bark cloth, *Awam Green* which developed a focus on traditional medicines, and *Karsa* which supports the establishment of cooperatives for coffee production.

[62] When asked to specify, the only concrete example forwarded was a certain amount of money for road building, however.

Clandestine Aristocratic Revival? – The Case of Toro

The case of Toro does not fit neatly into the overall picture of "democratic" versus "aristocratic" leaders. Here, we found a strong demonstration of democratic values, combined with a somewhat rigid insistence on the leader's own ideas and a hidden slant towards favouring the old elites. Under their leadership the present village had been established after the flight from their original village due to the attack of a spirit who wanted revenge for a killing committed by the villagers, and a subsequent time of wandering – as their oral history tells the story.[63]

The present *Kepala Desa* is the nephew of a previous one who remained in office for 17 years, starting in 1972. The long-standing link of this family with official functions was visible in an old photograph showing the old *Kepala Desa's* grandmother together with a Dutch missionary. Similar to the present village head, and another former village head we talked to, he proudly told us about the achievements during his term of office, which included the building of a road which could be used by vehicles, where there had only been a footpath before. He attributed his popularity and his achievements not least to his willingness to personally contribute to the village affairs out of his private means. Honesty and openness were for him essential qualities for a village leader – which he thought were sadly missing in many present office-holders. The present *Kepala Desa*, who is reported to have been reluctant at first to stand for election, has strong views about the faults of the government machinery and an equally strong tendency for local autonomy, which has won him acclaim as a "rebel" in a newspaper article (Irvan and Sangadji 2004), but has also led some to wonder whether the village was still part of the Indonesian state. Despite the reservations he himself expressed about the agenda of NGOs, one village functionary even asked whether village governments were now run like NGOs. This was, of course, a rather wild statement. At the same time, it reflected in an extreme form the general perception of the village head's good relationships with NGOs which stems from the time before he became village head and contin-

[63] Sohibuddin describes in detail the different versions of the myth and their importance for the communal identity (2003).

ued through his two re-elections. It is also connected to his long-standing interest in conservation issues which led to this village being one of the first to have a written voluntary conservation agreement in return for which the Park Authority granted certain limited use rights on park territory. He clearly demonstrated his commitment both to local independence and to forest conservation when he used the opportunity of a gathering called by the agricultural extension officer[64] to introduce a programme for the distribution of free *durian* seedlings, to give an extemporised speech about the need to rely on people's own resources for their livelihoods, particularly in agriculture – and without destroying the forest.

Nowhere else had the commitment to forest conservation been linked so explicitly to *adat* as in *Toro* (see particularly Toheke and Pelea 2005). Nowhere else was the cooperation between the village administration and the *Lembaga Adat* as close, and nowhere else did the *Lembaga Adat* so confidently declare that it was the village's supreme body for decisions concerning land matters and the use of natural resources.[65] Nowhere else had the unity of the village "community" under the guidance of the rules of *adat*[66] been projected so successfully as in this village, which had been a pioneer in openly supporting traditional values by building a traditional assembly hall even during Suharto times, using exclusively traditional building materials and designs.[67] Nowhere else had a village meeting in *musyawarah* fashion agreed on something like a constitution which contained principles according to which these institutions should function and cooperate with each other. These principles were:

1) Democracy: everything is decided together;

2) Participation: everything done together;

[64] The agricultural extension officer had invited us to this meeting and confirmed that the village head had a habit of giving similar speeches whenever the opportunity arose.

[65] For the reasons of this unusally strong claim in the village's struggle against resettlement see also Sunito, this volume.

[66] In one interview, the most senior member of the *Lembaga Adat* called the possession of rules the hallmark of civilization.

[67] The two German LAGSUS researchers had been privileged to attend a visit of a group of students of architecture from all over Indonesia to this meeting hall in 2003.

3) Transparency: everything is known by all;

4) Accountability: all components carry the responsibility together.[68]

Finally, nowhere else did we find an agreed-on organigram of the village institutions which differed from the standard set-up by ensuring positions for women in the *Lembaga Adat* and including an organization for *adat* women.[69]

These achievements, combined with contacts to NGOs, earned the village some acclaim in newspaper reports (e.g. Hartiningsih 2005) as well as in the eyes of researchers, culminating in the participation in an international competition for a model of a "traditional community" successfully combining development and conservation issues,[70] a participation which was supported by BTNLL, whose assistant director agreed to be the English voice of the delegation at the international event of the presentations of the finalists in Kuala Lumpur in 2004.[71]

---

[68] *1) Demokratis: segala sesuatu diputuskan bersama.*
2) Partisipatif: segala sesuatu dikerjakan secara bersama.
3) Transparansi: segala sesuatu diketahui bersama.
4) Akuntabilitas: semua komponen bertanggung jawab bersama.
We obtained these formulations from the pastor who had at the time been involved in their formulation. Later, we learnt that the meetings and constitution were at least partially initiated by the suggestions of an Indonesian researcher present in the village at the time, who - at that time already - noticed the tensions between the different village institutions which surfaced in our own research (M. Sohibuddin, personal communication).

[69] The official version of that organigram (published in Toheke and Pelea 2005: 73) contains – according to the pastor involved in the deliberations – an essential mistake which led to some of the problems observed during our research: while the women's organization should have been represented as part of the Lembaga Adat, it was in fact drawn at the same level as the village administration and the Lembaga Adat, thus formally according it more authority than warranted by the verbal agreements.

[70] It seems that the suggestion for participating in that competition originally came from researchers present in the village and impressed by these achievements.

[71] As one of finalists, the *Ngata Toro Community* is described on the Equator Prize website with the following words: "This forest community has revitalized traditional laws and institutions to support sustainable resource use in 18,000 hectares of Lore Lindu National Park. They have boosted incomes through activities such as the sustainable harvesting of forest products, low-impact agriculture, and ecotourism. Much of their great success results from the fact that their rights, knowledge

This, in turn, led to UNDP becoming involved as a donor, awarding a grant to the village and sponsoring the participation of the chairperson of the women's organization[72] and the village head in international conferences, thus spreading the reputation of the village as a model for how the revitalization of *adat* could go together with and even support community development, women's emancipation, and conservation.

During the early phases of our own research, this esteem seemed well-earned (see also Fremerey 2005): in the interviews with virtually all chairpersons of the farmers' groups the leadership of the village head was praised. The interviewed also emphasised the groups' good relationship with the *Lembaga Adat,* whose importance they stressed, particularly with respect to forestry issues, and they described the unity of the village as exemplary. Then we started hearing remarks from members of village institutions, that this village was in fact no different from all the other villages in the area, a remark which was echoed by the extension officer, and a school teacher who at first did not want to be interviewed at all. Another Indonesian researcher who happened to do field work at the time of our stay there, informed us that despite the unusually large expanse of wet rice fields, the village did not produce enough rice for consumption, and that in fact the majority of the villagers had accumulated debts with the rice mill owned by a Chinese. We also learnt about another peculiarity unique to *Toro*: in order not to lose the legal right to the title of "customary community" provided for in the Indonesian forestry laws, the villagers were discouraged by the village head to exercise their legal right to have their land registered and to acquire written titles. Without formal titles, the land could not change its owner through sale, making it impossible to have a situation as in *Watumaeta*, where the land is legally owned by immigrants.[73] It also made it more difficult to obtain credit from

---

and customary laws are recognized by the Park Authority and have been incorporated into the park's management strategy." (http://www.equatorinitiative.org/equatorprize/EquatorPrize2004/2004-finalists.htm#indonesia1)

[72] It should be noted that the chairperson of the women's organization is a daughter to a previous long-term Kepala Desa and a cousin of the present one.

[73] There was, in fact, a sizeable proportion of immigrants from Rampi in this village, who had come as refugees in the 1950s and been granted the right to settle down in the village with the understanding that the land they cultivated would revert back to village ownership once the refugees would return home – which never happened.

the banks, which was reflected in the substantially lower lending rates of the banks to inhabitants of this village, as compared to the inhabitants of neighbouring villages.[74] These facts partly explain the high incidence of poverty in the village,[75] and hence, the hopes placed in the creation of a credit union.

There were discussions between members of the various institutions in the village of using some of the UNDP grant as a capital stock for this credit union, but unfortunately, the purpose of that grant had been fixed as "capacity building". Therefore *training* for the setting up and running of a credit union could be financed, but the money could not be used for lending purposes. Even though the ultimate goal of capacity building was proclaimed to be the improvement of the economic conditions of people to a level which would make it easier for them to also abandon excessive use of forest resources, the money provided by the grant could not be used for economic purposes, such as a capital stock for the lending operations of a credit union. The *Kepala Desa* bemoaned this fact, but other villagers, even the members of village institutions, seem to have been ignorant of it. This resulted in the peculiar situation that after the first training course run by members of a well-known credit initiative in Kalimantan, the trainees (mainly the members of the BPD) waited for further steps from the village head, according to whose view they should not have waited, as it was *their* task to "educate" the people in their respective *dusun* (sub-section) with what they had learnt.

Towards the end of the 2005 research period, we also heard a story about someone whose ancestors had come to the village as people captured in a war with Rampi, and then continued to live here as *budak* (the lowest category of slaves). Later they had been given some land for cultivation by their *maradika* owners, land which the main character of the story himself had developed into a cocoa plantation.

[74] We are grateful to Günter Burkard for this information.

[75] This was confirmed by the extension officer, who also complained about irregularities in the distribution of the *raskin* (beras miskin): it was sold at a higher price than the government-stipulated below-market rates. A *dusun* head also talked about the high incidence of poverty – and that the government aid to the poorest did not reach some of those to whose condition the village head had been specifically alerted. In two other interviews with members of that group they confirmed that they had never received any assistance, even after formal registration.

After the trees had started producing, a descendant of the former owners, and therefore a member of the *maradika* class himself, claimed that according to *adat* regulations land belonged to those who had first cultivated it, and hence the plantation was his by right. Being a descendant of a former slave family, the developer of the plantation did not dare to complain and bring the matter to the *Lembaga Adat* whose declared official function it is to settle any land matters: he was afraid that he might be chased from the village which he felt to be his home.[76] We began to understand this fear, which knowledge of Indonesian law should easily dispell, because the law grants everybody who has been legally registered in any village for a stipulated time, the right to continue living there,[77] when we heard how members of the BPD had experienced rebukes in the form of "sharp glances" any time they attempted to fulfil their duties as "representatives of the people" (*wakil masyarakat*) by tabling controversial issues or criticising decisions they considered wrong. This had had the effect of destroying their courage to speak up and say the truth in the name of the people who had elected them into office, consequently silencing precisely the "voice of the people" in the village administration, which the reformation era of the Post-Suharto period had sought to strengthen with the new law No. 22 (1999).

Because of such rumours, particularly about more than one case of land-grabbing involving senior members of the village elite and using *adat* regulations as a justification, we decided to interview as many members of village institutions as possible, with ears open for stories relevant for these allegations, but aware that we were in no position to distinguish fact from rumour. We felt, however, that the existence of rumours of this nature in itself is a fact which throws considerable doubt on the image of village unity presented to the outside world. In the process we were surprised about the extent of dissatisfaction with the village administration and the *Lembaga Adat* among virtually all members of the BPD, and about the lack of internal support for the

[76] We presented the case of land-grabbing to one of the older members of the *Lembaga Adat* in the form of a story we pretended to have heard in a different village. His response was: "Such a case would be a sin. It cannot be justified by *adat*."

[77] Despite this, the village head was heard to exclaim on occasion that whoever was not willing to comply with the rules of the village, might as well leave (Günter Burkard, personal communication).

village's most famous "export article".[78] the "Organization of the *adat* women of Toro village" (*OPANT*). Even the village head said he was still "waiting for them to present a clear programme". The lack of a clear programme was the most common complaint from male and female members of both the BPD and the *Lembaga Adat*. They contrasted this with the clear ten-point programme of the *PKK,*[79] the place of which OPANT took in the set-up of village institutions. This move was justified by pointing to *PKK's* alleged role in maintaining women's subservient position. The "confusion" about the programme and role of the women's organization was compounded by a professed ignorance about the amount and use of outside funding granted to that organization by UNDP and other NGOs. This directed resentment most of all towards the lack of transparency of the women's organization, linked to the autocratic style of leadership of the chairperson,[80] who presented herself to us as being strongly motivated by a spirit of "struggling for the people". The discontent was also fuelled by the perceived contrast between the wealth of externally funded activities in other villages and the dearth of activities in Toro itself.

---

[78] We call it "export article" here because we know of at least two programmes to spread the village's model of an organization for *adat* women to other village – funded in part also by UNDP.

[79] The *PKK* (*Pembinaan Kesejahteraan Keluarga* = Building family welfare) had been set up in the Suharto period in all Indonesian villages, with a highly ideological ten-point programme in the service of the Suharto regime's version of national development, which was taught to all girls at school, and which can still be found on public display. Nonetheless, the organization seems to have provided the women involved in it with both self-esteem and a small but self-managed financial support, resulting, for example, in group managed gardens and savings associations, things the interviewees in that village bemoaned as lost. This was in stark contrast to the still functioning PKK in Kaduwaa (see above), where we witnessed a meeting of an active and self-confident organization whose members deliberately did not aspire to the kind of open political influence they felt had been achieved in Toro. For a comparison of *PKK* with other women's organizations see Marcoes (2002).

[80] One interviewee remarked that the chairperson's public exhortations for transparency stood in stark contrast to her actual behaviour. There are confirmed cases of treasurers of that organization being sacked when they wanted to engage in accurate bookkeeping, and there is at least one instance of committee members being given money to keep silent about one such case (Günter Burkard, personal communication).

With respect to the relationship between the *Lembaga Adat* and the BPD, most members of the BPD claimed to feel somewhat pressured by the presence of two members of the *Lembaga Adat* in the BPD, which in our view was not provided for by law No. 22 (1999), and marginalized by what they saw as the closeness of the *Lembaga Adat* to the *Kepala Desa*. He, in turn, found them wanting in their ability to discharge their functions properly despite several government-sponsored trainings they had attended: "They do not understand their job." In our interviews, however, most of them displayed a rather clear understanding of their function as carrying the "hopes of the people" to the village administration, and to supervise the financial management of the village administration, as stipulated by law No. 22 (1999) and Amendment 32 (2004). The chairman of the BPD had even managed to get ahead of the officials of the sub-district office by obtaining a copy of Amendment 32, which had not yet been officially distributed. Comparing Law No. 22 and Amendment 32 in a discussion with him, we found that the functions of financial supervision had remained virtually unchanged in the new regulation, which lent official support to a grievance voiced in some of the discussions: that the polite request to be informed about the use of the UNDP funds, which had been officially announced in a public meeting, was countered by members of the *Lembaga Adat* with the remark: "You do not need to know about that."

One of our respondents had a simple explanation for behaviours such as this: The "family factor" (*Faktor kekeluargaan*). Because we found this explanation cryptic at first, he elaborated that the members of the *Lembaga Adat* and the *Kepala Desa* all belonged to a small group of families closely related to each other and forming an elite network which directed all available benefits to their members. In the light of allegations about land-grabbing,[81] about possible misappropriation of funds concerning the handling of assistance to the poor, and also about real manipulations of funds,[82] this statement sounded

[81] To protect the identities of our informants we cannot give details of these cases, which in our view might lead to court proceedings if the victims both knew the right channels and were prepared to face the social tension associated with such action.

[82] One of our informants, who openly supported the present village leadership, explained that the members of the BPD had not received the compensation for attend-

like a reasonable and clear-sighted analysis of the situation to us. Because we also heard more than one indication of a growing mood of anger[83] of preparations for more open opposition,[84] and of intentions to step down from official functions[85] or to stop supporting existing village regulations,[86] we thought it fair to write a short and polite summary of the surprisingly widespread discontent within the larger circle of village leaders in *Bahasa Indonesia*, avoiding the most compromising statements and emphasising the also existing unanimous hope of our respondents for a village gathering in the manner of the earlier 2003 exercise. They saw this as a way to resolve and clarify all the issues mentioned using the process of *musyawarah* and in the spirit of openness. Circulation of this report was restricted to the village head and the chairperson of the women's organization. After the final discussions with only these two recipients of our report, in which we expressed our hope to be able to observe and report on indegenous mechanisms of conflict resolution, we left with the promise of being

---

ing meetings stipulated in the government regulations because this money had been used by the village head to cover for the shortfalls in land-tax. The *Kepala Desa* himself later prided himself in having been able to cover the land-tax in full for several consecutive years, which earned him a motorcycle in acknowledgement of good service. We found a parallel case in Kaduwaa, where the compensations for BPD members had been meticulously paid out, however.

[83] Thus, some of the leaders we interviewed spoke about threats villagers had issued, and about more aggressive actions they had dissuaded other villagers from taking. One instance concerned the intended refusal to welcome visitors to the village in the traditional costumes, another the threatened burning down of the old traditional meeting hall – the *lobo.* The latter case had to do with the building of a new meeting hall by the chairperson of the women's organization and the alleged mishandling of a grant from BTNLL for the renovation of the old one.

[84] Two of those one might call informal leaders stated that they were quietly preparing for launching action aimed at redressing the imbalances and incongruities in the present handling of affairs: they were just waiting for the "right moment" for a successful move.

[85] This concerned one member of the BPD who said he was only held back by the chairman's plea to continue, and one member of the village administration who complained that there was too much lying (*banyak bohong*) going on.

[86] One former official told us that he would soon stop admonishing villagers to follow the forest regulations promoted by the *Lembaga Adat* regulations which had been praised as exemplary by officials of BTNLL. Rather, he would encourage them to cut trees and clear forest for cultivation as they saw fit – if the situation in the decision-making bodies of the village did not change to the better.

invited to attend this *musyawarah* which the leaders, having been aware of the discontent for some time, claimed to have planned for the coming months anyway. Instead of the expected invitation, however, a few weeks later a fax arrived in Germany, asking us to abstain from further research in the village.[87]

We were therefore unable to observe whether and to what extent the rift between the official discourse maintained with the outside world and a parallel discourse in the village was healed, and to what extent the credibility of and loyalty to the village leadership was restored. Clarifying this question in some independent research would also assist in testing the hypothesis which emerged from the results of our interviews in all the villages: A leadership which misses out on the essential qualities of openness, justice, and attending to the needs of the people – qualities which are essential according to traditional values (as expressed by a former *Kepala Desa* in Toro), according to modern conceptions of democracy and human rights, and also according to ideals of participation and empowerment management – loses credibility and trust, and hence legitimacy. When that happens, the leadership becomes incapable of ensuring solutions to problems of resource management which can be considered sustainable.[88]

We believe that this hypothesis merits testing in different research contexts as well. Toro is simply a particularly striking case because of the discrepancy between the public claim that the guardianship and revitalization of traditional values exercised by the *Lembaga Adat* did in fact promote village unity, greater adherence to rules of forest conservation including a greater role for women (see Toheke and Pelea 2005), and the statement of disagreement with this view which one of our respondents, himself a former functionary, expressed: "If the

---

[87] This is not the first attempt by the Toro leadership to restrict researchers' access to information. A previous case is the attempt to demand that interviews with villagers only be conducted in the presence of a person with proven loyalty to the present leadership (Günter Burkard, personal communication)

[88] After the completion of the text we found that the concern expressed in this hypothesis has already been addressed as the "moral dimension of sustainability" by Beckmann and Pies (2004: 2): "Institutions, however, can be fully effective in the long run only if they are considered fair, legitimate, and mutually advantageous. This *moral dimension of institutional legitimacy* presents the missing link to empowering sustainability."

situation in the village continues like this, I will soon allow people to resort to cutting trees and clearing the forest for cultivation at will" – instead of sanctioning such behaviour by applying the *adat* regulations which made the village famous.

It should also be made clear, however, that it is precisely the fame and the concomitant external funding that have led to a situation which has spun out of control of the present leadership. The first UNDP grant was announced in a public meeting, hence expectations were raised in the whole village. These expectations could not possibly be met because UNDP rules governing the spending of the grant precluded uses whished for by both villagers and leadership, but these rules were not transparent to the villagers. For later assistance awarded specifically to the women's organisation, neither the amount nor the purpose became transparent to the villagers, leading, for example, to wild rumours concerning the new meeting hall which was in the process of being built while we were there. While the intention was to have an appropriate place for a national seminar (i.e. one fitted with electricity), people suspected the new hall was to replace the old one – and some of them interpreted the death of one of the helpers after being stung by a bee as an expression of the wrath of the spirits.

The leadership finds itself in a situation where it can neither control the flow and purpose of external assistance, nor the reactions of the villagers to these events. What remains is the attempt to control the interface between the village and the outside world, in order to retain a measure of control over the image of Toro in the outside world. Asking researchers not to come back by sending a fax is probably one aspect of this image management.

Conclusion: Subversion of Democratic Rights Through Revitalization of Adat?

We hope that the preceding case histories have given enough detail to support the overall conclusion of our research: the revitalization of *adat*, which local power holders claim to be supporting both democratization and conservation of forest resources, in fact often, but not always, serves mainly the interests of the descendants of the former aristocratic class, in the process contributing further to the impoverishment of the "commoners" and former slaves. At the same time the recourse to *adat* as a device to protect forest resources seems to be an invented one, if one goes by the one example of a written form which aims to become something like a general code for a whole *Kecamatan*: natural resource management does not figure in the *Bolapapu* codex. Still *feeling* in a subordinate position, the poorer members of the local societies are too afraid to transgress forest boundaries, except for individually manageable activities such as illicit hunting and collection of butterflies and plants for medicinal purposes, but also for supplying the regional furniture industry with *rotan*. The larger transgressions are initiated by more wealthy members of the society. Whether they are of aristocratic origin or come from the economically active immigrant merchant class, they use the services of the poorer segments of the population for clearing the forest, establishing plantations, and tending them, but retain ownership and control over the produce. It would certainly be interesting to follow the lead provided by the example of *Toro* to establish whether the practice of descendants of former aristocrats claiming already established plantations from less privileged groups is more widespread: whether the continuity of privilege can be observed in more cases.

We are not suggesting that there is any conscious design behind these developments. Taking into account the increasing role played by NGOs, we rather think that their quest for spheres of influence, most often motivated by genuinely democratic intentions, sometimes combined with conservation concerns, combines with a lack of understanding of the local history in such a way that they fail to observe that their participatory efforts are usurped and misused by powerful local interests. To the extent that they *need* to be accepted at the local level in order to develop and support local economic and conservation initiatives, and to be able to translate their enthusiasm for change for the better into action, they have little choice but to accept this tacit cooptation – even if they see it. From various conversations with members of NGOs we believe that "turning a blind eye" is more characteristic of foreign NGOs such as CARE, who are reported to sometimes follow a deliberate policy of keeping a low profile as a necessary prerequisite for being able to continue working in the country, and who do whatever they can to initiate changes "under the radar screen" of established interests.

If this chapter has succeeded in transmitting in a small measure a sense of an ongoing real struggle, it has achieved its purpose: we wanted to show that in such situations trying to be objective, to "tell the truth" as it looks from a scientific point of view, may become part of the struggle. It becomes part of the struggle because in cases like the ones presented in this chapter, there tend to be different versions of the "truth", each version supporting a different course of action, and supported by a different group of people, often with a power differential between them. "Truth" is contested not only in order to establish facts, but also because what is accepted as truth affects who will benefit from the course of action chosen: accepting one particular truth also means accepting the legitimacy of the course of action following from that truth, and with it accepting the distribution of benefits following from that legitimised course of action: accepting as fair what would have looked unfair in other versions of the truth. In this situation the scientist's version of the truth will necessarily become involved – and present him or her with a particular dilemma: telling the truth as s/he sees it is part of the scientific ethic, but doing so *locally* interferes with the local situation because the global reputation of science attaches a particular weight to the scientific version of the truth. It does not help to declare every scientific opinion a preliminary one: this inner-scientific ethos is not part of the public esteem of scientific results as being "truer" than other versions because the use of scientific methods "guarantees" objectivity. Objectivity in the usual scientific understanding, however, means *not* interfering: to restrict the scientific activity to observation. The solution to "tell the truth" only in publications normally not accessed by those who provided the data does not work when the local population explicitly asks for information about research results and reports – as consistently happened in the Lore Lindu area, where the hopes for a positive contribution from the scientists were also expressed rather strongly.

But compromising the very objectivity, which the scientific ethos calls for, is not the only dilemma. The second dilemma arises when the scientist's truth contradicts the truth of the more powerful local groups, who to a large extent control access to the research area: telling one's *truth* as demanded by these same powerful local groups can mean saying something which contradicts their self-image, *their* truth. And they may then regard the scientific truth as simply "biased" and feel that they have every right to ban researchers who are less than objective and only meddling in "internal affairs" from entering their area: By telling what the researcher according to his or her own ethic believes to be the truth *may* not only compromise objective observation of what is happening without the presence of the researcher, it may in extreme cases prevent any observation at all. The dilemma is as old as the social sciences, but to encounter the dilemma in the research situation in person can be a shocking experience – for all concerned. At the same time, the experience sharpens the researcher's sensibility to the existence of the dilemma. It may be this added sensibility which leads us to believe that there is a greater likelihood of finding oneself on the horns of the above dilemma in all research concerning sustainable development and natural resource management. The experience also makes it seem likely to us that the researcher's experience of the "sharpness of horns" of the dilemma in itself is an indicator of the quality of leadership – or power relationships, a the local level: leaders who follow old or new models of "good leadership" are likely to feel secure enough in their positions not to feel threatened even by a slanted scientific result, because they can simply expose and discuss it openly with all concerned. They may even welcome an opportunity to have an open discussion about an aspect hitherto overlooked. Because the scientific discourse usually downplays the role of feelings and intuitions, we would like to put on record the conclusion from our own experiences: That according the *feeling* of the researcher may be a valid measuring rod for "social facts". We also believe that where the scientist's feelings call attention to a particularly acute dilemma, there is a great likelihood that the situation itself *calls* for a redress of injustices. Not simply because of a humanly understandable feeling of sympathy with those individuals or groups who are wronged in that situation, but because, according to current knowledge about leadership and management, such situations are unlikely to remain stable for long and are unlikely to secure the kind of stable solutions for the management of resources which are needed for sustainability in local as well as a global sense.

References

Adiwibowo, S. (2005): Dongi-dongi – Culmination of a Multi-Dimensional Ecological Crisis: A Political Ecology Perspective. Doctoral Dissertation, Kassel.

Bearth, Thomas, (2008): Language as a key to understanding development from a local perspective. A case study from Ivory Coast. In : Henry Tourneux (Ed.): Langues, cultures et développement: 35-116. Paris.

Beckmann, M. and Pies, I (2004): Sustainability By Corporate Citizenship. The Moral Dimension of Sustainability. Martin-Luther-Universität Halle-Wittenberg Discussion Paper No. 04-11). available online from http://papers.ssrn.com/sol3/papers.cfm?abstract_id=991323 (accessed 20 May 2008)

Borrini-Feyerabend et al. (2004): Indigenous and Local Communities and Protected Areas. Towards Equity and Enhanced Conservation – Guidance on Policy and Practice for Co-managed Protected Areas and Community Conserved Areas. Gland, Switzerland.

Chambers, H. E. and R. Craft (1998): No Fear Management. Rebuilding Trust, Performance, and Commitment in the New American Workplace. Boca Raton.

CBNA (2000): Capacity Building Needs Assessment for Local Governments and Legislatures: Decentralization in Indonesia – The Framework for Local Governance. Discussion Paper; available online at: http://www.gtzsfdm.or.id/dec_in_ind.htm [click on: CBNA Discussion Paper "Decentralization in Indonesia (May 2000)].

Döbel, R. (2006): Good Leadership and Sustainability – Evidence from Three Countries. Paper presented at the annual LAGSUS conference, Tadulako University, Palu, September 5-6, 2006.

Fremerey, M. (2005): Local Communities as Learning Organization: The Case of the Village of Toro, Sulawesi. In: Neef, A. (Ed.): Participatory Approaches for Sustainable Land Use in Southeast Asia: 253-275. Bangkok.

Hunowu, M. A. (2004): Forest Resource Management: Battlefield of Local Knowledge among Local Leaders. A Case Study of Local Leadership in the Kulawi Community of Bolapapu, Central Sulawesi. MSc Thesis, Bogor.

Irvan N.R. and R. Sangadji (2004): Naftali, a Rebel with a Cause. In: *The Jakarta Post* (Saturday, 5 June 2004); quoted in: Community Forestry E-News No. 2004.06 (30 June 2004); online at: http:/www. recoftc.org/site/fileadmin/docs/e-letter-documents/eletter2004/CF_E-News2004.06_June_30_2004.htm (accessed 28 Oct.2007).

Hartiningsih, Maria (2005): *Konservasi Berkerangka Kultural Komunitas Toro* (Conservation in a cultural framework in the Toro Community). In: *Kompas* (15 September 2005). Online at: http://64.203.71.11/kompas-cetak/0509/15/humaniora/2050994.htm (accessed 19 May 2008)

IUCN (2002): Local Communities and Protected Areas. *The International Journal for Protected Area Manager*s. Vol.12 No. 2.

Majelis Adat Kecamatan Kulawi (April 2003): Watu Tambaga Momepanimpu (Togetherness and Helping Each Other). Unpublished Manuscript. Bolapapu.

Mappatoba, M. (2004): Co-Management of Protected Areas. The Case of Community Agreements in the Lore Lindu National Park, Central Sulawesi – Indonesia. Doctoral Dissertation, Goettingen.

Marcoes, L. (2002): Women's Grassroots Movements in Indonesia: A Case Study of the PKK and Islamic Women's Organizations. In: Robinson, K. and S. Bessell (Eds): Women in Indonesia. Gender, Equity and Development: 187 – 197. Singapore.

SMERU (2001): Otonomi Daerah dan Iklim Usaha; (Regional Autonomy and Investment Climate). Online at http://www.smeru.or.id/report/field/otdaiklimusaha/otdaiklimusaha.htm

Sunito, S. (2005): Continuation and Discontinuation of Local Institutions in Community Based Natural Resource Management. Doctoral Dissertation, Kassel.

Toheke, R. P. and Pelea, Krispus (2005): Perempuan dan Konservasi. Revitalisasi Kultural Peran Perempuan dalam Pengelolaan Sumberdaya Alam die Komunitas Toro Sulawesi Tengah (Women and Conservation. Cultural Revitalisation of the Role of Women in the Management of Natural Resources in the Community of Toro in Central Sulawesi). OPANT & CARE International Indonesia. Palu.

Tokare, I. Ds. (1990): Sejarah Lembah Lore Di Kabupaten Dati II Poso, Propinsi Sulawesi Tengah (The History of the Lore Valley in Kabupaten Dati II Poso, Province of Central Sulawesi). Unpublished Manuscript.

Tokare, I. Ds. (1999): Sejarah Datu Tindarura alias Gumangkoana dan Sejarah Penduduk Asli Tanah Perkurehua (The History of Datu Tindarura alias Gumankoana and the History of the Original Inhabitants of the Pekurehua Area). Unpublished Manuscript.

TNC (2004): Kesepakatan Konservasi Masyarakat Desa Watumaeta (Conservation Agreement of the People of the Village of Watumaeta – Appendix Sejarah Desa = Village History). The Nature Conservancy. Palu.

TNC (2002a): Kesepakatan Konservasi Masyarakat Desa Kaduwaa (Conservation Agreement of the People of the Village of Kaduwaa – Appendix Sejarah Desa = Village History). The Nature Conservancy. Palu.

TNC (2002b): Kesepakatan Konservasi Masyarakat Desa Sedoa (Conservation Agreement of the people of the village of Sedoa – Appendix Sejarah Desa = Village History). The Nature Conservancy. Palu.

TNC-LLFO (2001): Usulan Rencana Pengelolaan Partisipatif Wuasa (Proposal for the Management of a Participatory Planning Exercise in Wuasa). The Nature Conservancy. Palu.

Weber, R., Faust H. and W. Kreisel (2002): Colonial Interventions on Cultural Landscape of Central Sulawesi by "Ethical Policy" – Impacts of the Dutch Rule in Palu and Kulawi Valley 1905-1942. STORMA Discussion Paper Series No. 2. Bogor.

Chapter VII

From Rice to Cacao, from Women to Men: Farmer's Struggles in the Domination of Market Valued Knowledge

Laksmi Adriani Savitri & Michael Fremerey

Abstract

The shift from subsistence to petty commodity production which, in the case of "Revolusi Cokelat" (Cacao Revolution), is the shift from rice to cacao farming, provides the basic context for the constellation of gendered local knowledge in the community of Bolapapu. Rice farming, which is women's domain and women's knowledge, is in the process of being marginalized by the introduction of cacao as a commercial commodity. Market attributes have defined the "valued knowledge" which has then become a knowledge dominated by men. The option for commercial commodities (cacao) at the cost of rice farming is an individual farm household decision. It is these decisions which threaten the community's socio-economic security as well as the stability of the rain forest margin.

Gendered Agroecological Knowledge and Power Inequality

The incorporation of local knowledge into natural resource management has been increasingly accepted, since a hitherto centralized, expert-led control over nature had not succeeded in stopping the progress of natural resource depletion. According to *Global Forest*

Watch,[1] Indonesia is losing nearly 2 million ha of forest every year, through legal and illegal logging, clearance for plantations and agricultural estates, and fires. Acknowledging the importance of local communities as resource managers, the contemporary resource management strategy emphasizes grassroots approaches, sustainability, empowerment, and participation. The strategy, however, often takes an uncritical approach to local knowledge. One area where this is apparent is the lack of attention paid to the gendered nature of local knowledge (Wayland 2001). Resource development decisions are often produced without acknowledging gender differences.Gender-oriented research indicates that men and women tend to possess different kinds of knowledge because they have different roles in production activities (Wayland 2001; Kelkar and Nathan 1991; Agrawal 1994). In many cultures, women have specialized knowledge about farming, healing, and resource management that is unique to them. These differing bodies of knowledge are linked to culturally constructed gender roles. As men and women perform their daily activities, they accumulate knowledge about various topics and develop expertise in differing areas.

However, gender differences are rarely neutral – in almost all societies, gender is a form of social stratification. The prevailing division of labor has led to men and women assuming unequal positions in terms of power, prestige and wealth (Giddens 2001). This fact is also reflected and reinforced by a knowledge divide. Even in relatively egalitarian communities in South East Asia, Li (1998) evidenced gender-related inequality since women stayed behind men in terms of power and prestige. This differences are often derived from the activities women engage in or fail to do so. The productive activities of women signify rather weakness than power. For instance, the influence of women gained from their active role in the economic domain (e.g. as traders) cannot match the power which men acquire from their exclusive access to the spiritual domain.

The possession of certain knowledge paves the way to status and authority, and certain bodies of knowledge come to be privileged over others (Foucault 1976). As far as agricultural communities are concerned, agro-ecological knowledge is a privileged or valued knowl-

[1] http://www.globalforestwatch.org/english/indonesia/overview.html

edge, since people's livelihood largely depends on agricultural activities and the ecological stability of their living environment. Empirical findings further elucidate that local agro-ecological knowledge is characterized by a gender divide which puts men in a noticeable power advantage (Shiva 1992; Kelkar and Nathan 1991). Jewitt's findings on the unequal distribution of agro-ecological knowledge in Ambatoli, India (Jewitt 2000), are of special interest in this context. This divide has put women outside the decision making arena in production since the degree to which Ambatoli's women can develop agro-ecological knowledge is dependent on how far their husbands and parents-in-law allow them to contribute to field-based agriculture. Such circumstances automatically put the control of decision-making in production in the hands of men because it is men who have direct access to land and resources. This, in turn, opens the opportunity to accumulate agro-ecological knowledge. Having the relevant knowledge, men gain the expertise that gives them authority in making decisions in the agricultural production process.

The authority to decide goes hand in hand with the power to exclude or include others. In the context of agro-ecological knowledge as a valued knowledge, the decision either to plant or not, which commodity to plant, how much labor to use, where to sell etc. are made by that member of a farm household who has the relevant expertise and, therefore, the authority to decide. In the process of decision-making at the household level, as an expert, he or she will have the power to include or exclude other household members. As Guyer (1980) points out, the household is not an undifferentiated grouping of people with a common production and consumption function, but is better understood as a system of resource allocation where patterns of decision making vary according to culture.

Gender interaction is always an arena where face-to-face encounters of social actors – differing in gender, interests, resources and power – occur. As Foucault (2002) asserts "every human relation is to some degree a power relation. We move in the world of perpetual strategic relations." Then, when looking for power, Villarreal (in Long 1992) suggests that we should focus our attention on struggle, negotiation, and compromise. Power is not only about the capacity of an individual to impose his or her will upon others, but also the capacity of the 'victim' to respond to such pressure. Foucault (in Luke 1999) explains that, as there are no relations of power without resistance, power is inseparable from contestation and struggle. This is what Long (1992) had in mind when referring to his concept of the 'knowledge interface'. Departing from the interrelation between gender, knowledge and power concepts, the focus of this chapter is on social relations as they develop in the context of rice farming and cacao cultivation. This is the arena where the interface of gendered agro-ecological knowledge in the forest margin zone of the Lore Lindu National Park, Central Sulawesi, occurs. The village of Bolapapu in the Kulawi valley is depicted as a place where the shifting mode of production from subsistence (rice) to petty commodity (cacao) is setting off a shift in valued local knowledge and hence a gender-related change in power concepts within the individual households as well as on the community level as a whole. This interface includes negotiation, compromise and strategies in order to gain legitimacy, authority, and power over resource use (Long 1992), and it finally relates to the issue of sustainability of such resources.

During our field research, in-depth interviews as well as participant observation were the main methods applied to develop a deeper understanding of the ongoing process of change which may yield far-reaching socio-economic and ecological consequences.

Aristocracy of the Past and Present: The Base of Land-ownership Structures in Bolapapu

Land holding and ownership are the main bases for knowledge accumulation. In Bolapapu, the local aristocracy determined one's access to land. From the past until the present, the village of Bolapapu is perceived as the core center of Kulawi culture and the regional governance system. As a part of the overall Kulawi homeland, the community of Bolapapu is more or less strictly divided into three strata[2] until today, e.g. the *Maradika*, *Todea* and *Batua*.

The noble *Maradika* constitute the recruitment base for local leadership. Kulawi leadership manifests itself in a collective entity of *Totua*

[2] Shohibuddin found the same stratification scheme in the community of Toro (Shohibuddin, this volume)

Ngata, that is individuals acknowledged for their wisdom and charisma. Traditionally, the *Totua Ngata* chose a person from the *Maradika* strata to become their headman or headwoman. In the past, every village in Kulawi had its own leader with several villages constituting a sort of federation. Later in the Colonial period, the Dutch changed this governance system, and crowned only one person as 'the King', or *Magau* of Kulawi.

The majority of the ordinary people were called *Todea*, whereas the term *Batua* denotes the lowest section of Kulawi society. Originally, *Batua* were the slaves of the *Maradika*. Nowadays, the existence of this class is sometimes denied due to feelings of shame among the *Maradika* about their ancestors' involvement in slavery. Bolapapu village profile of 2003 shows that out of a total of 1244 persons in productive age (14-64 years), 79% are farmers. Unlike Sanderson's (2000) generalization of a unitary class of farmers, Bolapapu farmers represent not a repressed homogeneous group at the same time constituting the lowest socio-economic and political level. Instead, their social and economic status is depending on the degree of control and ownership over land and other productive means, including aristocratic descent in some cases. Thus, to a high degree, socio-economic status depends on what kind of capital is available.[3]

In Bolapapu's past social system, economic, cultural and social capital played an important role. Cultural and social capital such as the position of *Maradika*, could be automatically converted into economic capital, since being a *Maradika* included the privilege of owning and distributing productive properties. *Maradika*, for instance, held dominant control of land distribution and ownership. If a commoner wanted to open a new plot outside the family land, he had to obtain the consent of the *Maradika* and *Totua Ngata*. Thus, economic capital was linked to the social and cultural capital of *Maradika* and was strengthened by large land ownership, aristocratic descent, higher

---

[3] As described by Bourdieu (in Turner 1998), the social world is a sphere with different dimensions based on differentiation and distribution, and the actor is defined by his/her position in this sphere. Furthermore, the actors' positions are determined by two dimensions: (1) the degree of capital ownership and (2) the weight of their capitals' compositions. There are four types of capital, e.g. (1) economic capital; (2) social capital; (3) cultural capital; and (4) symbolic capital.

education attainment, positions in government service, etc. *Maradika*, then, became the most dominant class at that time or – in Bourdieu's terms – a class characterized by extensive possession of various forms of capital, including the effective means to produce and reproduce social positions.

Traditional Socio-economic class Structure			**Present Socio-economic Class Structure**		
Capital's elements	Types of capital	Vertical positions	Capital's elements	Types of capital	Vertical positions
Aristocratic descent Wealth Political position Extensive social relations	Cultural Economic Social Symbolic	Landlord (*Maradika*) Headman/headwoman (*Maradika*) and advisory council (*Totua Ngata*)	Aristocratic descent Education Wealth Political position Extensive social relations	Economic Cultural Social Symbolic	Landowner (*Maradika, Totua Ngata*) Government officer (*Maradika, Totua Ngata*)
			Wealth Education Political position Social relations	Economic Social Cultural Symbolic	Big retailers, big-scale farmers, large-land owners (mostly *Todea*) Government officer with higher education (mostly *Maradika*, some *Todea*) Government officer in general
Wealth Social relations	Economic Social	Small-medium traders and medium farmers (*Todea* or migrants)	Wealth Social relations	Economic Social	Small-medium retailer traders, medium farmers (*Todea, Maradika*)
Social relations	Social	Small farmers or farm labor (*Batua*)	Social relations	Social	Small-scale farmers, farm labor (*Todea, Maradika)*

***Table 1**. Socio-economic classes in Bolapapu*

Today, since land has become privatized and is valued first of all as a commodity, distribution and redistribution of land are regulated by the market. Migrants or outsiders may acquire land simply by purchasing it. The status of *Maradika* does not anmore automatically include control of economic capital. The allocation of land, for instance, is decided by the village head – a representative of modern village governance. In other words, aristocratic descent has lost its ability to be converted into other forms of capital. A wealthy trader on the other hand may have all the abilities to symbolize his/her economic, social, and cultural capital, but he may not access the cultural capital of the former aristocracy. Such compositions characterize Bolapapu's present-day social map, where the dominant class is dispersed, power centers are not anymore clustered in a single point, and where the power of the aristocracy as the most important capital element is eroding (see table 1).

This description of Bolapapu's social structure may lead us to the assumption that the concentration of economic power in the hands of the *Maradika* has been displaced by a more redistributive socio-economic structure characterized by a more equal allocation of land ownership. At present, the large majority of landowners in Bolapapu (93.5%) own less than 1.0 ha, while the maximum land owned by one person is only 3.0 ha.[4] Farm sizes of 0.25 - 0.5 ha are most common. In general, there are three types of land: wet-rice fields (*sawah*), dry-land (*kebun*), and home lots (*kintal*).

The distribution referred to above, however, does not include land ownership outside Bolapapu. The two biggest local land owners, who are not of *Maradika* descend, have accumulated substantial pieces of land outside the village. One of them owns about 60 ha of cacao estates in Laone, a hamlet in nearby Namo village, the other owns 20 ha of land in Marena, a hamlet in the Kulawi village of Oo Parese. These are the only actors who are generally recognized as commercial cacao farmers (*petani pengusaha*) in the Kulawi sub-district. Thus, land

4 Data source: Daftar Himpunan Ketetapan Pajak dan Pembayaran Tahun 2003. Desa Bolapapu, Kec. Kulawi, Kab. Donggala, Provinsi Sulawesi Tengah. (Report on Tax Assesment and Payment in Bolapau Village 2003). Issued by: Ditjen Pajak, Dep. Keuangan, KanWil XIII, Sulawesi Tengah. Kantor Pelayanan PBB Palu (Fiscal Departement XIII, Palu, Central Sulawesi, Service Center).

ownership in present Bolapapu displays its most significant inequalities between commercial cacao farmers on the one hand and the majority of smallholders on the other. This structure gives privilege to large landowners to accumulate "cacao knowledge".

From Rice to Cacao: From Subsistence to Commercialization

The Green Revolution entered Kulawi only in the 1980s, fairly late in comparison to the western parts of Indonesia where the Green Revolution started already in the 1960s.

With an area of 449.5 ha (72.6%) dry-land farming is still the dominant cultivation system in Bolapapu; 170 ha (27.4%) are wet-rice fields.[5] The majority of farmers (64%) own only dry-land, while 28% till exclusively wet-rice fields. A small portion of (mixed) farmers (8%) own both types of land.[6]

In former times, rice farming used to be an inseparable part of Kulawis' cultural and belief system. At each step of the cultivation cycle, a ritual was performed acknowledging the integrative nature of human beings and their natural environment. It was believed that both, men and nature, belong to one cosmic system, and the maintenance of harmony between them was sought by the creation of a direct communication with nature by the use of specific 'mantra' and offerings. However, since the Salvation Army introduced Christianity in 1913, followed by Islam in 1935, which was followed in turn by Protestant missions in 1947 again (Garang 1985), all practices associated with animism and spiritism were forbidden (Schrauwers 2002) in a process of ritual secularization. Being a target of agricultural modernization, many aspects of the traditional rice farming technology, such as the *ani-ani,*[7] the *baparuja*-sstem[8] and various indigenous varieties[9] have

---

5 Data source: Bolapapu Village Profile 2003.

6 Data source: *Daftar Himpunan Ketetapan Pajak dan Pembayaran Tahun 2003. Desa Bolapapu, Kec. Kulawi, Kab. Donggala, Provinsi Sulawesi Tengah.* Issued by: *Ditjen Pajak, Dep. Keuangan, KanWil XIII, Sulawesi Tengah. Kantor Pelayanan PBB Palu* (see above).

7 A manual rice stalk cutter. The cutter operates like scissors, so that it can only be used to cut one stalk at a time, which makes harvesting a rather tedious work.

vanished from Bolapapu. A*ni-ani* was replaced by the *sabit*[10] about 20 years ago. *Baparuja* was also not anymore practiced since the plough and hand tractors entered the village at the same time. The change in rice varieties from the traditional long stalk types to the modern, hybrid short stalk varieties has turned the *ani-ani* into an unsuitable tool for harvesting. Moreover, water buffaloes had became scarce and expensive due to a high consumption rate in traditional

life-cycle rituals. Modern tools and techniques become a affordable alternative for local farmers.

Monetization plays an important role in transforming the economic and socio-cultural value of rice farming. For farmers whose income depends solely on agricultural activities, rice cultivation is still a necessity since having a stock in a barn provides subsistence security. Under this logic, conversion of rice fields in to cacao plantations is rare, except for the plots that have been ruined by flood and sedimentation. This type of 'security first farmer' is still dominant in Bolapapu. On the other side, for upper-class farmers with various sources of income, rice farming is considered to be a non-profitable business. Hence it is not important to own or cultivate a paddy field. The market price of rice is low and it is easily available. Therefore, cacao or any kind of cash crop is given priority, not only because it generates more cash but also because it increases the value of the land.

These 'rational' farmers are the entry point of cash crop cultivation. The elderly farmers in Bolapapu claim that cacao was first introduced by the Dutch, but never as a commercial commodity since it was never planted *en masse* and never produced for the market. It was not until the late 1980ies when the price on the world market started to increase that cacao caught the attention of the farmers. In 1990, when the price of cacao reached Rp 5,000 (€ 0.50) per kg, cacao planting started to spread widely. Most cacao farmers stated that they gained their knowledge of cacao cultivation techniques either by observing

---

8 A technique to soften the soil before planting. It is done by large numbers of water buffalos led by the herdsman. The buffalos are circling the fields several times until the soil becomes loose.

9 *Rakki*, *Gumbira*, and *Kamba* are examples of vanishing indigenous varieties, not to mention many varieties of sticky rice used to be planted in the dry land farms.

10 Another manual u-shaped stalk cutter, operates like a knife. It can cut a bunch of rice stalk at once.

other farmers or by listening (*dengar-dengar*) to conversations and discussions among fellow farmers. Cacao is not the first cash crop to enter the village economy. Agricultural commercialization in Bolapapu started in the 1930s with the introduction of coffee by the Dutch. Initially however, the villagers planted coffee trees only for their own needs. The planting of coffee as a commodity started not earlier than 1963, when local migrants planted it initially in larger scale plantations as a response to the rising market price. Nowadays, due to the even higher and more stabile prices, and triggered by a sevenfold price increase in 1997-1998, cacao is preferred to coffee.

The global market interest in Sulawesi's cacao is even manifested in Bolapapu. An international alliance of cacao industries, consisting of the US non-profit organization ACDI/VOCA, Masterfoods, and the World Cacao Foundation, decided to intervene in local cacao cultivation practices by deploying a training program for Bolapapu's farmers. They train farmers to apply frequent harvesting, pruning, sanitation and fertilizing (PsPSP) methods with the aims to double production rates, to prevent pests and diseases and to keep the produce free from chemicals. These programs created some enthusiasm in the beginning, but when it comes to application, many farmers opted for the traditional practice since the 'new' knowledge required a total focus on cacao. This is just unaffordable for many small farmers, since their single and small plot is often designed for many crop varieties, including maize, cacao, vanilla, coffee and candle nuts. Monoculture is a cultivation technique that is still not fully accepted as a provider of total security. And the farmer will turn to vanilla, coffee, or clove, as additional sources of income instead of cacao, whenever the market price appears to be more favorable.

Transformations in Gender Relations

The traditional Kulawi norms of gender relations have undergone a noticeable transformation in the course of the described agro-economic processes. Historically, according to *Kulawi* norms and traditions, the public roles of women were recognized in the community. A woman was esteemed as *ova*, a symbol of wisdom, one who was able to level down tensions in a situations of conflict. Therefore, whenever there was an important meeting, women were always invited and sought in order to prevent that dissent turns into open conflict. Furthermore, women used to be ritual leaders, especially in the case of rites related to farming activities. In late 19th century, just before the Dutch entered in 1905, Kulawi was headed by a woman named Hangkalea. She was a person whose deep wisdom was praised, and who is remembered to have received the highest respect from her people.

Before monetization and cash crops entered Bolapapu's village economy, rice was the only means of exchange and its cultivation the dominant farming activity. Women held the authority to determine the planting season, to organize variety selection, and to lead the rituals that used to be performed prior to, and during the rice planting cycle. It was the women who had the exclusive access to the rice barn. Men were not totally excluded in the rice planting process: they assisted in land clearing, hoeing and plowing. However, in this period of time, when the rice cultivation was the dominant village economy, women held the knowledge of and power in rice farming. The transformation began with Christianity followed by state intervention as it was brought about by colonialism, until finally, the introduction of a patrilineal culture. This change found its formal expression in the crowning of Hangkalea's nephew, Tomampe as the king (*magau*) of Kulawi in 1917. The attribution of a family name in accord to the men's genealogical line (Aragon 2002) is another form of patriarchy. And Christianity induced the fundamental belief that the woman was created from man's rib: a belief that supported a total change in gender norms.

In effect, the matri-focused society, as expressed in the role of woman as the heiress of the house and the paddy fields, was no longer strictly applied. Because of the attribution of men's surname to the landownership, the right to sell lands became to be perceived as men's right. As Christianity forbids native rituals, the woman's role as a ritual leader has vanished as well. Nowadays, women are no longer involved in legal deliberations within the traditional law council (*Lembaga Adat*), which has become an almost exclusive forum of men. And concerning women's position at the community level, a significant transformation has occurred in Bolapapu as well. Bolapapu's women are excluded from the mainstream process of village decision-making. The PKK, as the only existing women's organization, is confined by the state[11] to domestic tasks and responsibilities. It is not considered as equal to other village institutions in the local decision-making process.

At the household level, the traditional Kulawi norms asserted an equal position of men and women in daily interactions and the division of labor. To balance intra-household power relations, a woman was placed as *Tua Tambi*, a 'place' to secure traditions, because she was the one who was supposed to maintain and teach local traditions to the younger generation. Therefore, women were in charge to manage the family's property. The eldest sister held the power to record and manage the family's wealth. In contrast, the present Kulawi custom conveys a marriage advice as follows: "*The husband should be responsible and hard working, eagerly seeking everywhere to provide for his family, and the wife is to wait for the husband and manage as best as she can all of her husband's earnings to provide for the family's needs.*" This advice no longer asserts the power traditional balance within the family, instead it places women in a position inferior to men.

The description of present *adat* norms in gender relations unequivocally depicts a segregation of men and women in the productive and reproductive domain, as it declares men as the breadwinners. It is also emphasizes that "*nowadays, men are the only power holders in the family, we – women – are powerless because they think that we don't know anything. When men want to sell the land, they tend not to involve us in the decision, although some women are involved. Men also intervene in managing family income and assets, such as land. Sometimes they sell land without telling the woman.*" This testimony of an interviewed woman shows that the previous role of a woman as the person who is responsible for managing family property – a position that was supposed to balance the control and access in household decision-making – is not being applied anymore in consistent manner. Men have partly taken over control over family assets too.

[11] The state has a main role in women domestication, since PKK is an organization formed by the state and imposed on the village level (see also Burkard, Wild West….., this volume).

Transformations in Social Relations of Production

Rice farming activities require high labor inputs. Hoeing, planting and harvesting absorb at least 4 to 5 laborers in a plot of 0.5 hectares. This need was usually met by labor exchange. *Mapalus*, literally means "working together and helping each other", was the most preferred option. People form a group of ten or more, working together successively in each member's farm. If one member is not able to perform the task because of sickness or other reasons, she/he is expected to find a temporary substitute until she/he is able to re-join the group. But, the person who was replaced should repay his/her substitute's labor (called *baku balas tangan*) based on needs, which is not going to be articulated but tacitly understood, either in the form of cash or labor. Strict exact reciprocity is the basis of this exchange.

Nowadays, *mapalus* has been partly transformed into a group of individually paid labor (*buruh upahan*) or group-paid labor (*borongan*). Payment for planting and weeding is usually on a daily basis, with working hours from eight in the morning to three in the afternoon, including a one-hour break for lunch. The wage is Rp 15,000 (€ 1.50) per person/day without meal or Rp 10,000 (€ 1.-) per person/day with meal provided by the farm owner.[12] Harvest work is always paid in-kind, which is one bucket (*blek*) of paddy (*gabah*) per day, and the tasks range from *basabit*[13] to packing the rice in sacks.

However, labor rationalization through forms of paid *mapalus* has not yet eliminated solidarity and reciprocity. Mostly it is the better-off farmers who opt for such arrangements, since, as government officials or traders they don't have enough time to care for their own farm or engage in group activities. For small farmers, *baku balas tangan* is still a preferred option. The personal and reciprocal nature of the social relation of production is also expressed in the non-existence of fixed-rent and share tenancy in the land tenure system. Landless farmers can borrow other farmers' land for cultivation (*mehabi* or *pinjam garap*), and if the permission is given, the borrower will promise to share part of his/her harvest to the landowner as an expression of good will (*momirara* or *kemanisan hati*). There is no formal condition to

---

[12] Labor wage in 2004

[13] *Basabit* = cutting rice stalks

share, because it is a manifestation of the norm of reciprocity. Thus, in the context of shifting modes of production toward crop commercialization, the fact that reciprocity remains in the present social relation asserts the persistence of non-market values against commercialization.

The present division of labor in rice farming, technological transformations and the introduction of modern varieties have altered the quality of certain tasks. For example, hoeing (used to be done by women) was replaced by the plough (operated by men), and the *ani-ani* (used to be a women's tools for harvesting) was substituted by the sickle (*sabit*) that can also be used by men. Yet, women in Bolapapu, to some extent, are able to retain their parts in the labor market, since women still perform many of rice farming tasks such as planting and weeding exclusively. Owners of rice fields, who are usually women, also still control variety selection and the determination of the planting season. Therefore, despite some replacements caused by mechanization, women are still the main actor in rice farming. Their roles may not be as dominant as they were in the past when all rituals and manual work demanded women's leadership, but women are still recognized as "the experts" in rice farming. This means that the ownership of knowledge is theirs, and they tend to pass it to their daughters too, because most women refer to their mothers when questions about their source of information on rice farming are asked. Clearly, rice, from the field to the table, is a domain of women's knowledge.

Unlike in rice farming, a high labor input requiring to work in a group or *mapalus* is rarely necessary in cacao growing, except for land clearing and manual weeding (which is mostly paid, the wage being similar to rice farming activities). Furthermore, in contrast to rice farming, cacao cultivation is clearly the men's domain. Almost the entire decision-making in production is controlled by men. It may not be exclusively the man who handles the work in a cacao plantation. Of course women can help in planting, weeding, pruning and harvesting, taking care of the drying process, and sometimes selling. However, men decide when and how to buy fertilizer, pesticides and labor. In many cases of woman's full involvement due to her ownership of the cacao plantation, production decisions were still done by the man.

There are cases of women's non-involvement in cacao farming too, which occur in upper-class farm households. The household division of labor puts the wives either in the domestic setting, or in productive work other than cacao farming, such as the wife of the biggest cacao plantation owner who is in charge of managing the sundries shop, but never cares for the cacao business. Marketing is the area where women usually take a relatively large part in decision-making. They know when an immediate household need must be met, such as paying the children's school fees. Therefore, setting the time for selling is often a matter of women's decisions. From time to time retail traders who collect farmer's produce will pay a visit, and they are not reluctant to buy cacao beans even only after a day of drying. Farmers would sell their cacao to this retail trader to care for their immediate needs. However, for a higher level of expenditure, such as wedding preparations, when and where to sell is a man's decision or compromised between wife and husband. For such purposes, farmers would bring their produce directly to the wholesale trader in order to get a higher price.

Actually, most farmers come to the wholesale trader because they have borrowed money, fertilizer, or pesticides from the trader, so there is an obligation to sell their produce only to this particular trader in order to repay the loan. In such a setting, the wife will take part in weighing the beans, and making sure that the weight in the trader's store is not less. Most farmers are actually confined to this type of socio-economic relation in terms of marketing of their produce. They know very well that they may receive a better price if they sell cacao directly to the provincial capital, but they are attached to the wholesale trader because of the never-ending debt cycle. Despite shared-participation in the decision-making process of cacao marketing, men control the other decisions about cacao production, as concluded in table 2.

From Subsistence to Commercial Agriculture: The Shift of Domination

In Bolapapu, subsistence and commercial agriculture are coexistent. Rice farming is struggling to maintain its status, with all its cultural connections, while commercial commodities – not only cacao but also coffee, clove, and vanilla – are gradually becoming more rooted in the agricultural and socio-economic context of Bolapapu. While the commercial crops started to obtain a good market price, rice remains primarily important as the staple food. It has never been commercialized because its exchange value is much lower than any of the export commodities.

Progressive farmers perceive wet-rice fields in a rational way. They decide to sell or rent out the fields to other farmers, because rice farming is not profitable for them. Others prefer to maintain their ownership of paddy fields as a source of staple food, but focus on cacao for their income. As the standard of value is shifting towards market measurements, the 'importance' of each commodity is classified in terms of *rupiah* value. This indicates a shift of dominance not only in economic terms. Following Bourdieu's perspective, the values are changing, based on the capital to be used as a standard of valuation. Thus, when there is a transition from subsistence to a commercial mode of production, it then becomes a context where the value is shifting from its root in the social and cultural capital, to the economic sphere where productive forces are the essence of such capital.

The head of the Lore Lindu National Park sub-office in Matawe stated that Park encroachment along the border adjacent to Bolapapu in 2004 was about 60 ha planted with cacao and coffee ; and 700 ha for the whole Kulawi Sub-district. This is just another indication that the shift of domination from subsistence to the commercial mode of production is in progress. The fact that rice farming in Bolapapu never leaves its subsistence nature implies that the economic value is never attributed to this knowledge. Therefore, when a value is in transition (which is actually a process of solidifying economic capital), the value of knowledge is also shifting from a non-market to a more market-valued knowledge. In other words: A transition of subsistence to a commercial mode of production becomes a context where knowledge about cacao as a commodity is more valued than knowledge about rice farming.

Gendered Knowledge and Household's Struggles for Socio-Economic Security

In the past, women's role and status in rice farming in Bolapapu was highly valued because production-for-subsistence was the dominant type of economic activity. As outlined above, this situation is changing profoundly, accompanied by a marked rearrangement of gender-related issues. Since knowledge about subsistence-oriented rice farming is embedded in the women's domain, the described process asserts women's subordination in their reproductive work. At the same time, men's exclusive knowledge in commercialized cacao production substantially adds to their socio-economic, political and cultural domination. This materializes, first of all, in decision-making processes on the household level. Decision-making can be translated into control and access. By control, we mean the power to decide when and how a task should be performed and a resource should be used. By access, we mean the freedom or permission to perform tasks and to use resources. A decision-maker certainly has the power to control, which always includes access. But having access does not imply automatically control over resources.

As shown in table 2, women dominate the decision-making and task execution in rice farming. In cacao cultivation, however, almost all production decisions are taken by men, except for determining the types of commodity planted on the farmland and cacao selling. Women give input based on their household's subsistence needs. Usually, before planting cacao seeds, the farmers plant faster cash-earner crops, such as corn, because cacao takes at least two to four years to yield. Later, in the post-harvest period, farmers face many options of buyers. Therefore, only in these two types of decisions women push their involvement in the decision-making process, since those are periods with a direct relation to the household's income management. Table 2 depicts the gender-divide in terms of decision-making and task execution in the present day farm households. However, the actual variations are quite high, depending on the form of communication and the type of relationship within each family (see table 3). One decisive element that can elevate a woman's bargaining position is her ownership of land. This is where gender analysis reveals the actual process of negotiation within households.

Based on seven household cases, the emerging pattern of decision-making in Bolapapu is distinguished by the aristocracy and economic classes. Interestingly, the aristocracy and lower economic classes relate to a better position of women in intra-household decision-making. The pattern of decision-making among the *Maradika* shows a relatively equal position of women compared to non-*Maradika* in the decisions regarding 'high expenditure' types, such as household asset allocation and farm production. Even, in the lower economic class, a*Maradika* family places women as equal partners in the household decision-making process. On the contrary, women in the upper economic class of non-aristocratic households must comply with men's domination in the decision-making process.

Organization of production	**Decision maker(s) (Control and Access)**		**Task executor(s) (Access only)**	
	Woman	Man	Woman	Man
Rice Farming:				
Type of variety planted	✓		✓	
Labor recruitment	✓	✓	✓	✓
Use of inputs (fertilizer, pesticide)		✓		✓
Types of inputs (types of fertilizer/pesticide, the volume, the frequency of application)		✓		✓
Types of task performed:				
– Put fences		✓		✓
– Spread seeds	✓			
– Plowing or hoeing	✓	✓	✓	✓
– Transplant the seedlings	✓		✓	
– Weed	✓		✓	
– Apply fertilizer and/or pesticide		✓		✓
– Harvest	✓	✓	✓	✓
To sell the produce or to store	✓		✓	
If to sell: where to, how much the best price is	✓		✓	

Organization of production	Decision maker(s) (Control and Access)		Task executor(s) (Access only)	
	Woman	Man	Woman	Man
Cacao cultivation:				
Type of commodity planted prior to cacao		✓	✓	✓
Labor recruitment		✓		✓
Use of inputs (fertilizer, pesticide)		✓		✓
Types of inputs (types of fertilizer/pesticide, the volume, the frequency of application)		✓		✓
Types of task performed:				
– Plant the seedlings		✓		✓
– Weeding, pruning		✓	✓	✓
– Apply fertilizer and/or pesticide		✓		✓
– Harvest	✓	✓	✓	✓
– Drying (how long it should be dried: depending on how soon cash is needed)	✓	✓	✓	✓
Marketing (when and where to sell: depending on when cash is needed, and which buyer gives the best price)	✓	✓		✓

Table 2. *Agricultural intra-household decision making and task execution*

In all cases, eventually, women comply with men's decision in cacao production due to their lesser access and control over cacao knowledge. Nevertheless, women did not give their compliance without resistance. An extreme case presents a resistance of a wife to sell her rice field or help her husband in his cacao plot. At the end, she complied with her husband's argument that their household's socio-economic security is better guaranteed by cacao production than rice.

Socio-economic class	Land-ownership	Household structure	Type of relationship	Type of decisions	Pattern of decision making	
					Single decision maker (husband only)	Shared decision making (husband and wife)
Maradika and upper class	Man Woman	Male headed household with grown-up children: Th-SM case	Senior-junior partner	Asset allocation (land transfer, expansion, inheritance)		✓
				Production	✓ *(female)*	
Maradika and lower class	Man Woman	Male headed household with grown up children: Dh-DT case	Equal partner	Asset allocation (land transfer, inheritance, expansion)		✓
				Production		✓
Non-maradika and upper class	Man only	Male headed household with grown up children: Kk case Ys case	Male dominated	Asset allocation (land transfer, inheritance, expansion)	✓	
				Production	✓	
Non-maradika and middle class	Man only	Male headed household with young children: Is case	Senior-junior partner	Asset allocation (land transfer, inheritance)	✓	
				Production		✓
Non-maradika and lower class	Woman only	Female headed household with grown up niece and nephew: Rg case	Senior-junior partner	Resource allocation (income utilization for production and consumption)	✓	
				Production		✓
Non-maradika and lower class	Man only	Male headed Household with grown up children: Dm case	Senior-junior partner	Resource allocation (income utilization for production and consumption)	✓	
				Production		✓

Table 3. *Gender relations and patterns of decision making in Bolapapu farm households*

Conclusion: Repercussions on the Stability of Forest Margins

It is a fact that some women in Bolapapu who have the main responsibility for maintaining rice farming are well aware that the intensive need of continuous water supply for rice farming requires an effective conservation of the forest. But, possessing such knowledge does not always imply that this knowledge is or can be applied in the negotiation process of whether to cut the forest for cacao plantations or to conserve it for rice farming purposes, because most women are rarely involved in such decision-making. Moreover, Kulawi women in Bolapapu rely on earnings from cacao in the management of their households, despite their ownership of rice field. So, a tendency to comply with and conform to a 'cut the forest for cacao' decision is likely to appear.

The interface of gendered knowledge in the agro-ecosystem constituted by forest, rice and cacao manifests itself as a diverging process. Cacao knowledge now dominates the decisions on the household as well as on the community level. Negotiation and compromise to converge the knowledge of rice and cacao do not occur. Cacao expansion to the forest evidences the weak existence of negotiation and compromise. The persistence of rice in the present agro-ecological landscape, however, may not provide a strong base to defend forest existence, as long as the link between the forest and cacao farms is not revealed.

It is important to ensure women's access to cacao knowledge accumulation. The number of women who participate in cacao cultivation training is not adequate to ensure their access to the field-base work and decision-making. Types of adequate empowerment needed are a "collective knowledge exchange" between male and female farmers, and a collective learning process on sustainable cacao farming. Equally important is a revitalization of the traditional norms that ensure women's access to landholding and ownership, as well as a more frequent involvement of women in the community decision-making process.

References

Agarwal, B. (1994): A Field of One's Own: Gender and Land Rights in South Asia. Cambridge.

Arragon, L.V. (2002): Migrasi, Komoditas Ekspor, Dan Sejarah Perubahan Hak Pemakaian Tanah di Sulawesi Tengah (Migration, Export Commodity, and The History of Land Use Change in Central Sulawesi). In: A. Lounela and R.Y. Zakaria, 2002: Berebut Tanah: Beberapa Kajian Berperspektif Kampus dan Kampung (Struggle for Land: Several Studies with Campus and Kampong Perspectives). pp 271-281. INSIST Press. Yogyakarta.

Bourdieu, Pierre (1993): The Field of cultural Production. 322 p. Columbia University Press.

Denzin, N. K. and Y. S. Lincoln (2000): Handbook of Qualitative Research. London.

Dey, I. (1993): Qualitative Data Analysis. London.

Foucault, M. (1976): The Archeology of Knowledge. New York.

Foucault, M. (2002): Power/Knowledge (Wacana Kuasa/Pengetahuan). Jogjakarta.

Garang, J. (1985): Dunia Kulawi: Masyarakat, Budaya dan Gereja di Sulawesi Tengah (The

Kulawi World: Communities, Culture, and Church in Central Sulawesi). Jakarta.

Giddens, A. (2001): Sociology. Fourth Edition. Polity. UK.

Guyer, J.I. (1980): Household Budgets and Women's Incomes. Working Paper No. 28. African Studies Center, Boston University. Boston.

Haryatmoko (2003): Menyingkap Kepalsuan Budaya Penguasa: Landasan Teoritis Gerakan Sosial Menurut Pierre Bourdieu (Uncover the False of the Power Holder Culture: Theoretical Foundation of Social Movement According to Bourdieu). BASIS No.11-12, Tahun ke - 52, November-December 2003. Yogyakarta.

Jewitt, S. (2000): Unequal Knowledges in Jharkand, India: De-Romanticizing Women's Agroecological Expertise. *Development and Change*, Vol.13: 961-985.

Kelkar, G and D. Nathan (1991): Gender and Tribe, Women, Land and Forests in Jharkand. New Delhi.

Li, T. M. (1998): Working Separately but Eating Together: Personhood, Property and Power in Conjugal Relation. *American Ethnologist*, Nov 1998: 25, 4.

Li, T. M. (2002): Local Histories, Global Markets: Cacao and Class in Upland Sulawesi. *Development and Change*, 33 (3): 415-437.

Long, N. and A. Long (1992): Battlefields of Knowledge: The Interlocking of Theory and Practice in Social Research and Development. London.

Luke, T.W. (1999): Eco-managerialism: Environmental Studies as Power/ Knowledge Formation. In: Fischer, F and M.A. Hajer. (Eds): Living with Nature. Oxford.

Sanderson, K.S. (2000): Makro Sosiologi: Sebuah Pendekatan Terhadap Realitas Sosial (Macro Sociology: An Approach to Social Reality). Jakarta.

Schrauwers, A. (2002): "Itu Tidak Ekonomis": Sifat Ekonomi Moral Yang Berakar Pada Ekonomi Pasar di Dataran Tinggi Sulawesi, Indonesia ("It's Not Economical": The Nature of moral Economy rooted to a Market Economy in Upland Sulawesi, Indonesia). In: Li, T.M (Ed): Proses Transformasi Daerah Pedalaman di Indonesia (Transformation Process in the Upland Indonesia), pp 105-130. Jakarta.

Shiva, V. (1992) : Bebas Dari Pembangunan. (Free from Development). Jakarta.

Sitorus, MT F. (2002): "Revolusi Cokelat": Social Formation, Agrarian Structure and Forest Margin in Upland Sulawesi, Indonesia. STORMA Discussion Paper No.9, Bogor.

Turner, J. (1998): The Structure of Sociological Theory. London.

Villarreal, M. (1992): The poverty of practice: Power, gender and intervention from an actor oriented practice. In: Norman Long and Anna Long, 1992: Battlefields of Knowledge: The Interlocking of Theory and Practice in Social Research and Development. pp 247-267. London-New York. Routledge.

Wayland, C. (2001): Gendering Local Knowledge: Medicinal Plant Use and Primary Healthcare. *Medical Anthropology Quarterly*. Washington.

Chapter VIII

DONGI-DONGI – BATTLEFIELD OVER ACCESS, CONTROL AND MEANINGS OF FOREST RESOURCES

Soeryo Adiwibowo

Abstract

The incident of Dongi-dongi, where more than 1,500 ha of forest in the Lore Lindu National Park have been cut down within a few days by land-seeking local people is investigated under the perspective of "political ecology". The analysis focuses on the context of state in gaining international reputation and support have influenced the man-environment relationship on the local level. It concludes that the source of environmental degradation in Indonesia is, to a large extend, rooted in the political condition and nature of the state. As another external intervener, the complex NGO community is identified as a power in itself in the battle on bio-diversity and natural resource management. Individual NGOs appear as self-appointed mandate holders which act on behalf of local communities on the basis of questionable legitimacy.

Introduction

Around 8 p.m. on December 13, 2003, after an hour of torrential rainfall, thousands of cubic meters of soil slid down from the steep mountainsides of the Dongi-dongi area. Dongi-dongi, part of the slopes of Mount Nokilalaki in Central Sulawesi, is one of several areas that have been appointed as "core zones" of the Lore Lindu National Park (LLNP). Located 1,100 m above sea level within the upper watershed of the Sopu-Gumbasa river system and covering approximately 3,500 ha of forested land, the topography of Dongi-dongi is plain to undulating with land slopes less than 20 percent. The massive flood of December 13, 2003 was the outcome of ecological changes initiated on June 19, 2001, when residents of the villages Kamarora A, Kamarora B, Kadidia and Rahmat opened and encroached into the Dongi-dongi forests. These villages had been established under official resettlement programs in the 1970s, the majority of its relocated inhabitants being of the local Kaili Da'a ethnic group. Within the next two years the villagers' activities dramatically changed the landscape and biodiversity of the Dongi-dongi area. Until the present, the farmers still remain in Dongi-dongi. Leaders of the "Independent Farmers' Forum" (*Forum Petani Merdeka*: *FPM*) claim that around 1,030 households from the four villages mentioned are involved in farming activities in Dongi-dongi. In contrast to the first two months after the encroachment, when only 63.5 ha of forest had been cleared (Setyo and Neville 2002), one year later the area of forest opened had already reached 3,400 ha (*Kompas* August 7, 2002).

Is the occupation of Dongi-dongi merely the product of a failure of government policy and a corrupt political-economic system, which regards natural resources as a source of revenue to be exploited for political ends and individual gain as pointed out by Global Forest Watch (2002) or, is it, as indicated by Peluso (1992:19-20), a symptom of resource scarcity and agrarian change as well as an indicator of complex social conflicts over territory and access, surplus appropriation and the distribution of surplus? In Indonesia, the resistance of local peoples to state forest control has a long history (Peluso 1992, 1995; Bryant and Bailey 1997). Albeit this history indicates conflict over access, use and control of resources between the state and local communities, the turbulence surrounding the Dongi-dongi affair reveals that competition over access to land and trees is not simply a diametrical conflict between the state and the peasantry.

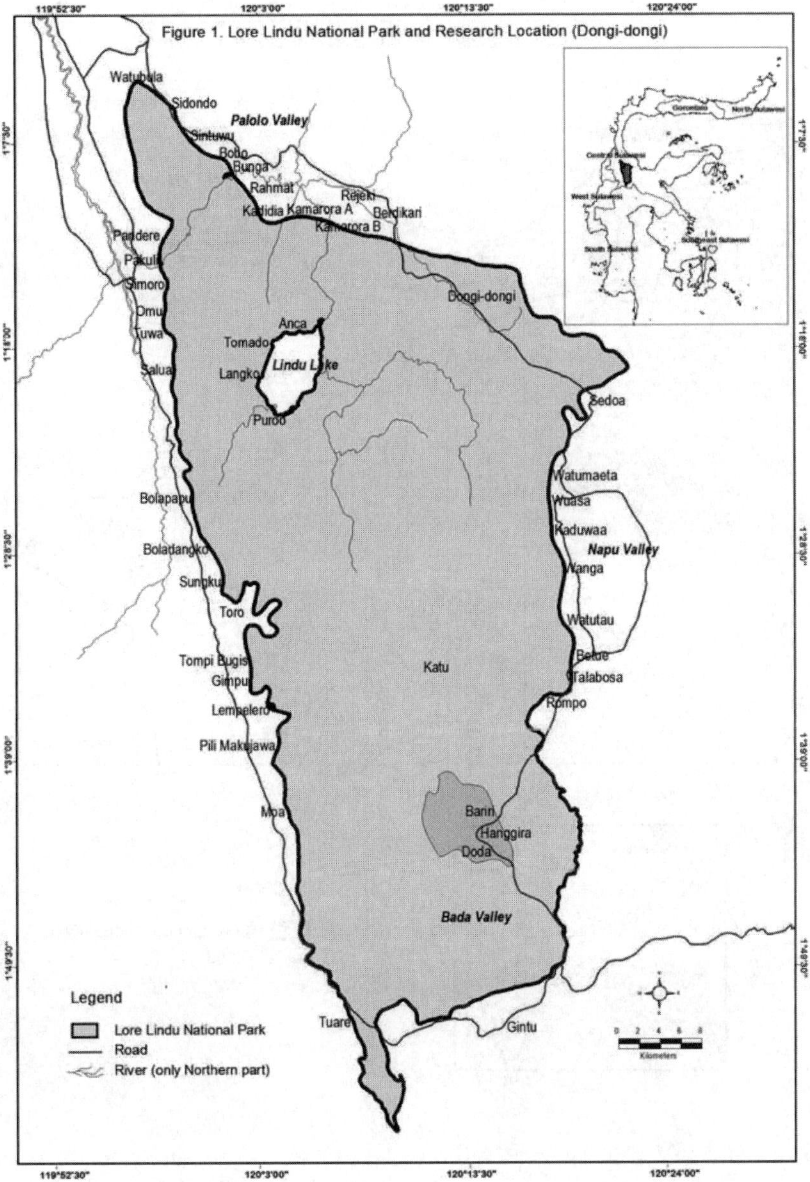

Figure 1. *Lore Lindu National Park and research location*

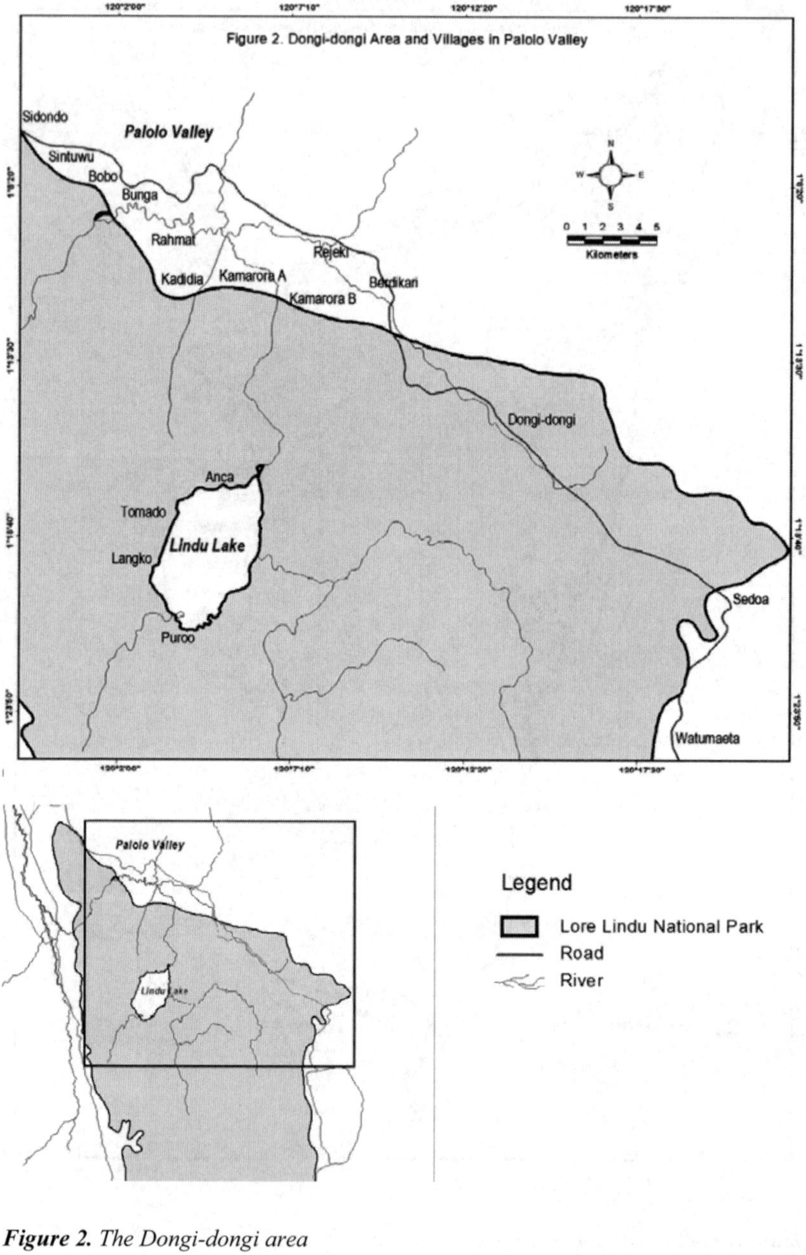

Figure 2. *The Dongi-dongi area*

The purpose of this paper is an attempt to examine why the villagers of Kamarora A, Kamarora B, Kadidia and Rahmat decided to reside in the Dongi-dongi forest even though they recognized that it was a protected area. Further, it examines the reasons the peasants were able to resist their more powerful counterparts and could continue their activities in the core part of the National Park And, last but not least, it tries to analyze the various ways and forms through which the various actors involved seek to exert their power of access and control over the Dongi-dongi resources and the Lore Lindu National Park in general? Hopefully, this study can deepen our understanding of the complex socio-economic, cultural and political settings and events that might contribute to the stabilization and destabilization of the conservation forest and its margins. The data were collected in a period stretching over two years from February 2002 to February 2004 using in-depth interviews, group discussions, direct observations and the archival study of documents and records as well as secondary data.

Political Ecology: Knowledge, Power and Actors

There are three main reasons in favor of a "political ecology" approach in analyzing the Dongi-dongi case. First, the Dongi-dongi case is an outcome of the interaction of actors operating within a context of unequal power relations. Second, the case of Dongi-dongi is not merely a reflection of political and market failures, but a manifestation of broader political and economic influences and forces. Third, it represents a struggle over the discursive meanings of bio-diversity conservation as it is deeply rooted in scientific knowledge and inscribed on the imaginations and material lives of the objects of development (Blaikie 1985, 1999; Bryant and Bailey 1997; Peluso 1992). An actor-oriented approach was adopted due to its potential to facilitate our understanding of how actors with diverse interests and characteristics, being socially, politically or economically inter-linked at local, regional, national and international scales are framed and locked into a series of battles over resources, meanings, institutional legitimacy and control (Long 2001). In this context, "power" cannot be described in terms of rigid hierarchical categories and hegemonic ideologies that "oppress passive victims" (e.g. Blaikie 1985, Peluso 1992,

Bryant and Bailey 1997) Instead, "power" is perceived as the outcome of complex struggles and negotiations over authority, meanings, status, reputation and resources, and necessitates the enrollment of networks of actors and constituencies (Long 2001).

Linked to the embeddedness of power in social relations is the creation and transformation of knowledge that emerges as a product of interaction, dialogue, reflexivity and contest over meaning and identity. This process of knowledge production is deep-rooted in every day social life (Long 2001). As knowledge processes lead to the establishment of common perceptions, interests and intentions, they do not only reflect, but also contribute to conflicts between actors (ibid: 17, 183). Therefore, one of the challenges in this study is to examine by which means and processes actors with diverse aims, interests and characteristics are able to create, generate and transform "knowledge" into "power".

The Global Politics of Bio-diversity Conservation

The Convention of Bio-diversity (CBD) of 1992 has been adopted by countless governments and multilateral organizations, with millions of dollars being rushed into bio-diversity conservation. International conservation NGO's have long been engaged in the bio-diversity discourse. WWF, The Nature Conservancy (TNC), and Conservation International (CI),[1] as well as IUCN / World Conservation Union, and World Resource Institute (WRI) have become big international conservation enterprises that have strongly shaped the global discourse and our knowledge on bio-diversity conservation.

Modern conservation management established a precedent of preserving habitats and ecosystems by the exclusion of resident people from the exploitation of natural resources. In this model, the protection of physiographic regions, biotic communities, genetic resources

---

[1] Conservation International (CI), The Nature Conservancy (TNC) and World Wide Fund for Nature (WWF) are the three leading environmental NGOs in the world. They all contribute significantly to the production and formation of knowledge and discourses of bio-diversity conservation, have extensive field actions for the protection of nature throughout the world, gain enormous revenues, and employ large numbers of human resources (see Chapin 2004).

and species represents the highest priority over other, conflicting interests. One of the commonly used management tools of the exclusionary mode is zoning or “zonation” by which a sanctuary or Park is divided into zones in terms of their biological functions and management criteria, e.g. core zone, wilderness zone, activity zone and tourist zone. The core zone as the heart of and the unique feature of the Park is entirely dedicated to bio-diversity preservation. Whereas other zones allow limited access for purposes of research, tourism, and limited traditional resource extraction, no human activity is allowed in core zone areas. Hence, mapping Park zonation as well as Park boundaries is an intrinsically political act, which excludes or includes people within the same spaces (Menzies 1992 in Peluso 1995). Not surprisingly, newly declared National Parks often create conflicts with local communities because the state often ignores the traditional customary rights over natural resources, imposes limitations and exclusions on traditional users, and even moves them out of the protected areas (Peluso 1993b, 1995; Bryant and Bailey 1997; Kramer and Schaik 1997; Colchester and Erni 1999; Lynch and Harwell 2002; Colchester 2004; Wulan et al 2004).

The notion of making protected areas relevant to local communities has been addressed since 1980 by IUCN, UNEP and WWF in the World Conservation Strategy. Thus, in 1985, the WWF introduced the “Integrated Conservation and Development Project” (ICDP). Despite the diversity of terminology and variation in the scope of activities, ICDPs have a number of common features. Thus, bio-diversity conservation is the primary goal of all ICDPs worldwide. Further, ICDPs do not devolve control or ownership of protected areas to local communities. Receiving funds from external sources, i.e. from bilateral or multilateral donors and international conservation organizations, ICDPs are always externally motivated (Brandon 1997; MacKinnon 1997; Hughes and Flintan 2001). The larger scale ICDPs in Indonesia have been implemented in the Kerinci-Seblat National Park (KSNP) in Sumatra and the Lore Lindu National Park (LLNP) in Central Sulawesi. In case of LLNP, the ICDP project started at 1997 under the Central Sulawesi Integrated Area Development and Conservation Project (CSIADCP), funded partly through a loan from the Asian Development Bank and partly through grants from USAID and The Nature Conservancy (TNC).

Environmental Non-Governmental Organizations

Since the mid-1980s, the global environmental movement transformed into transnational NGOs, controlling multi-million-dollar budgets with extensive representations in various countries (Jamison 1996; Bryant and Bailey 1997; van der Heijden 2002, Chapin 2004). Two of them, TNC and Friends of the Earth (FoE), exerted direct or indirect influence on the Dongi-dongi case.

In June 2000, TNC in addition to membership dues and contributions that generated $357.4 million in the fiscal year of 2000, TNC earned $60 million from government awards, $14 million from private contracts and $161 million from investment income (Cohen 2001). Today, it has become the "world's richest conservation organization" with assets of an estimated $3 billion (Ottaway and Stephens 2003; Chapin 2004). TNC describes itself as being science-based, non-confrontational, collaborative and focused, with an emphasis on creative business strategies. In Indonesia, its activities started in 1991 at two major protection sites, e.g. the Lore Lindu National Park and the Komodo National Park.

Friends of the Earth (FoE) International, on the other hand, is the world's largest grassroots environmental network uniting 68 national member groups, 15 affiliated organizations and around 5,000 local activist groups (FoE International 2004). It describes itself as a global federation of national environmental organizations. One of its leading members in Indonesia, the "Indonesian Forum for the Environment" (*Wahana Lingkungan Hidup Indonesia* = WALHI) joined FoE International in 1989. FoE works on various environmental and social issues including the preservation of the earth's ecological, cultural and ethnic diversity and increasing public participation. FoE challenges the unjust current model of economic and corporate globalization, patents of life, and the abuse of human rights and indigenous peoples (FoE 2004). FoE works closely with local communities, as well as alliances with indigenous peoples, farmers' movements, trade unions and human rights groups.

Due to differences in organizational patterns and values, TNC and FoE differs sharply in their policy, their scope of activities as well as their forms of political activity. FoE developed a confederation-like organizational structure in which members could carry out their activities autonomously, whilst the organizational structure of FoE International remained light and flexible making FoE an organization with a "quasi-anarchic character" (Wapner 1996: 124) whereas TNC developed a more centralist, non-confrontational type of organization. In indirect manners, both types of NGOs colored and intensified the controversy around the occupation of Dongi-dongi.

The environmental NGO network in Indonesia was established in 1980 under the name of "The Indonesian NGO Forum for the Environment" uniting more than 450 environmental NGOs throughout Indonesia. WALHI perceives environmental problems in Indonesia as manifestation of larger issues that may only be improved by social justice and a more democratic political life. The second characteristic is the development of networks, collaborations, coalitions, and alliances among environmental NGOs as well as other organizations and groups such as the Anti-Debt Coalition (*Koalisi Anti-Utang* – KAU), the Coalition of Non-Governmental Organizations (*Koalisi Organisasi Non-Pemerintah* – KORAN), and the National Alliance of Customary Communities (*Aliansi Masyarakat Adat* – AMAN). The third characteristic is the tendency to exert political influence through well-published campaigns, most notably through the mass media. The circuit of power is thus accumulated via the various discourses and narratives on democratization, social justice and equity, which – by complementing the pro-democracy movement in Indonesia – transformed WALHI into one of the strongest powers outside the state apparatus and political parties.

The Lore Lindu National Park and the Dongi-dongi Area

Within the last 30 years, the forest regime of the Dongi-dongi area has been altered at least three times. In 1976, the Directorate of General Forestry (under the Ministry of Agriculture) appointed 25,000 to 50,000 ha of forest surrounding the Sopu and Gumbasa rivers, including Dongi-dongi, as concession area of PT. Kebun Sari. The concession to exploit *Agathis lorantifolia* was granted for 20 years (Sondakh

2002: 72).[2] Due to its strategic position, a helicopter paddock, maintenance and workshop facilities, oil and fuel supplies as well as a non-permanent housing complex for field operators were erected in Dongi-dongi.[3] Under the production forest regime, the communities of the nine surrounding resettlement villages (Sintuwu, Bobo, Sigimpu, Bakubakulu, Rahmat, Kadidia, Kamarora A, Kamarora B and Tongoa) still had access to the forest and were not prohibited from collecting rattan and cultivating coffee, cacao, maize and cassava (Syahyuti, 2002: 66; Sondakh, 2002: 73). As the intensity of forest exploitation by PT Kebun Sari was decreasing, the Department of Forestry (DoF) converted the concession area into the "Lore Lindu Wildlife Reserve" (*Suaka Marga Lore Lindu*) in 1981. In 1993, the forest concession permit of PT. Kebun Sari was terminated, and the area was returned to the government.

Although it was an extension of the existing Lore Kalamanta Wildlife Reserve, the Lore Lindu Wildlife Reserve was perceived as a probable means to protect the upper water catchment area of the Gumbasa and Sopu irrigation schemes. Despite it had been designated as a wildlife reserve with the rules of government being more restricted in comparison to National Parks, no boundary demarcation had been made for more than a decade and no management plan for the reserve had been drafted. Although the Lore Lindu National Park was declared as early as 1982, the National Park Regime began to exert its power only after 1993 when the forest administration integrated two nature sanctuaries and a nature conservation area which were formerly separated. Those two natural reserves were the Lore Kalamanta Wildlife Reserve (131,000 ha) and the Lore Lindu Wildlife Reserve (67,000 ha), whereas the nature conservation area included the Danau Lindu Tourism/Protected Forest (31,000 ha) (Helmi, 2001; The Nature Conservancy, et al, 2001a: 181-191).

---

[2] PT Kebun Sari obtaned a permit (*Hak Pengusahaan Hutan:* HPH) to exploit specific forest resources within an area of 74,000 ha in Donggala District on the basis of Forest Agreement No. FA/N/032/III/1977, issued on March 28th, 1977 (later extended on the basis of Forest Agreement No. FA/N-AD/008/1/1980, issued on January 26th, 1980), and Decision Letter No. 41/Kpts/Um/1/1978, issued on January 24th, 1978 (Dinas Kehutanan Propinsi Sulawesi Tengah 1986: 132).

[3] The remains of the base camp and helipad structures are located approximately 200 meters from the Palolo-Napu road.

The government also resettled the local communities of the Dodolo and Katu villages who lived in the depths of the forest. In 1998, the Dodolo villagers were resettled to Toe Jaya, located between the villages of Wanga and Kaduwaa in Poso District. The community of Katu was supposed to be resettled, but the community resisted and remained inside the Park. In 1999, the Katu people were to be resettled again, this time via the "Central Sulawesi Integrated Area Development and Conservation Project" (CSIADCP). However, this effort failed due to the successful advocacy of the "Free Land Foundation" (*Yayasan Tanah Merdeka,* YTM) (Sangaji, 2000a, 2001b: 8). The villages of Kamarora A, Kamarora B, Rahmat, and Kadidia (as well as Sintuwu, Bobo, Sigimpu, Bakubakulu, and Tongoa) faced a similar situation when the residents were no longer allowed to collect rattan and to cultivate coffee, cacao, maize, dry rice and cassava in the forest. Those still utilizing the forest were accused of being encroachers (Syahyuti, 2002: 66; Sondakh, 2002: 73; Forum Petani Merdeka et al., 2001a: 2).

After establishing the boundary demarcation, the Park administration prepared the Lore Lindu Management Plan and the Park zonation. With support of the Natural Resource Management (NRM) II-USAID, the Nature Conservancy (TNC) formulated the Management Plan. In this draft management plan, the Park was designed to comprise eight zones (The Nature Conservancy, et al. 2001b: 99-102)[4] e.g. the core zone, wilderness zone, activity/utilization zone, intensive utilization zone, traditional utilization zone, rehabilitation zone, social and historical zone and tourist zone. According to the principles of the National Park Management, the core zone is the heart of the Park and represents its unique feature. Therefore, any activities leading to the alteration of the function of the core zone, such as influencing the unique features of the ecosystem, its natural aesthetic and physical

[4] The first outlines for the zoning of the National Park were laid out in the Lore Lindu National Park Management Plans of 1981 and 1995. Zoning was reassumed again in 1997 (Ministerial Letter No. 135/BKSDA/VI/1997) but was put on hold due to proposed changes regarding the Park's boundaries. Zoning was made possible again with the issuance of an information letter from the Minister of Forestry and Plantations (Menteri Kehutanan dan Perkebunan No. 464/Kpts-II/9), (The Nature Conservancy, *et al,* 2001b: 95).

phenomena, or the collection of forest products are strictly prohibited.[5] The zoning process designed by TNC however, entailed several problems and weaknesses.

First, the Management Plan of Lore Lindu National Park, including zoning, was initially designed to utilize the information gathered during a survey carried out by TNC and the Central Sulawesi Integrated Area Development and Conservation Project (CSIADCP).[6] In fact, however, the management plan was based only on the ecological and bio-diversity survey carried out by TNC. This is due to the following two inter-locked factors:

1) The survey funded by CSIADCP was hopelessly delayed.[7]
2) TNC, being responsible to the trust fund of NRM II/USAID, had to complete the management plan before December 30, 2001; the deadline for the extended contract between TNC and NRM II/USAID (Juliani and Buchori 2001: 16).

Second, the emergence of the Dongi-dongi occupation motivated TNC to accelerate the accomplishment of the management plan immediately. Thus, TNC finished the first draft of the management plant at the end of August 2001, or less than five months after Juliani and Damayanti (2001) found that on 1 March 2001 it had covered only ten percent of the entire area. TNC held its first public consultation for the

[5] Government Regulation No. 68/1998 on Nature Sanctuaries and Nature Conservation Areas (Kawasan Suaka Alam dan Kawasan Pelestarian Alam)

[6] According to the Project Administration Memorandum of CSIADCP (ADB, 1998: 3), USAID/TNC and CSIADCP collaborate to strengthen the capability of the Park Authority in managing bio-diversity conservation and user-friendly services for Park visitors, as well as in dealings with the local community. Those capabilities are (1) the development and implementation a Five Year Park Management Working Plan; (2) the training of Park rangers, (3) the development of visitor programs, (4) conducting inventories and ecological studies as a guide for the zoning of the Park, (5) the construction of guard posts, staff housing and a visitors' center as well as the provision of field equipment and vehicles for monitoring purposes, and (6) the construction of new and rehabilitation of existing Park trails. The item components (1), (2) and (3) were funded by USAID/TNC, whereas item components (4), (5) and (6) were funded by CSIADCP.

[7] TNC started the surveys in March 2000 and finished them in June 2001. Meanwhile, CSIADCP did not issue the contract for the bio-diversity study until November 2001 (The Nature Conservancy, 2002a: 12).

draft management plan and zoning of Lore Lindu National Park, on August 29-30, 2001. In this public consultation, Dongi-dongi was announced as a core zone due to its function for the protection of the Sopu and Gumbasa watersheds. The core zone, then, soon became a "contested discourse".

Third, there are no strong grounds to determine Dongi-dongi as a core zone as the strategic all-weather road connecting the provincial city of Palu with its hinterlands cuts across the area. The physical attributes of Dongi-dongi, hence, do not meet the core zone requirements as stipulated in Government Regulation No 68/1998 on Nature Conservation Areas. The present Head of LLNP and as well as the General Directorate for Nature Protection and Nature Conservation in Jakarta pointed out to this fact.[8]

Fourth, the draft management plan was based almost exclusively on biological and ecological data. It makes only little reference to the social, economic, and cultural aspects of the forest margin communities, particularly their agrarian relations with the surrounding forest. For example, Volume 1 of the draft comprises 203 pages, but only 15 of the pages elucidate the cultural and socio-economic aspects of the communities with only one page explaining population structures and livelihoods around the Park.

Fifth, even though the draft management plan states that the zoning of the Lore Lindu National Park was prepared through public consultation involving local NGOs, government officials, and village communities (The Nature Conservancy, et al. 2001b:107), the zoning process was criticized by various parties, including the local government, environmental activists and local NGOs in Palu. They pointed out that the zoning process did not at al involve the local communities affected by the plan.[9]

---

[8] Personal communication with the present Head of Lore Lindu National Park (June 2006).

[9] Sources: Position Statement Rejecting the Zoning of the Lore Lindu National Park (issued by Forum Petani Merdeka, September 12, 2001), commentary and views in a paper presented by Sangaji (2001a, 2001b), letters from the Central Sulawesi Executive of WALHI to the Director General of Forest Conservation and Nature Conservation and to Duncan Neville at the TNC Lore Lindu Field Office, as well as a letter from People's Education Foundation (Yayasan Pendidikan Rakyat) to the Governor of Central Sulawesi.

Thus, the changes in the forest property regime of the Dongi-dongi area are more than a technical issue only. Rather, they represent a political act in which an international conservation NGO plays a significant role in assigning and arranging access and control of the area by inscribing its mission and interests in the management plan of the Park. Hence, the contemporary politics of protected area management involves not only modifications of the property regime; it also seeks to control acts, narratives, discourses and attitudes as well as the modes of production of the local peoples living inside and adjacent to the protected areas.

The Farmers, the Villages and the Dongi-dongi Area

The Demographic and Agrarian Features of the Four Resettlement Villages

During the 1970s and 1980s, the resettlement of so-called "community tribes" (*suku terasing*) was conducted throughout Indonesia. The program included provisions of guidance in regard to cultivation techniques, cultural expressions, formal education, health care, religion, and the interaction within the wider Indonesian community. The program targets were the isolated tribes in various upland areas who – in the eyes of many bureaucrats and academicians – were viewed as ignorant, disordered, subsistence oriented and threatening the sustainability of the Indonesian forests (Li 1999). In the Lore Lindu area, the resettlement program begun in 1974 with the establishment of Rahmat Village. Further, around 500 households (2,437 people) were resettled to Kamarora Village in 1979 and another 100 households resettled to Kadidia sub-village in 1983. In the next two decades, 1998, Kamarora was expanded into two villages namely Kamarora A and Kamarora B. A decade after the resettlement program ended, Palolo Valley attracted many newcomers due to its relatively fertile soils and its excellent road connections. Massive increases in prices of cacao and coffee during the monetary crisis (*krismon*) in 1997/1998 accelerated the inflow of migrants. Statistical data indicates that the population of Palolo grew 8.8 percent in 1998-1999 in comparison to the average district rate of only 1.63 percent for the same period. Data released by the Legal Aid for the People Foundation (YBHR) indicates that the number of families had doubled the original number of residents in the 1970s. Every family that received between 0.8 and one hectare of land at the time of resettlement had meanwhile split into two or three families (Suara Pembaruan, September 28, 2001).

Subsequently, one of the most predominant characteristics of the agrarian structure of the four-resettlement villages was the large numbers of family heads that did not have their own land to cultivate. As noted above, when the settlers arrived at their new resettlement location, each family head received only between 0.8 and one ha of land instead of the promised two ha. This discrepancy however, was not initially felt as people still enjoyed free access to the forest. Thus, research conducted by Li and Sangaji (2003) in Rahmat village indicates that in the period of 1975-1977 villagers initially had permission to cultivate land in the forested areas bordering their village. More important for the changes in the agrarian structure than the resettlement scheme was the spontaneous immigration of Bugis farmers from South Sulawesi in combination with the "cacao boom" during the economic crisis when large tracts of land shifted hands through purchase (Li and Sangaji 2003, Sitorus 2002c). Thus, from 1997 onwards, a significant number of plots had fallen into the hands of the Buginese migrants who used those plots almost exclusively for cacao cultivation. A sizable segment of these plots were procured from government officials (village heads and village administration staff, as well as other local officials), who had come into possession of them through both legal and illegal methods. In Rahmat for instance, the land sold to the Bugis migrants was actually part of the land initially allocated for the re-settlers, which had been misappropriated by local officials. In the villages of Watumaeta and Sedoa, the sale of land by local officials was even the main motivating factor for the migration of the Bugis into those villages. (Li and Sangaji 2003: 19, see also Burkard, this volume). In the early of 2000s, the arrival of the Buginese migrants had fundamentally changed the rural socio economic conditions of the villages around the Lore Lindu National Park. The cacao boom had radically changed the rural social formation from a subsistence mode of production previously practiced by the indigenous population to a market production predominantly practiced by the Buginese. Hence, the "cacao revolution" upgraded the Buginese from "landless" to "landed", whereas the local so-called Kaili-groups were downgraded from "landed" to "landless" (Sitorus 2002c).

Increased population pressure in the resettlement villages, the closure of access to the forest by the establishment of the National Park as well as the transfer of significant amounts of land to the hands of the Buginese migrants resulted in an acute land crisis in the resettlement villages. Often, the younger generation has to live in cramped quarters at their parents' homes and to work as coolies on the cacao plantations. Data gathered by Firdaus (2003) indicates a disastrous situation as can be seen in table 1.

Village	Total House-holds	Households owning Housing		Landless Farmers	
		Households	%	Households	%
Kamarora A	479	-	-	220	46
Kamarora B	313	298	95	261	83
Rahmat	600	400	67	325	54
Kadidia	160	100	62	125	78
TOTAL	**1,552**			**931**	**60**

Table 1. *Number and percentage of land less farmers in the Kamarora A, Kamarora B, Rahmat and Kadidia villages. Source: Firdaus (2001), Forum Petani Merdeka,* et al. *(2001a: 4-5) (- : no data)*

To sum up, over the last three decades the agrarian structure of four resettlement villages was formed as the product of various influences as e.g. the government resettlement policy (1970s), changes in the state defined forest property regime (1980s) and forest zonation (late 1990s), changes in demographic characteristics (in-flux of migrants) and last but not least the "cacao boom" (2000s).

Dongi-dongi – A Contested Terrain for Access and Control

Before the major occupation in June 2001, Dongi-dongi was partly occupied by some dozen individuals at least four times, namely in 1981, 1982, 1988 and 1999. Shortly after the fourth occupation, the Donggala District Head held a coordination meeting to prevent a recurrence of the Dongi-dongi occupations resulting in the plan to develop a social forestry system, to identify the cultivated areas in the Park and to map the degraded forest areas. It was agreed that the "National Heritage Foundation" YPAN (*Yayasan Pusaka Alam Nusantara*)[10] should facilitate the activities. After YPAN facilitated a series of data collection activities, village mapping, and a workshop for the formulation of a "Village Plan and Community Agreement" for three of the resettlement villages, the outcome[11] was presented in a multi-stakeholder dialogue in order to communicate the plan to the decision makers of CSIADCP.

Finally, in a follow-up meeting in September 2000, the local government declared the proposed programs as accepted and said that they were in line with the mission of CSIADCP. Since CSIADCP was designed to protecting the Lore Lindu National Park by increasing the livelihood quality of the communities surrounding the Park, the villagers were quite optimistic that CSIADCP could disentangle the problems. After more than six months, however, no follow-up action from the CSIADCP management had materialized and the villagers become more and more frustrated and distrustful. In the meeting in September 2000, the leaders from Kamarora A and Kadidia had already anticipated the situation:

> *"I don't want to hear any more promises ... We have been trained to make proposals, and in March 2000 we submitted a proposal to the Lore Lindu National Park Authority for which we still have not received a response. Are you gentlemen not embarrassed that the Lore Lindu National Park is indeed a fine*

---

[10] YPAN was established in 1995 by The Nature Conservancy (TNC) as an Indonesian sister organization. In 1997, YPAN established its branch office in Palu

[11] The "Village Plan and Community Agreement" documented and presented the socio-economic condition of the villages, community need assessment, rattan collector need assessment, village land use map, and a draft for a community conservation agreement.

place, while its people remain ignorant and backward?" (Notes from dialogue between villagers of Kamarora A, Kamarora B and Kadidia with local government officials, September 6, 2000).

The failure of CSIADCP and the Provincial Government to find a way out of the land crisis in the four resettlement villages eradicated any hope among local people to regain access to and control over their surrounding lands and forests. Their future becoming increasingly uncertain, the villagers began to feel as if they were truly a "marginal people". The construction of "marginality" through processes of state impinged "territorialization" is a well known phenomenon in the Indonesian uplands (see Li 1999).

Marginality, Power and Resistance

The marginalization and dis-empowerment of the residents of the four resettlement villages attracted the sympathy of two local NGOs in Palu, YBHR (Legal aid Foundation for the People) and Central Sulawesi WALHI. The approach of the two NGOs was different from the one previously adopted by YPAN in the sense that YBHR and WALHI chose the method of "advocacy through mediation" which implicated a fundamental change in power relations between the communities and the state. A number of meetings held in the villages as well as in the city of Palu ended with the conclusion that action had to be taken in order to reestablish the people's rights to the land after the various efforts to mediate the conflict had failed. The residents of the four-resettlement villages, motivated to unify and to organize their activities, joined together to form the Independent Farmer Forum (*Forum Petani Merdeka*, FPM).

The long lasting process of marginalization resulted in the adoption of a hard, overly sensitive and myopic attitude among the villagers. Therefore, it is not surprising that a large number of seemingly innocuous issues arising in the villages often ended up in turmoil and violence. As it is well known from the literature, long felt "coercive power" may not only produce resistance, but may also lead to processes of alienation (Etzioni 1961 in Pierce and Nestrom 1995). In such a situation, any individual, who seems courageous enough to stand up against the "coercive power", will become the focus of hope for the populace or group in question. In other words: the overall social situation exerted a great deal of influence on the decision of leadership. The question of "who will be chosen as leader" of FPM may also be influenced by the capability of an individual to access and control the resources needed by the group to overcome the pressing issues. The backing of the resettlement villages by YBHR and Central Sulawesi WALHI carried with it the implication that there must be sufficient resources to support the various advocacy activities directed by the two NGOs. Neither of the local NGOs however, were affluent organizations. Hence, the FPM leader should be able to provide the resources required for local meetings in the village, organizing lobbying and negotiation in Palu, and for the mobilization of villagers for demonstrations in that city. The formation of leadership in such a social situation will inevitably result in the emergence of leaders who are courageous, determined, cool-headed, and fearless and who never lack confidence in confrontations with anyone. At the same time, they have access to the required resources. These leadership characteristics were to be found in a local figure, named PGA, who was selected as the first leader of the FPM. The social movement set in motion by YBHR and WALHI was rooted in the resistance to the "coercive power" exerted by the state and its various institutions. Therefore, it is not surprising that the alliance of YBHR, Central Sulawesi WALHI and FPM succeeded, in a relatively short time period, in organizing a "field of resistance" that was almost equal in power to the existing coercive power of the state.

In pursuing its aims, the alliance had principally two major sources of power at its disposal. The first source of power was certainly the "power of discourse". YBHR and Central Sulawesi WALHI succeeded in transforming the "agrarian justice discourse" into a power as strong as the "conservation discourse" adopted by the state and TNC. Through this "agrarian justice discourse", other NGOs, who sympathized with the resettlement villagers, grew in number as well. The second sources of power were the extensive fiscal resources needed for various advocacy actions directed by YBHR and Central Sulawesi WALHI. The urging fiscal needs presented a dilemma that could only by resolved by courageous decisions, which only leaders of a type like PGA could make: encroach the Dongi-dongi forest and sell the logs, a decision PGA made deliberately with the support of several villagers and his closest companions. Illegal logging was definitely not an agenda of YBHR and WALHI. However, with these resources at hand, PGA achieved the power to facilitate all kinds of advocacy activities undertaken by the two NGOs and FPM while he simultaneously created a strong network with police and government officers engaged in illegal logging.

These two sources of power functioned within their own channels and worked in such a way that the totality resulted in a cumulative strength that rivaled that of the "coercive power" flaunted by the state. Thus, the unequal power relations between the state and the resettlement villagers had shifted, which enabled the resettlement villagers to continue their residence and farming activities in Dongi-dongi. It was not long, however, until the practice of illegal logging and the ensuing environmental damages brought negative criticism to bear on WALHI. The activities of PGA and his followers caused WALHI's support of FPM's struggle to become only half-hearted. Besides, a number of FPM members, including leading female figures, also disapproved of the illegal logging practices undertaken by PGA although no one had the courage to stand up against PGA. At the end of 2002, when a number of the Dongi-dongi female leaders displayed banners in the village in order to protest against the ongoing illegal logging, PGA did not only threaten them, but also approached female NGO activists in Palu and threatening them with sharp weapons.

The Conflicts

The controversy surrounding the occupation of Dongi-dongi exhibits a typical pattern of an escalating conflict. The social process, which started with acts of protest directed at the government by local farmers, ended anticlimactically with conflicts among the various components of the civil community itself. In order to understand how this could have happened, it is important to analyze the dynamics of the political processes occurring during the period of controversy, the interactions of the people and organizations involved as well as the different phases and forms of the unfolding controversy.

The Emergence of Conflict

The Occupation

On June 19, 2001, 200 people from the four-resettlement villages, along with the Central Sulawesi WALHI and YBHR, demonstrated at the Central Sulawesi Provincial Parliament Building asserting their intention to move into Dongi-dongi if the government did not fulfill their demands to provide agricultural land and to use the plots already established inside the Lore Lindu National Park. Both NGOs supported the demands of the farmers from the outset, although their initial statements of support aimed at the control of the cultivated lands in the forest rather than unilaterally encroaching into Dongi-dongi. Although WALHI and YBHR as well as other environmental NGOs tried to solve the problems through intensive dialogues with the provincial and district governments, the farmers one-sidedly occupied Dongi-dongi a few days after the demonstrations.

The NGO's support was an effective political resource for the FPM in its fight to establish its political interests. Press releases, campaign plans, debates in various forums and seminars, the use of newspapers and electronic forums as well as street rallies all represented new types of power exertion that local farmers, now grouped under FPM, had neither used, nor known in the past. Due to the facilitation by the two NGOs, the movement to occupy Dongi-dongi was immediately dispersed across provincial and district boundaries. Given the high degree of support and the widespread public knowledge of their struggle, the villagers started to mobilize efforts in defense of their settlement and land should the police or security forces take actions to move them out to other places.

From Dialogue to Polarization of Stances

When the occupation of Dongi-dongi first gained public attention, all parties agreed to take a persuasive approach in order to find a solution to the problem. As a result, the initial approach to resolve the controversy surrounding the Dongi-dongi case involved intensive discussions among the various interested parties, which in the course of time led to a polarization of stances and even open conflict among them. It is worth noting here that during this conflict, the strategy adopted by the state was quite different from the strategy adopted by the NGOs.

The actions taken by the Central Sulawesi government and legislature during this phase of conflict reflect a serious lack of understanding of the socio-economic and historical background of the agrarian situation in the villages. Instead of following up the "Village Plan and Community Agreement" that had been formulated by various government agencies on September 2000, the Central Sulawesi Governor preferred to establish a new committee, the "Special Committee for the Solution for Dongi-dongi". The committee was a purely state-based institution with no single NGO or local representative being involved.[12] One week after the Dongi-dongi declaration, the Committee reported its task to the Central Sulawesi legislature as well as to the Ministry of Forestry in Jakarta. In the report, the Committee refused to convert the function of the lands in Dongi-dongi to that of agricultural and residential status as demanded by FPM. Instead, the government would provide land in Palolo Subdistrict, specifically in Manggalapi, Uenuni and Lemban Tongoa.[13] The committee conducted several coordination meetings and arranged a site visit to Dongi-dongi in order to collect detailed data and to obtain a better picture of the local situation. Unfortunately, the site visit never happened. The committee stopped at Tongoa village, some 20 km north of Dongi-dongi.[14] This incident once again disappointed the farmers who had prepared themselves to meet with the government officials. From their point of view, it was becoming increasingly apparent that the motivation of the government to solve the problems they were facing was not at all serious.

In contrast to the state at this initiative step, the NGOs adopted a more comprehensive approach with the NGOs hosting a series of meetings and discussions between the various actors such as WALHI,

---

[12] The Special Committee consisted of the Head of the Provincial Forestry Agency (Leader of the Committee), the Head of the Lore Lindu National Park, the Head of the Provincial Community Service Agency, the Head of the Forestry and Estate Agency of the Donggala District, and the Head of the Development Planning Board of the Donggala District.

[13] In 1999, the Provincial Government offered Manggalapi as an alternative land for Dongi-dongi. However, the farmers rejected this solution.

[14] In Tongoa, the Committee immediately summoned the Head of Palolo Subdistrict and the Village Heads of the four problem villages in order to present pertinent data and provide information.

YBHR, government representatives, and private company associations. This initiative was mainly endorsed by the "Partnership Forum for the Lore Lindu National Park" (Forum Kemitraan Taman Nasional Lore Lindu, FKTNLL), an umbrella organization dedicated to the protection of the Lore Lindu National Park. At this stage, all parties agreed on the formation of the "Tongoa Dongi-dongi Solution Committee" (*Komite Solusi Tongoa Dongi-dongi*).[15] During this time, the conservation NGOs, like FKTNLL and the advocacy type of NGOs (YBHR and WALHI and its networks), developed an intensive cooperation in order to increase the effectiveness of the committee as well as to facilitate the dialogue between the government and the occupiers. However, this cooperation lasted only a week. The process of polarization of stances began to become apparent when all parties involved, including members of Provincial Parliament, the Lore Lindu National Park Authority, and the traditional leader of the Sedoa/*Pekurehua* customary community met in Dongi-dongi on June 28, 2001. During the meeting, the FPM rejected the alternative land in Manggalapi, Uenuni and Lemban Tongoa proposed by the Provincial Government. According to the farmers, the land provided was too steep and not fertile. They further complained that the location was already fairly populated and had no transportation facilities. For those reasons, the farmers insisted on remaining in Dongi-dongi. Their demands, however, met with a strong rejection on behalf of the leader of the Sedoa/*Pekurehua* customary community from adjacent Napu valley (albeit he could understand the situation of the farmers). His rejection was simply because the area in dispute was, historically, part of the *Pekurehua* forests, the traditional customary lands of his people. The statements of the *Pekurehua* leader introduced a new issue to the problem, the claim of the lands in dispute as "customary area" (*wilayah adat*). In the meanwhile, although the government's plan to develop the relocation area of Manggalapi, Uenuni and Lemban Tongoa obtained approval by the regional legislature, the villagers occu-

[15] Tongoa is an area in the lower regions of Dongi-dongi near the residential communities of Lemban Tongoa and Kamarora. Encroachment in this area has been sporadic an unorganized, and seems to be motivated more by illegal logging rather than by land shortages. Due to the close distance between Tongoa and Dongi-dongi, the NGOs agreed to solve the Dongi-dongi problems simultaneously with those of Tongoa.

pying Dongi-dongi apparently no longer had any interest in it. The public's confidence in the government's commitment and sincerity had reached its nadir. Instead of cooperation, two distinctly conflicting processes had emerged: the continued agitation and mobilization on the part of the farmers on the one hand, and the government's continued preparations to move out the farmers on the other. It was clearly at this stage of conflict that the polarization of stances had reached an irreversible level. By the end of June 2001, there was no sign of a convergence of thinking, but rather signs of increased polarization among the actors opposed to and in support of the Dongi-dongi occupation.

EXERTION OF POWER THROUGH DISCOURSE AND VERBAL ABUSE

With this increased polarization in stances, the controversy over the occupation escalated into a new phase of discourse involving a more sophisticated argumentation. The increasing dynamic in this phase of conflict was most apparent in a public debate hosted by the National Park Management on July 5, 2001, in the provincial capital of Palu. In this public debate, Arianto Sangaji, the Director of the "Free Land Foundation" (YTM), presented a "Moratorium for Lore Lindu National Park" which suggested a moratorium on the National Park based on three points. First, land designated as National Park must return to its previous function before being re-designated for conservation purposes. Second, once the moratorium was in effect, all exploitation of natural resources in the contested area must end, including the activities of plantation companies, and third, the ownership structure of all the lands surrounding the Lore Lindu National Park must be re-designed anew. Only then would it be possible to designate new conservation areas through public discussion and consultation with the concerned communities. Within a short time period, the moratorium narrative expanded into a discourse that pushed the parties into a real "zero sum game". This moratorium narrative represented the peak of the "agrarian justice discourse" supported by WALHI Central Sulawesi and YBHR. For a long time WALHI and YBHR shared the opinion that the Dongi-dongi problem could only be resolved through efforts to reform the existing agrarian structure, i.e. by redistributing land declared as conservation area. Thus, the moratorium narrative suggested by Sangaji triggered even more confrontation and hardened the stances of the parties concerned with the case.

On the other hand, the Park Authority reiterated its intention to retain Dongi-dongi as a core zone of the Lore Lindu National Park on the meeting on July 2, 2001, with the Central Sulawesi Legislature, the Provincial Governor, the Regent of Donggala District and the Regent of Poso District. The stipulation of Dongi-dongi as part of the core zone of the Park placed emphasis on the ”conservation discourse” advocated by the Park Authority to counter the “agrarian justice discourse” set forth by WALHI, YBHR and FPM. Through this discourse, the Park Authority garnered the powerful support of the provincial government and legislature, the central government in Jakarta as well as conservationist NGOs.

An important actor that supported the conservation discourse was of course The Nature Conservancy (TNC). TNC, which was in the midst of finishing the Management Plan of the Park, with zonation as one of its major outputs, played an important role in this matter. Delays in the completion of the Management Plan provided the opportunity for TNC to classify Dongi-dongi as part of the core zone of the Park. The “Dongi-dongi core-zone narrative” acted as a strong magnet to draw together the strengths of several NGOs that disapproved the occupation of Dongi-dongi, including FKTNLL and a number of nature lover’s organizations. Thus, it was not long before the occupation that the NGOs in Palu split into two opposing poles as laid out in table 2. This polarization among the local NGOs then escalated into a reiteration of discourses supporting their individual positions, dashing any hope for a constructive open dialog. Instead, the polarization created a win-lose situation in which every party felt driven to fight to the finish. Along with the escalation of the debate, the defense and construction of these discourses became a part of the campaigns waged by the NGOs supporting the occupation. For example, following the moratorium on National Parks, the adherents of the “agrarian justice” point of view (YBHR, WALHI etc.) suddenly brought up other related issues, including what they call the “eco-fascism” approach in the management of National Parks. The choice of the term “eco-fascism” is very interesting because it constituted a counter-discourse directed unequivocally towards the policy of “eco-populism” introduced by the

Head of the Park Authority.[16] Although the Head of the Park Authority did exhibit the courage to recognize a number of customary communities' rights as the manifestation of his "eco-populism" approach, the "National Park policy" itself however, was deemed as the representation of a typical "eco-fascist" approach because it was based on the classical paradigm of exclusionary preservation.

With the emergence of a moratorium against National Parks and the discourse on eco-fascism, the perception of the problem expanded to include the question of the status and legitimacy of National Parks as such. Both of these controversial topics became immediately popular, with the terms "moratorium", "eco-fascism" and "eco-populism" used as a kind of jargon within a number of debates and negotiations. In several heated discussions on the internet, metaphorically dubbed "star wars",. these topics became a source of never-ending criticism and even threats to be exchanged between and among the various NGOs. In the meantime, the local people involved in the occupation exhibited their determination to remain in Dongi-dongi. The villagers made their "do or die" statement to the Head of Donggala District and representatives of the Donggala legislative council during their visit in Palolo on July 7, 2001. In response to the Government Committee idea conveyed by the Regent, the farmers firmly rejected the offered solution and reiterated that they would remain in Dongi-dongi either alive or dead. The farmers conveyed similar sentiments to the Central Sulawesi Provincial Legislature, as well as the provincial and Donggala district administrations, on meeting July 9, 2001.

Thus, one month after the occupation, the Dongi-dongi controversy was even further away from a win-win solution, colored by the occurrence of a pro and con polarization within a zero-sum atmosphere. This situation led to the elimination of the intermediate position adopted by a number of parties making efforts to find a balanced solution that could satisfy everyone involved in the dispute.

[16] The eco-populism concept was first introduced by the Head of the Lore Lindu National Park in a coordination meeting to solve the problem of encroachment into the National Park held at the Donggala District office on June 3, 1999.

Actors	Supportive of Dongi-dongi Occupation (Pro-Occupation)	Opposed to Dongi-dongi Occupation (Con-Occupation)
International Based NGOs	–	The Nature Conservancy (TNC) CARE Indonesia – Central Sulawesi
Local NGOs and Student Organizations	People's Legal Aid Foundation (YBHR) Free Land Foundation (YTM) People's Education Foundation (YPR) Palu Bay Fishermen Union (SNTP) National Farmers Union (STN) Donggala People's Legal Aid Institute (LBHR) League of National Democratic Students (LMND) National Front for Indonesian Labor (FNBI) People's Solidarity (SORAK) The Solidarity for the Struggle of Agrarian Reform (SPRA) The Red White Foundation (YMP) Evergreen Indonesia Foundation (YEI)	LBH Bantaya Katopassa Foundation Green Advocates Jambata Foundation Tadulako Student Nature Lovers (MAPATALA) University of Tadulako Member of BKKPA Member of KPA Central Sulawesi

Associations/ Alliances	WALHI Central Sulawesi Central Sulawesi Customary Community Alliance (AMASUTA) Kamalisi Customary Community Alliance (AMAK) Katuvua Alliance Student Nature Lovers Regional Information Center – Central Sulawesi	Lore Lindu National Park Partnership Forum (FKTNLL) Nature Lovers Group Coordinating Body Central Sulawesi (BKKPA-ST) Nature Lovers Community Central Sulawesi (KPA-ST) Kabeloata Singgani Alliance Tolelembunga Alliance (a mysterious organization)
Local Communities/ Village Administrations/ Community Organizations/ Customary Communities	Independent Farmers Forum (FPM) Farmers from the villages of Tuva, Pakuli, Bulili, Panimbo, Vatutela, Pandere, Lindu, Sirenja and Dodolo	Poor Farmers Solidarity Forum (FSPM) Pekurehua Customary Community Urban Farmers Community (KOMPAK). Farmers and Leaders of Rahmat village (Head and Secretary of Village, Chief of Customary Council, etc.) Leaders of Sintuwu Village (Head and Secretary of Village, Chief of Customary Council, and other formal and non-formal leaders) Urban Farmers Community Forum Pitu Nggota Ngata Kaili Customary Council

Table 2. *NGOs, associations/alliances and community organizations involved in Dongi-dongi controversy*

The Escalation of the Conflict

MOBILIZATION OF THE MASSES AND THE CLAIMS OF CUSTOMARY COMMUNITIES

As the dispute increasingly heated up, the concerned parties felt themselves forced to make efforts to mobilize the public in order to strengthen their own positions. These efforts took the form of exploiting the jargon of customary communities, the creation of alliances among various organizations as well as the creation of new counter-organizations. The first mass action after the occupation of Dongi-dongi involved FPM itself and a number of associated NGOs gathered under the umbrella of the *Katuvua* Alliance. One month after the Dongi-dongi declaration, this alliance, consisting of FPM, YBHR, WALHI, YTM, LMND, YPR, SNTP, STN, FNBI, and AMASUTA, held a demonstration in front of the Governor's office also The action taken by the *Katuvua* Alliance immediately drew both pro and con reactions from a number of sources. A new alliance – the *Kamalisi* Traditional Community Alliance (*Aliansi Masyarakat Adat Kamalisi*: AMAK) and a group of Central Sulawesi People's Youth Movement (*Gerakan Pemuda Kerakyatan Sulawesi Tengah*) was set forth and issued a statement supporting the actions of FPM and its alliance. As a response to this situation, a new group called "The Poor Farmers Solidarity Forum" (*Forum Solidaritas Petani Miskin*: FSPM) was formed. FSPM was a local community organization in Berdikari village and had a strong affiliation with the "Legal Aid Institute Bantaya" (*Lembaga Bantuan Hukum Bantaya*, LBH). FSMP openly urged opposition to the occupation for the reason that a local lumber syndicate was exploiting the farmers' movement in Dongi-dongi.

The con-occupation networks then increased its pressure on the pro-occupation side by organizing the "*Pekurehua* Customary Community Congress". The congress condemned the occupation and urged the government to "remove all settlements and residences from Dongi-dongi, and to provide alternative land for settlement outside the National Park". Further, the congress "strongly rejects the efforts of the parties who are facilitating and supporting the people who have been settling in Dongi-dongi and its surroundings". As a counter-reaction to the latest development, the pro-occupation network and FPM organized two strategic moves. The first was the traditional land clearing ceremony held in Dongi-dongi. Through this ritual, attended by more than 1,000 people according to claims, where four water buffaloes were sacrificed, the farmers stated their ties to the land in what they called "The Village of Hope" (*Ngata Katuvua*). By such a (cultural) statement of commitment to the land, and through the ritual procession performed, the farmers determined themselves to fight to death in order to hold onto the land in Dongi-dongi. "Because," as one participant explained, "it is like we have buried our heads here."

Second, in response to the concerns expressed by the con-occupation parties in regard to ecological degradation, the pro-occupation side issued the "Dongi-dongi Unified Charter for the Management of Natural Resources" (*Piagam Kesepakatan Dongi-dongi untuk Pengelolaan Sumber Daya Alam*). This Charter, issued on July 26, 2001, constituted a number of regulations on the management of the region including the implementation of traditional sanctions. The regulations dealt with matters like the distribution of land and its allocation, the management of water, rivers and forests, as well as the settlement of land and the administration of settlements. The Charter further stated that land should not be sold, that trees larger than 30 centimeters in circumference should not be cut down; that nothing should be cultivated along river banks, that rice fields should not be planted in the area, and that only tree crops such as coffee, pepper and candle nuts should be planted. After the announcement of the Charter, the YBHR leader sent an e-mail encouraging all parties to support the desires of the FPM farmers settling in Dongi-dongi. His reasoning was that since the traditional ceremony was already held, no efforts to move out the settlers would be effective but rather would give rise to new problems. Thus, the only feasible focus at this point of time could be "how to guarantee that their management of the area in question did not deviate from the principles of conservation". At the end of July 2001, the pro and the con NGOs achieved an agreement on the following three points: First, no further community assistance should be given by any NGO in the Lore Lindu National Park area while the full sovereignty to make decisions should be returned into the hands of the local people (the "moratorium on assistance"). Second, the farmers should be given the time they need to discuss the Dongi-dongi case among themselves. Third, the NGO activists should no longer discuss the Dongi-dongi controversy on the internet, unless there is a decision made by the people concerned.

The agreement to withdraw from assistance in Dongi-dongi turned out as a appropriate means to reduce the tension for the time being, although it had no impact on the resolution of the conflict itself. According to a senior NGO activist in Palu, it was during this break in assistance that the incidence of illegal logging in the Dongi-dongi area increased immensely.[17] However this may be for the parties involved, this period of cooling down was used to consolidate their own individual positions and strengths, instead of seeking a compromise between them. As such, the break became no more and no less than a period of preparation for the next phase of conflict. In this way, both parties created a new battle arena for even greater conflict and controversy. Meanwhile, the government simply disappeared from the unfolding process and took no initiative in order to find a solution. This "impotence" or "lack of presence" on part of the state apparatus in the midst of a heated community conflict became another factor giving way to the escalation of the conflict. Instead, the government used this period of time with the coordination of the efforts to realize the implementation of its relocation policy. Experts and government officials were preoccupied with taking steps against illegal logging. However, even in this effort, the coordination failed due to an overlap of interests related to illegal logging which reached its peak at that time.

STRIKE-COUNTER STRIKE, EFFORTS TO REPRESS AND TO DEFEAT OPPONENTS

Facing a situation in which all attempts to facilitate a dialog were defeated by the hard stance taken by the farmers grouped in FPM, the government began to feel it had no other option than to enforce the law strictly and forcefully. Therefore, this phase was marked by several repressive moves on part of the government as well as by efforts among various groups within civil society to negate the existence of others with opposing views. The Regent of Donggala District, supported by the Department of Forestry, the Governor of Central Sulawesi as well as the Central Sulawesi Regional Police ordered its

[17] Interview with Nasir Abas on March 19, 2002.

apparatus (1) to summon the area residents who had occupied Dongi-dongi, (2) to clear all squatters out of Dongi-dongi, (3) to put an end to the clearing of land there, (4) to enforce the status of Dongi-dongi as a water catchment zone and (5) to carry out all of the mentioned orders within a period of no more than three days beginning from August 20, 2001.

While the state authorities were preparing repressive action, the Lore Lindu National Park Authority and members of the public made efforts to submit a formal complaint to the police in which the farmers occupying Dongi-dongi were portrayed as being stirred up by certain NGOs. In mid-August 2001, in a letter to the Central Sulawesi Regional Police, the Head of the National Park requested that the Regional Police "investigate YBHR… and to enforce the law in line with all existing regulations and rulings". In response to the steps taken by the government, the National Park Authority and the con-occupation NGOs, the pro-occupation NGOs did not wait long to take counter actions. The Central Sulawesi WALHI and its network demanded the Regional Police to restrain from imposing a security approach to the situation and urged the authorities "not to view the actions as a criminal case just because the land in question is designated as National Park". Instead, the Dongi-dongi case "should be viewed as an agrarian conflict caused by an imbalance of land ownership in the area surrounding the National Park". Further, WALHI perceived the recent steps taken by the National Park Authority as a proof of the true militaristic and eco-fascist nature of this agency. The rejection of an oppressive approach raised by WALHI Central Sulawesi gained support from its NGO network members such as YPR, STN, SNTP and AMAK (see glossary).

As both parties solidified their power, the NGOs against the occupation of Dongi-dongi increased their pressure through new demonstrations. The first protest rally held was on end of August 2001 in front of the Provincial Parliament (DPRD) and was organized by the Central Sulawesi Nature Lovers (*Komunitas Pecinta Alam Sulawesi Tengah*: KPA-ST). Within the next three weeks, a second mass demonstration was set forth at the same place, this time organized by the Alliance of *Kabeloata Singgani* (*Persekutuan Kabeloata Singgani*), an organization made up of residents of the villages of Rahmat and Sintuwu, KPA-ST, FSPM, and the Urban Farmers Community *(Komuni-*

tas Petani Kota or KOMPAK). These two protests demanded the disbandment of WALHI in Central Sulawesi as well as the prosecution of the organization. Other issues raised during the protests were the rejection of the use of force in the resolution of the Dongi-dongi case, provision of alternative land for the landless farmers, and the conversion of the status of Dongi-dongi to its original function. Due to this pressure, Central Sulawesi WALHI and its pro-occupation network started to organize counter actions. The difference however, was that WALHI focused its actions primarily on issues of human rights. Further, it did not target the NGOs in Central Sulawesi, but focused deliberately on the foreign NGOs. The counter strike began on September 12, 2001, when around 300 members of FPM closed down and sealed the doors of the Nature Conservancy (TNC) and CARE offices. They legitimated their action with the argument that the two foreign NGOs had misled the public during the process of Park zonation. During the strike, the farmers stated their rejection of the zoning plan developed by TNC. The second counter strike began on September 24, 2001. Using the momentum of the commemoration of the nation's 40th Agrarian Day, a number of NGOs, community and university student organizations, including AMASUTA, AMAK, FNBI, LBHR, LMND, SPRA, SNTP, SORAK, STN Donggala, WALHI, YBHR, YEI, YMP, YPR, and YTM held a protest rally in Palu. During the event, claimed to be the largest demonstration taken in the history of protest in Central Sulawesi, the protestors demanded that the stop of the abuse of human rights and the appropriation of customary lands through the government. Further, they urged the reformation of the management of the Lore Lindu National Park in the direction of more rural participation in the decision-making process. Further, WALHI demanded the rescinding of a Memorandum of Understanding (MOU) between TNC and the Directorate General for Forest Protection and Nature Conservation in Jakarta. The argumentation of WALHI was that first, the income generating projects carried out by TNC did not provide any economic benefit for the local communities. The approach adopted by TNC was not at all sustainable and did not touch on the basic problems the people living in the forest margin faced. Second, from the very beginning, TNC had opposed the idea of a "community based conservation management". Instead, the activities of TNC rested entirely on the assumption that the ones who damage and plunder the conservation area are exclusively local people. Third, in preparing the management plan, TNC made a fatal mistake in issuing the zoning draft without involvement of the local communities.

The conflict then shifted from the local to the national level when TNC Indonesia negotiated with the National Executive of WALHI in the capital. During these negotiations, both agreed to bridge the gap of communication and to calm down any conflicts between the two NGOs. They further agreed on a "cooling off period" of two weeks beginning on October 1, 2001. In the meantime, however, the TNC office in Palu requested protection from the police after it received six anonymous threatening letters. This respite phase marked the end of the open conflict among the various NGOs in Palu. Even so, it did not end efforts to defeat opponents. The head of the National Park put the Central Sulawesi Police under pressure in order to resolve the Dongi-dongi case immediately through legal channels because the police were already in possession of the names of the people who were responsible for illegal logging. However, even though it had strong support from various parties to take firm action, the police never took any concrete action. The reluctance of the Regional Police to take any risks placed the Dongi-dongi area under the *de facto* control of the local community.

ILLEGAL LOGGING

From June 27 through December 31, 2001, the Lore Lindu National Park management, along with regional Police officers, investigated the illegal logging activities in the Park area. As a result, 302.1 m^3 of wood and 55 trucks were confiscated.[18] The highest incidence of illegal logging occurred between June 27 and July 16, 2001, with the confiscation of 92 m^3 of wood and 20 trucks. In this three weeks period, the amount of wood and number of trucks confiscated reached approximately 30 percent of the total amount confiscated in the entire six months of the surveillance. There appeared to be a close connection between the commencement of the Dongi-dongi occupation on June 19, 2001 and the intensity of illegal logging. There are several

[18] Slide presentation of the head of the Lore Lindu Park Authority on February 2nd, 2002.

reasons for the widespread incidence of illegal logging. First, there is a high level of demand for wood in the regional sawmills and wood shops most of which are operating illegally in the Palu valley. A survey carried out by the Park Authority, the Central Sulawesi Regional Police, and FKTNLL on 5-15 November 2001 found that (a) of the two molding factories investigated, one was operating without a permit; (b) of 58 investigated sawmills, 40 had no permits, and (c) of the 77 wood shops investigated, 65 operated illegally (Laban, 2002: 3). Second, the Lore Lindu region is characterized by a low level of weak law enforcement and monitoring. Thus, the legal process in prosecuting log shipments without the proper "Certified Forestry Products Document" *(Surat Keterangan Sahnya Hasil Hutan,* SKSHH) is extremely slow. For example, of the 40 cases of illegal wood shipment monitored from June 27 through December 31, 2001, the Police sent only three suspects to trial, five cases shifted to the Donggala Police station, and the Military Police Detachment received two suspects (Laban, 2002: 7). Third, the issuance of SKSHH by the Provincial Forestry Agency is not selective enough in the investigation of receivers (Press Release Katopassa Foundation, May 11, 2002). This dark portrait of illegal logging in Central Sulawesi is just a miniature picture of the overall extent of this problem as it unfolded throughout the Indonesian archipelago in the wake of the economic crisis. The imbalance in supply and demand of raw materials and forests products in combination with an over-capacity of the wood industry, a serious weakness in law enforcement, as well as the continuance of corrupt practices, collusion and nepotism, represent the main causes of illegal logging in Indonesia today (Kartodihardjo, 2002).

Human Victims and Ecological Damage

The illegal logging carried its victims when on October 8, 2002, two forest rangers patrolling under the "Operation for the Eradication of Illegal Logging" came into conflict with illegal residents transporting illegal logs from the Dongi-dongi forest to Palu. One Dongi-dongi farmer fell dead, fired on by the forest ranger while one ranger was injured and hospitalized. The next day, on October 9, 2002, a crowd burned down the facilities/branch office of Lore Lindu National Park in Kamarora village. Villagers occupying Dongi-dongi burned down an information center building, four guesthouses, a Motorola radio communication unit, and two office structures. Meanwhile, in Bobo village, a crowd burned an office and destroyed a Motorola communication radio. In reaction to these incidents, Central Sulawesi WALHI issued a press release that condemned the violence and requested the Head of the Park to resign from his post. Central Sulawesi WALHI was of the opinion that the area residents had run amuck and burned the Park Authority's property in response to the forceful actions taken by the forest rangers (Nuansa Pos, October 14, 2002).

One year after the shooting incident, the degradation of the forest ecosystem in Dongi-dongi continued as more and more people moved to Dongi-dongi. Whereas in December 2001, there were only 390 new houses in the Dongi-dongi area, by October 2003, a little less than two years later, Dongi-dongi counted 1,148 buildings. In the same time, the number of houses with zinc roofing, an indicator for an increase in construction quality, had significantly increased.[19] In the two years after the occupation, the physical landscape and ecosystem of Dongi-dongi had changed dramatically. This has a severe impact on Dongi-dongi's stability as a water catchment area for the downstream regions. The extreme result of these ecological changes was the landslides and massive flooding on December 13, 2003 referred to in the initial section of this paper. The flooding not only damaged Dongi-dongi, but also caused severe damage in the downstream villages of Ranteleda, Tongoa and Berdikari. A seven-km section of the Tongoa - Sedoa access road fell out of service due to landslides and fallen trees. The flooding damaged three bridges, and washed away a hanging bridge in Pakuli (Biromaru Sub-District). Electrical power disrupted in Tongoa, and no fewer than 40 houses in the Tongoa, Dongi-dongi, and Karava (Pakuli) villages inundated, while four homes crushed. Besides, among 50 ha of rice fields and cacao plantations were flooded. Not long after the flooding, the Coordinator of FPM firmly stated in a talk show broadcast of a private radio station in Palu, that he was actively involved in illegal logging and that he was not afraid

[19] This data was collected through observation and a rapid assessment conducted by Christian Schulze, a co-researcher of the STORMA-research project from December 2001 until September 2004.

of arrest by the police. Instead, he explained that in case of arrest, he would disclose the names of the police, military and government officials involved in and supporting illegal logging in Dongi-dongi.

Although history seemed to be repeating itself, the whole disaster seemed to be over in a flash. The Tongoa - Dongi-dongi - Sedoa route started functioning again, the Dongi-dongi residents went back to their previous activities of clearing forest, growing cacao and coffee or collecting rattan and logs from the forest. A small number of Dongi-dongi residents have even found another additional source of income. They are now charging fees from every vehicle crossing the emergency bridge built to replace the main bridge washed away by the flood. As if they became exhausted finally, the NGOs have settled back into their own routine activities.

Conclusions

Our research revealed several factors that caused the residents of the four resettlement villages to continue their occupation of Dongi-dongi until the present. One factor is certainly that the Post-Suharto Indonesian state is still in a transition from an authoritarian regime to a democracy. A weak legal system, poor law enforcement, a notorious lack of commitment on behalf of the state and a feeble state bureaucracy continued to exist in most governmental institutions. Therefore, the various initiatives put forth by the central government, the provincial administration, and the Central Sulawesi legislature, all ended up in failure. These repeated failures produced costly results in terms of a lack of confidence in the capabilities of the government.

Another factor is a strong overlap of economic interests among the residents of Dongi-dongi, the police, the military and government officials in the business of illegal logging. Thus, since the advent and rapid expansion of illegal logging in Dongi-dongi, FPM's actual power base was no longer the support of the NGOs. The Coordinator of FPM rather preferred that FPM ally itself with the movers and shakers of the illegal logging business, who also wielded increasing economic and political power in Central Sulawesi. This shift in the power base of FPM brought WALHI, YBHR and the other NGOs supportive of the Dongi-dongi occupation into a difficult situation.

Should they withdraw their support from FPM with the risk that WALHI, as the leading NGO in Indonesia, would "lose face" in the eyes of its constituents, or to continue to support FPM with the risk of being labeled as an environmental NGO that was inconsistent with is stated vision and mission. In order to hold their position in Dongi-dongi, the occupiers shrewdly manipulated two types of power for their own interests and survival: the power of the support provided by the NGOs, and the power they gleaned from their relations in the illegal logging business. FPM succeeded in leveraging both types of power becoming a "coercion-struggle of power", that is, a form of opposition that carried strength equal to that of the "juridical-dominance of power" exerted by the state. This lends further credence to Foucault's statement that – "resistance is never in a position of exteriority in relation to power" (Foucault, in Gordon 1980:95).

So what is the solution to the Dongi-dongi dilemma? There are two alternative solutions. The first alternative (Alternative I) would be to change the function of Dongi-dongi from its initial status of a National Park core zone to that of an utilization zone allowing legal access and exploitation by villagers who do not have land of their own. The second would be to enforce Dongi-dongi's initial core zone status and to move the occupants out to another area. In Alternative I, Dongi-dongi would remain state property, although local people could access and utilize it. The area residents would only enjoy usufruct rights but no individual or private property rights without rights to transfer the land by purchase. In Alternative I, Dongi-dongi would eventually become an agro-forestry landscape with a spatial planning and utilization scheme determined in a participative manner by the local community. As a consequence, a strong social contract between the Park Authority and the local community would be required concerning the rights to access, control, withdrawal and management of the area, as well as exclusion and alienation from resources. Included in this social contract should be the local community's obligation to replant and cultivate trees in areas agreed upon among all the parties concerned.

Our research revealed several factors that caused the residents of the four resettlement villages to continue their occupation of Dongi-dongi until the present. One factor is fore sure that the Post-Suharto Indonesian state is still in a transition from an authoritarian regime to a democracy. A weak legal system, poor law enforcement, a notorious lack of commitment on behalf of the state and a feeble state bureaucracy continue to exist in most governmental institutions. Therefore, the various initiatives put forth by the central government, the provincial administration, and the Central Sulawesi legislature, all ended up in failure. These repeated failures produced costly results in terms of a lack of confidence in the capabilities of the government.

Another factor is a strong overlap of economic interests among the residents of Dongi-dongi, the police, the military and government officials in the business of illegal logging. Thus, since the advent and rapid expansion of illegal logging in Dongi-dongi, FPM's actual power base was no longer the support of the NGOs. The Coordinator of FPM rather preferred that FPM ally itself with the movers and shakers of the illegal logging business, who also wielded increasing economic and political power in Central Sulawesi. This shift in the power base of FPM brought WALHI, YBHR and the other NGOs supportive of the Dongi-dongi occupation into a difficult situation. Should they withdraw their support from FPM with the risk that WALHI, as the leading NGO in Indonesia, would "loose face" in the eyes of its constituents, or to continue to support FPM with the risk of being labeled as an environmental NGO that was inconsistent with is stated vision and mission. In order to hold their position in Dongi-dongi, the occupiers shrewdly manipulated two types of power for their own interests and survival; the power of the support provided by the NGOs, and the power they gleaned from their relations in the illegal logging business. FPM succeeded in leveraging both types of power becoming a "coercion-struggle of power", that is, a form of opposition that carried strength equal to that of the "juridical-dominance of power" exerted by the state. This lends further credence to Foucault's statement that "resistance is never in a position of exteriority in relation to power" (Foucault, in Gordon 1980: 95).

So what is the solution to the Dongi-dongi dilemma? There are two alternative solutions. The first alternative (Alternative I) would be to change the function of Dongi-dongi from its initial status of a National Park core zone to that of an utilization zone allowing legal access and exploitation by villagers who do not have land of their own. The second would be to enforce Dongi-dongi's initial core zone status and to move the occupants out to another area. In Alternative I, Dongi-dongi would remain state property, although local people could access and utilize it. The area residents would only enjoy usufruct rights but no individual or private property rights without rights to transfer the land by purchase. In Alternative I, Dongi-dongi would eventually become an agro-forestry landscape with a spatial planning and utilization scheme determined in a participative manner by the local community. As a consequence, a strong social contract between the Park Authority and the local community would be required concerning the rights to access, control, withdrawal and management of the area, as well as exclusion and alienation from resources. Included in this social contract should be the local community's obligation to replant and cultivate trees in areas agreed upon among all the parties concerned.

There is a strong probability that Alternative I would be the best solution to the Dongi-dongi problem. Alternative II – already tried by the local administration – did not receive a positive response from the occupiers due to the issues of accessibility and soil fertility. Clearly, any attempt to resolve the matter through application of Alternative II would encounter more severe constraints than would the implementation of Alternative I.

However, no matter which alternative applies, it is imperative that the government follows up with three important steps. First, in an initial step, the government would have to make a sincere acknowledgement of its failure in policymaking while exhibiting a concrete determination to improve the lives of the residents of the resettlement villages. Second, the government must rebuild the public's trust consistently fulfilling its promises and realizing mutually established agreements. Third, the government must develop an intense and continuing dialog and maintain this intensity throughout the process until the agreement of a final resolution. These three vital steps may seem extremely difficult for the government to take, but if implemented carefully, with patience, sincerity and honesty, they represent the only effective way to solve the complex problems that led to the occupation of Dongi-dongi.

References

Asian Development Bank [ADB] (1997): A Report and Recommendation of the President to the Board of Directors on a Proposed Loan to the Republic of Indonesia for The Central Sulawesi Integrated Area Development and Conservation Project. RRP: INO 29336.

Asian Development Bank (1998): Project Administration Memorandum: Central Sulawesi Integrated Area Development and Conservation Project. Loan No. 1605-INO in Indonesia.

Blaikie, P.M. (1985): The Political Economy of Soil Erosion in Developing Countries. Singapore.

Blaikie, P.M. and S. Jeanrenaud (1996): Bio-diversity and Human Welfare. United Nations Research Institute for Social Development. UNS-RISD, Switzerland.

Blaikie, P.M. (1997): Classic in Human Geography Revisited. *Progress in Human Geography*, 21.1: 75-80.

Blaikie, P.M. (1999): A Review of Political Ecology-Issues, Epistemology and Analytical Narratives. *Zeitschrift für Wirtschaftsgeographie,* Jg. 43 (1999), Heft 3-4: 131-147.

Bryant, R.L. (1991): Putting Politics First: The Political Ecology of Sustainable Development. *Global Ecology and Biogeography Letters* 1: 164-6.

Bryant, R.L. (1992): Political Ecology: An Emerging Research Agenda in Third-World Studies. *Political Geography,* Vol. 11, No.1, January 1992: 12-36.

Bryant, R.L. (1997): Beyond the Impasse: The Power of Political Ecology in Third World Environmental Research. *Area* 29.1: 5-19.

Bryant, R.L. and M. J. G. Parnwell (1996): Introduction: Politics, Sustainable Development and Environmental Change in South-East Asia. In: Parnwell, M.J.G. and R. L. Bryant (Eds): Environmental Change in SE-Asia: People, Politics and Sustainable Development: 1-19. London.

Bryant, R.L. and S. Bailey (1997): Third World Political Ecology. London.

Bryant, R.L. (1998): Power, Knowledge Political Ecology in the Third World: A Review. London.

Central Sulawesi Executive for Wahana Lingkungan Hidup Indonesia [WALHI] (2001): Laporan Hasil Monitoring Taman Nasional Lore Lindu, Provinsi Sulawesi Tengah (Monitoring Report of Lore Lindu National Park, Central Sulawesi). Palu.

Chapin, M. (2004): A Challenge to Conservationist. Nov./Dec. 2004 World Watch magazine. WorldWatch Institute, see: www.worldwatch.org.

Cohen, B. (2001): The Green Land-Grabbers: It's Not Just the Feds Who Are After Your Land. See: http://www.sierratimes.com/archive/files/nov/28/arbc112801.htm.

Colchester, M. and L. Lohmann (1993): The Struggle for Land and the Fate of the Forests. The World Rainforest Movement. Kuala Lumpur.

Colchester, M. and C. Erni (1999): Indigenous Peoples and Protected Areas in South and Southeast Asia. International Work Group for Indigenous Affairs (IWGIA). Copenhagen.

Colchester, M. (2004): Conservation Policy and the Indigenous Peoples of the Commonwealth. Paper presented to the Conference on 'Indigenous Peoples of the Commonwealth and the Millennium Development Goals', Commonwealth Policy Studies Unit, Institute of Commonwealth Studies, University of London.

Colfer, C.J.P. and I.A.P. Resosudarmo (2002): Which Way Forward? People, Forest, and Policy Making in Indonesia. Washington.

Eksekutif Daerah Wahana Lingkungan Hidup Indonesia (WALHI) Sulawesi Tengah (2001): Laporan Hasil Monitoring Taman Nasional Lore Lindu, Provinsi Sulawesi Tengah (Monitoring Report of the Lore Lindu National Park, Central Sulawesi). Palu.

Fauzi, N. (1999): The Farmer and the Ruler – the Dynamic of the Politic of Agraria in Indonesia. Yogyakarta.

Fauzi, N. (2003): Bersaksi untuk Pembaruan Agraria – dari Tuntutan Lokal Hingga Kecenderungan Global (Appeal for Agrarian Reform – from Local Demand to Global Trend). Yogyakarta.

Firdaus (2001): Memahami Aksi Pendudukan Petani di Kawasan Dongi-Dongi (Understanding the Dongi-Dongi Occupation by the Farmers). Articles in Palu Local Newspaper.

FoE [Friends of the Earth] International (2004): Annual Report (2003): Friends of the Earth International. Amsterdam.

Forest Watch Indonesia/Global Forest Watch [FWI/GFW] (2002): The State of the Forest: Indonesia. Forest Watch Indonesia and Global Forest Watch. Washington, Bogor.

Forum Petani Merdeka [FPM], Yayasan Bantuan Hukum Rakyat [YBHR], Yayasan Tanah Merdeka [YTM], Yayasan Pendidikan Rakyat [YPR], ED WALHI Sulteng (2001a): Kesepakatan-Kesepakatan antar Masyarakat di Wilayah Dongi-Dongi: Ngata Katuvua Ngaripuri (Agreement among Community Members of Dongi-Dongi: Ngata Katuvua Ngaripuri). Palu.

Forum Petani Merdeka [FPM], Yayasan Bantuan Hukum Rakyat [YBHR], Yayasan Tanah Merdeka [YTM], Yayasan Pendidikan Rakyat [YPR], ED WALHI Sulteng (2001b): Kronologi Kasus Dongi-Dongi: Sejarah dan Permasalahan yang Dialami oleh Desa Rahmat, Kamarora A, Kamarora B dan Desa Kadidia (The Chronology of Dongi-Dongi Affair: History and Problem of Rahmat, Kamarora A, Kamarora B and Kadidia Villages). Palu.

Gordon, C. (1980): Power/Knowledge: Selected Interviews and Other Writings 1972-1977. Michel Foucault. New York.

Helmi (2001): Sejarah Kawasan Taman Nasional Lore Lindu (The History of Lore Lindu National Park Area). Paper presented at the Workshop on the Lore Lindu National Park Management Plan, Bogor, July 30th-August 2nd, 2001. Jakarta.

Hughes, R. and F. Flintan (2001): Integrating Conservation and Development Experience: A Review and Bibliography of the ICDP Literature. London.

IUCN, UNEP, and WWF [World Conservation Union, United Nations Environment Programme, and World Wildlife Fund] (1991): Caring for the Earth: A Strategy for Sustainable Living. IUCN, UNEP and WWF. Gland, Switzerland.

IUCN (1994): Guidelines for Protected Area Management Categories. MCMC-IUCN. Gland, Switzerland and Cambridge.

Jamison, A. (1996): The Shaping of the Global Environmental Agenda: The Role of Non-Governmental Organizations. In: Lash, S. et al. (Eds): Risk, Environment and Modernity: Towards A New Ecology: 224-245. London.

Juliani, S. and Damayanti (2001): Report on Evaluation of the NRM II-USAID Grant to the Nature Conservancy.

Kabupaten Donggala (2000): Kabupaten Donggala dalam Angka – Tahun 1999 (Donggala Regency in Numbers – 1999). Donggala, Sulawesi Tengah.

Kartodihardjo, H. (2002a): Kondisi dan Kerusakan Hutan di Indonesia: Permasalahan dan Rekomendasi Kebijakan (The Condition and Degradation of Forest in Indonesia: Problems and Policy Recommendation). Background Paper. International NGO Forum on Indonesian Development (INFID). Jakarta.

Kartodihardjo, H. (2002b): Structural Problems in Implementing New Forestry Policies. In: Colfer, C.J.P. and I.A.P. Resosudarmo (Eds): Which Way Forward? People, Forest, and Policy Making in Indonesia: 144-160. Washington.

Laban, B. Y. (2000): Membangun Gerakan Komunitas Lokal menuju Konservasi Radikal (Empowering Local Community Movement for Radical Conservation). Palu.

Laban, B. Y. (2001): Tragedi Dongi-Dongi, "Q-Sruh" Pandang Komunitas (The Dongi-Dongi Tragedy: Disputes Perceptions among Communities). Paper presented at the Workshop on Integrating Forest Conservation and Economic Recovery in Indonesia.

Laban, B. Y. (2002): Prospek Negatif Penebangan Liar di Taman Nasional Lore Lindu (Negative Prospects of Illegal Logging in Lore Lindu National Park). Palu.

Li, T. M. (1999): Transforming the Indonesian Uplands Marginality, Power and Production. Singapore.

Li, T. M. and A. Sangaji (2003): Perubahan Agraria di Sekitar Taman Nasional Lore Lindu (Agrarian Change surround the Lore Lindu National Park). *Seputar Rakyat.* (06) Agustus 2003. Yayasan Tanah Merdeka: 18-27. Palu.

Li, T. M. (2003): Situating Resource Struggle. *Economic and Political Weekly* 2003 (forthcoming)

Long, N. (2001): Development Sociology – Actor Perspectives. London, New York.

Lynch, O.J. and E. Harwell (2002): Whose Resources? Whose Common Good? – Towards a New Paradigm of Environmental Justice and the National Interest in Indonesia. Lembaga Studi dan Advokasi Masyarakat (ELSAM). Jakarta.

Ottaway, D. B. and J. Stephens (2003): Nonprofit Land Bank Amasses Billions: Charity Builds Assets on Corporate Partnerships-Day One: Inside the Nature Conservancy. *Washington Post*, May 4, 2003.

Parnwell, Michael J. G. and R.L. Bryant (1996): Environmental Change in South-East Asia: People, Politics and Sustainable Development. London.

Peluso, N.L. (1992): Rich Forests, Poor People: Resource Control and Resistance in Java. Berkeley.

Peluso, N.L. (1993a): 'Traditions' of Forest Control in Java: Implications for Social Forestry and Sustainability. *Global Ecology and Biogeography Letters*, 3: 138-157.

Peluso, N.L. (1993b): Coercing Conservation? The Politics of State Resource Control. *Global Environmental Change* 3: 199-217.

Peluso, N.L. (1995): Whose Woods are these? – Counter-Mapping Forest Territories in Kalimantan, Indonesia. *Antipode* 27:4, 1995: 383-406.

Pierce, J.L. and J.W. Newstrom (1995): Leaders and Leadership Process – Readings, Self Assessment, and Applications. Chicago.

Sangaji, A. (2000a): Orang Katu di Taman Nasional Lore Lindu: Perjalanan Panjang menuju Rekognisi (Katu People in Lore Lindu National Park: A Long Way for Recognition). Paper presented for International Symposium and Workshop on "Entering the 21st Century: Regional Autonomy, Local Culture and Nation Culture", August 1-5, 2000. Makassar.

Sangaji, A. (2000b): Politik Kawasan Lindung dan Penghancuran Kekuasaan Adat atas Teritori Asli: Kasus Masyarakat Adat Behoa (The Politics of Protection Areas and Devastation of the Customary Power over

Indigenous Territory: A Case of Behoa Community). Paper presented at Policy Dialogue on Lore Lindu National Park and Behoa People, November 16th, 2000. Palu.

Sangaji, A. (2000c): PLTA Lore Lindu: Orang Lindu Menolak Pindah (The Lore Lindu Dam: the Lindu People Refuse to Move). Yayasan Tanah Merdeka, ED Walhi Sulawesi Tengah dan Pustaka Pelajar.

Sangaji, A. (2001a): Moratorium Taman Nasional Lore Lindu (Moratorium for Lore Lindu National Park). Paper presented at Public Debate on the Moratorium of Lore Lindu National Park, July 5th, 2001. Palu.

Sangaji, A. (2001b): Potret Taman Nasional Lore Lindu: Buruk Pendekatan, Rakyat Disalahkan [A Portrait of Lore Lindu National Park: An Awful Approach, Blaming the Local People]. Position Paper 03/YTM/2001. Palu.

Sangaji, A. (2001c): Konflik Agraria di Taman Nasional Lore Lindu: Tersungkurnya Komunitas-Komunitas Asli (Agrarian Conflict at Lore Lindu National Park: Displacement of the Indigenous Community). Paper presented at Policy Dialogue on Knowledge and Rights of the Customary Community around Lore Lindu National Park. Yayasan Tanah Merdeka and NRM/EPIQ. Palu.

Sangaji, A. (2002): Menuju Pengelolaan Taman Nasional Lore Lindu Berbasis Masyarakat (Toward Community Based Lore Lindu National Park Management). Wahana Lingkungan Hidup Indonesia [WALHI] Sulawesi Tengah. Palu.

Sangaji, A. (2003): Taman Nasional Lore Lindu Berbasis Masyarakat (Community Based Lore Lindu National Park Management). Wahana Lingkungan Hidup Indonesia [WALHI] Sulawesi Tengah. Palu.

Sangaji, A., M. Hamdin, Sugiharto, F. Lumeno, and S. Lahigi (2004): Masyarakat dan Taman Nasional Lore Lindu (Lore Lindu National Park and Community). Yayasan KEMALA dan Yayasan Tanah Merdeka. Palu.

Setyo, B. Y. and D. Neville (2002): Analisis Perubahan Hilangnya Tutupan Vegetasi di Taman Nasional Lore Lindu menggunakan Interpretasi Citra Satelit Landsat (Analysis of Vegetation Cover of the Lore Lindu National Park using Landsat Interpretation). Paper presented at the Workshop Seminar on the Qualitative Evaluation Performance of the Mid-Term CSIADCP, March 4-9, 2002, Kulawi, Central Sulawesi.

Sitorus, MT. F. (2002a): Agrarian Dynamics in Forest Margin Communities in Central Celebes, Indonesia. STORMA Discussions Paper Series No. 5, Bogor.

Sitorus, MT. F. (2002b): Land, Ethnicity and the Competing Power. Agrarian Dynamics in Forest Margin Communities in Central Celebes, Indonesia. STORMA Discussion Paper Series No. 5, Bogor.

Sitorus, F. (2002c): "Revolusi Cokelat" Social Formation, Agrarian Structure, and Forest Margins in Upland Sulawesi, Indonesia. STORMA Discussion Paper Series No. 9, Bogor.

Sondakh, J. O. M (2002): Pola Pengaturan Pemanfaatan Sumber-sumber Agraria serta Implikasinya pada Keberlanjutan Kelompok Sosial dan Kelestarian Alam: Studi Kasus pada Masyarakat di Dua Desa sekitar Kawasan Taman Nasional Lore Lindu, Kabupaten Donggala, Propinsi Sulawesi Tengah (The Management of Agrarian Resources and its Implication to the Sustainability of Social Groups and Nature Conservation: Case Study at Two Village Communities, Donggala Regency, Central Sulawesi Province). Master Thesis, Graduate Program of the Bogor Agriculture Institute. Bogor.

Syahyuti (2002): Pembentukan Struktur Agraria pada Masyarakat Pinggiran Hutan: Studi Kasus di Desa Sintuwu dan Desa Berdikari, Kecamatan Palolo, Kabupaten Donggala, Propinsi Sulawesi Tengah (The Creation of Agrarian Structure among the Marginal Rainforest Communities: Case Study at Sintuwu Village and Berdikari Village, Palolo Sub-district, Donggala Regency, Central Sulawesi Province). Master Thesis, Graduate Program Bogor Agriculture Institute, Bogor.

The Nature Conservancy [TNC] (2000): Building Conservation Capacity and Partnership at Lore Lindu National Park: Fourth Semi Annual Report, July – December 2000. Jakarta.

The Nature Conservancy, Directorate General of Forest Protection and Nature Conservation, and Lore Lindu National Park Authority (2001): Draft Management Plan 2002-2027: Lore Lindu National Park. Volume One and Two. Jakarta.

The Nature Conservancy [TNC] (2002a): Lessons Learned from Building Conservation Capacity and Partnership at Lore Lindu National Park. NRM II-USAID Grant to The Nature Conservancy. Jakarta.

The Nature Conservancy Lore Lindu Field Office [TNC LLFO] (2002b): Evaluasi Perkembangan Ancaman terhadap Taman Nasional Lore Lindu (Evaluating the Threat Level of Lore Lindu National Park). Paper presented at the Workshop Seminar on the Qualitative Evaluation Performance of the Mid-Term CSIADCP. March 4-9, 2002, Kulawi, Central Sulawesi.

The Nature Conservancy [TNC] (2002c): Monitoring Sumber Daya Alam Desa Rahmat di sekitar Taman Nasional Lore Lindu. In: The Nature Conservancy Building, Conservation Capacity and Partnership at Lore Lindu National Park – Final Report to NRM. TNC, Jakarta.

The Nature Conservancy [TNC] (2003): *Annual Report 2002*. The Nature Conservancy, Arlington.

van der Heijden, H-A. (2002): Environmental Movements, Ecological Modernisation and Political Opportunity Structures. In: Rootes, C. (Ed): Environmental Movements: Local, National and Global: 199-221. London.

WALHI Sulawesi Tengah (2001): Laporan Hasil Monitoring Taman Nasional Lore Lindu, Provinsi Sulawesi Tengah (Monitoring Report of the Lore Lindu National Park, Central Sulawesi). Palu.

Wapner, P. (1995): Politics Beyond the State-Environmental Activism and World Civic Politics. *World Politics 47*: 311-40.

Wapner, P. (1996): Environmental Activism and World Civic Politics. Albany.

Wulan, Y.C., Y. Yasmi., C. Purba, and E. Wollenberg (2004): An Analysis of Forestry Sector Conflict in Indonesia 1997-2003. *Governance Brief*, Number 1, October 2004. Center for International Forestry Research (CIFOR). Bogor.

WWF (2004b): *WWF Annual Review 2003*. WWF. Gland, Switzerland.

Yayasan Tanah Merdeka (YTM) (2003): Taman Nasional Lore Lindu: Perlu Paradigma Baru (Lore Lindu National Park: Need New Paradigm). *Seputar Rakyat*. Edisi 06, Tahun I, Agustus 2003. Palu.

Yulianto, O.A. (2003): Aktor Pro-Demokrasi Berbasis Jaringan (The Networks Based Pro-Democracy Actor). In: Priyono, A.E. et al. (Eds): Gerakan Demokrasi di Indonesia Pasca-Soeharto (Democracy Movement in Indonesia – Post Soeharto): 525-558. Jakarta.

GLOSSARY

Acronym	Indonesian	English Equivalent
ADB	--	The Asian Development Bank
AMAK	Aliansi Adat Masyarakat Kamalisi	Alliance of the *Kamalisi* Customary Community
AMAN	Aliansi Masyarakat Adat Nusantara	National Alliance of Customary Communities
AMASUTA	Aliansi Masyarakat Adat Sulawesi Tengah	Central Sulawesi Alliance of Customary Community
BKKPA-ST	Badan Koordinasi Kelompok Pencinta Alam, Sulawesi Tengah	Nature Lovers Group Coordinating Body, Central Sulawesi
CBD	--	Convention on Bio-diversity
CI	--	Conservation International
CSIADCP	--	Central Sulawesi Integrated Area Development and Conservation Project
DPRD Provinsi	Dewan Perwakilan Rakyat Provinsi	Provincial People's Representative Assembly
FKTNLL	Forum Kemitraan Taman Nasional Lore Lindu	Partnership Forum for the Lore Lindu National Park
FNBI	Front Nasional Buruh Indoensia	National Front for Indonesian Labor
FPM	Forum Petani Merdeka	Independent Farmer Forum
FSPM	Forum Solidaritas Petani Miskin	Poor Farmers Solidarity Forum
IUCN	--	International Union for the Conservation of Nature or World Conservation Union
KAU	Koalisi Anti Utang	Anti Debt Coalition
KOMPAK	Komunitas Petani Kota	Urban Farmers Community
KPA-ST	Komunitas Pencinta Alam Sulawesi Tengah	Central Sulawesi Nature-Lovers Community

Acronym	Indonesian	English Equivalent
LBH Bantaya	Lembaga Bantuan Hukum Bantaya	Legal Aid Institute of *Bantaya*
LBHR	Lembaga Bantuan Hukum Rakyat	Legal Aid Institute for the People
NGO	--	Non-Governmental Organization
SKEPHI	Sekretariat Kerjasama Pelestarian Hutan Indonesia	NGOs Networks for Indonesian Forest Conservation
SKSHH	Surat Keterangan Sahnya Hasil Hutan	Certified Forest Product Document
SNTP	Serikat Nelayan Teluk Palu	Palu Bay Fishermen's Union
SORAK	Solidaritas Rakyat	Solidarity of the People
SPRA	Solidaritas Perjuangan Reforma Agraria	The Solidarity for the Struggle of Agrarian Reform
STN Donggala	Serikat Tani Nasional, Donggala	National Farmers Union, Donggala
USAID	--	United States Agency for International Development
WALHI	Wahana Lingkungan Hidup Indonesia	The Indonesian Environmental Forum
WWF	--	World Wide Fund for Nature (formerly World Wildlife Fund)
YBHR	Yayasan Bantuan Hukum Rakyat	Legal Aid for the People Foundation
YEI	Yayasan Evergreen Indonesia	Evergreen Indonesia Foundation
YMP	Yayasan Merah Putih	The Red White Foundation
YPAN	Yayasan Pusaka Alam Nusantara	National Heritage Foundation
YPR	Yayasan Pendidikan Rakyat	Foundation for the Education of People
YTM	Yayasan Tanah Merdeka	Free Land Foundation

Chapter IX

Cacao and Rural Transformations: Processes of Displacement and the Evolution of Exclusionary Agrarian Arrangements in Sintuwu[1]

Günter Burkard

Abstract

This paper analyses the effects of the Central Sulawesi cacao boom and its associated patterns of migration on the rural labor market in the village of Sintuwu, Palolo valley. The domination of cacao in the dry land sector and the subsequent conversion of wet rice fields into cacao plantations caused a split between those with relative stable access to rice and work and those who have become almost excluded from agricultural work. Workers with "family-access" can rely on stable contracts with fix partners for many years, others instead have to look out for minimal contracts with many partners. The new institution of exclusionary work arrangements in wet rice cultivation is framed within a locally dominant ideology of kinship. Rather than being a local feature since times immemorial, an exclusive form of shared poverty has arisen as a result of capitalist penetration and a commodification of the land market. On the other hand, the "cacao revolution" has not led to a commercialisation of relations of production, neither the dry land nor the wet rice sector, but it has contributed to an ever-increasing social polarisation in Sintuwu.

---

[1] Data collection for this paper was conducted in March/April 2004. It was integrated into a number of research activities under sub-project A2: "Social Organization and Processes of Ecological Stabilization and Destabilization" of the STORMA Research Program, a collaborative research program of the German Universities of Göttingen and Kassel and the Indonesian Universities "Institute for Agriculture" (IPB) in Bogor and Tadulako University in Palu, Central Sulawesi. Fieldwork in Central Sulawesi was conducted in 6 villages during the time of March 2001 until April 2006. Within each village 30-40 households have been chosen as random sample.

Introduction: "Difference", "Change" and the Lowland-Upland Stereotypes in Indonesia

Informed by the impact of the "green revolution", studies on socio-economic change in rural Indonesia focused primarily on the wet rice growing lowlands. These studies were predominantly "quantitative" or "statistical" in nature, analysing first of all aspects of *agrarian change*, such as disparities in wealth and land ownership, technological and demographic issues, diversification of labor markets, problems of rural credit access etc.[2] On the contrary, the Indonesian uplands were mainly investigated in terms of either *environmental issues* such as forest and bio-diversity conservation, the ecological impacts of shifting cultivation, and community based resource management on the one hand, and in terms of *socio-cultural issues* such as indigenous rights, customary law, local knowledge, problems of marginality, and tribal identities on the other (Li 2002: 416). This division has far-reaching implications in that it implicitly relates the lowlands to the "agrarian" and the uplands to the "environmental" key questions of the social and natural sciences.

In Indonesia, the lowland-upland dichotomy is rather a cultural than a topographic or geographic distinction.[3] However, in representing upland societies, scientists, bureaucrats and environmental activists give a rather "diachronic" picture: On the one hand there is the notion of the backward and ignorant highland farmer whose inappropriate agricultural practices will inevitably deplete any upland resource, threaten environmental stability and who is therefore in special need of "guidance" and specific "hillside development". This is the general attitude adopted within the government's discourse of development and informs much of the national planning activities in the "outer islands". The other extreme is the image of the "moral economy" of

---

2 For an anthropological critique of this quantifying trend see Burkard 1999.

3 According to Spencer (1949: 28, cited in Allen 1993: 226 and in Li 1999: xvi), uplands „could be defined as containing a core of hilly to mountainous landscapes of steeply inclined surfaces and the table lands and plateaus lying at higher elevations". That such an old and vague definition is still valid today shows that lowland-upland distinctions in Indonesia are indeed "constructions" rather than "physical facts". Terms like "interior" or "upriver" (in Kalimantan) are often used synonymously.

upland communities which – in its emphasis on the sustainability and resilience of customary institutions and indigenous livelihood patterns – transforms upland farmers into environmental experts ingrained with ancestral wisdom. This is the dominant upland representation among many of the conservation, empowerment and indigenous rights NGOs in contemporary Indonesia. Both views tend to consider upland societies as isolated, bounded and territorially localised "micro-cultures", more or less closed systems which had evolved to be functionally stable.

A major consequence of this tendency is the perpetuation of a long established stereotype which locates the lowlands at the centre and the uplands at the periphery of social and economic change. Thus, in his influential book "Agricultural Involution", Clifford Geertz speaks of the uplands as regions of "essentially unchanged swidden making", and a "monotonous expanse of enduring stability" (Geertz 1963: 116, cited in Li 1999: 36). In Indonesia, upland populations are often portrayed in terms of primordial loyalties and self-containing ethnic categories (e.g. the *Batak*, or the *Toraja*) with the effect that "what might otherwise be understood as inequalities in access to resources and/or power are read instead as problems in intercultural relations" (Kahn 1999: 81). This image of stable, bounded ethnic communities emphasises isolation and separation rather than relations and exchange and is thus quite at odds with the more dynamic understanding of group identities anthropologists developed in other SE-Asian upland settings. Following Lehman's theory of social systems (Lehman 1967), we reject the idea of primordial differences between lowlands and highlands. Rather, we believe that upland and lowland group categories are formally like roles and are "only very indirectly descriptive of the empirical characteristics of substantive groups of people" (ibid: 107). As exemplified by Barth (1969:15ff), an adequate understanding of upland (and any other minority) population must shift the focus from groups as culture-bearing units to the ethnic and social boundaries that define certain groups in relation to others (and the state), not the trait inventories and cultural characteristics a boundary encloses. The approach adopted by Lehman and Barth entails an important implication in the sense that it perceives of highland people not only as victims or recipients of outer forces of power (supposedly located in the lowlands), but as active agents who may well identify and position themselves as "uplanders" within a regionally defined system of inter-group relations (see Lehman 1969: 106). As pointed out by Li (1999, 2001), there is ample evidence that Central Sulawesi interior groups positioned themselves as subordinate uplanders in relational alliances with lowland powers on deliberate reasons of their own, such as evading feuding, famine and slavery (Li 2001: 43). In the examples given by Li, people opted to become "uplanders" first of all for reasons of security. In a more actual example, Tsing (1993) shows how Meratus Dayak successfully use, manipulate and elaborate the notion of "marginality" in their interactions with state agencies in order to receive government attention.

Within the last decade or so, the long established notion of uplanders as isolated cultures has become the subject of critical scrutiny and debate. As pointed out by Dove (1983: 93), the notion of upland societies, whose livelihood are often portrayed as providing "subsistence, but little else" (ibid: 93), becomes even more grotesque in regard to their economic activities. First, in contrast to the subsistence economy often ascribed to them, by trading forest products to the coast, uplanders have often been integrated into world markets long before their lowland counterparts. Exchange of forest products with coastal and foreign powers predates the arrival of modern "cash crops" such as rubber and cacao in insular and opium in mainland SE-Asia (Dove 1983, Geddes 1976). Second, as already pointed out by Pelzer in the late seventies (Pelzer 1978, cited in Dove 1983), being embedded in inter-regional commercial networks, uplanders and swidden agriculturalists contribute more to the Indonesian export economy than the intensive wet rice cultivators of the Javanese agricultural heartland.

Whereas most of the literature cited so far deals with the general aspects of upland livelihoods as they relate to upland-lowland dichotomies and interactions, this paper pursues a slightly different purpose. In focusing on a specific feature of agrarian change which is usually associated with the "hard" facts of lowland wet rice cultivation, such as "population pressure", "agrarian commercialisation" and the "green revolution", namely the emergence of *exclusionary work arrangements*, it describes many of the aspects related to processes of socio-economic change in the lowlands such as social transformation, class polarisation and diversification of the labor market as being in full operation in the uplands. In Central Sulawesi however, the emergence of exclusionary work arrangements (*bapetak*) is not rooted in technological changes in wet rice cultivation or in imperfections of the regional market, but is shaped by global market forces beyond the wet rice sector. Here, changes in employment opportunities were induced by the Central Sulawesi "cacao boom" and its concomitant waves of immigration which resulted in a large scale conversion of wet rice fields (*sawah*) into perennial stands. Within the processes of change observed, the transfer of agricultural land to outsiders and the subsequent changes in land use turned out to be the major lubricant by which local people became displaced from their land and simultaneously from access to work. In contrast to wet rice agriculture, cacao cultivation is characterised by relatively low demands in labor with all major seasonal activities being almost exclusively performed by household members. With significant parts of the surrounding wet rice fields being converted into cacao groves, local people found themselves confronted with an increasingly "involuted" labor market. This labor market is dominated by institutions which provide job security for a selected group of workers at the same time they are excluding others.

In contrast to lowland Java, where such institutions are known since at least since 1922 (Harjono 1993: 207), exclusionary work arrangements are a recent innovation in the Central Sulawesi uplands. Despite the "involuted" character of the agreements, in Central Sulawesi references to notions of "shared poverty" or "moral economy" remain questionable (see Schrauwers 2000). Supported by a locally dominant ideology of kinship, exclusionary work opportunities are confined to certain types of relationships and do not operate at the level of the community as a whole. Thus, they are rather enhancing class polarisation than reducing it.

Sintuwu: The Dynamics of a Central Sulawesi Upland Community

The village of Sintuwu is located on the Gumbasa river banks of the Palolo upland plateau, Central Sulawesi, Indonesia. Being a classical "frontier area", Palolo was gradually occupied by spontaneous as well as state directed immigration since the sixties and has meanwhile become one of the most densely populated micro-regions in Central Sulawesi. Palolo valley is located on the eastern fringes of the Lore Lindu National Park (LLNP), which was established in 1995. Sintuwu was founded in 1961 by several Kaili-speaking families.[4] The first Buginese migrants from South Sulawesi entered Sintuwu since the early seventies, after taking up wage work in a nearby logging company. The third category of early immigrants were rattan collectors from the so-called Kulawi valley, which is now forming the south western border of LLNP. (Sitours 2004).[5] Whereas the regional Kaili and Kulawi population increased steadily through natural growth, population increase was fuelled first of all by the immigration of Buginese migrants who were attracted by the Central Sulawesi cacao boom and the availability of cheap land reserves since the eighties.[6] This means that the boom-bust cycles and migratory patterns usually associated with cacao expansion predate the economic crisis which hit Indonesia in late 1997.[7]

As can be observed all over Central Sulawesi, there is a tendency of the Buginese in Sintuwu to establish themselves as a large and economically powerful group in the Diaspora. In contrast to the various "mixed cropping strategies" and the delicate "subsistence-cash crop balance" adopted by the local population, the Buginese economy is dominated by the export sector with a preference for cacao production.

---

4 The different Kaili groups are generally believed to be the indigenous population of the Palolo valley.

5 The term 'Kulawi' is used as an ethnic as well as an geographic marker in addressing both, the Kulawi upland valley as well as its long established inhabitants.

6 In mid 2001, the Bugis made up almost 1/3 f the population of Sintuwu (see Sitorus 2004: 107). Other onlv very marginally represented ethnic groups, such as one Toraja and some Sundanese families are not considered in the sample.

7 During the currency crisis 1997/1998 the US dollar tied returns to cacao increased seven fold (Ruf and Yoddang 1999: 248) leading to an unprecedented immigration of Buginese farmers to Central Sulawesi. With regard to cacao, Li's general remarks on upland cash crops that "even when prices are high, … the marketing chain reflects power imbalances which ensure that upland farmers receive only a small percentage of the gains" (Li 1999: 31), needs some rectification. Instead, the almost tax free Indonesian market system ensured that – at least during the economic crisis in 1997/98 – producer prices reached up to 85% of the export figure (Ruf 1999: 248).

Supported by high cacao prices, fertile forest soils (which may substitute for initial agrarian inputs) and the availability of cheap land, Buginese migrants became not only the most wealthy peasants in Sintuwu, but also the biggest landowners. Within our sample of 43 households, Buginese migrants owned 2.8 ha (SD = 2.2) of land on average, of which 2.3 ha had been reserved for cacao cultivation covered by an average number of 2031 (SD = 816) adult cacao trees. Local Kaili and Kulawi people instead own 1.6 ha (SD = 1.5) on average,[8] with an average of 1.1 ha (SD = 1.2) being reserved for cacao (accounting for a mean of 631 trees per household). There is a clear relationship between household size and the area of a holding that is devoted to annual-perennial mixed stands (r = 0.40, statistically significant with 0.01).

Size of land holding	Local groups (Kaili, Kulawi)	Buginese migrants	Total
Land less	7 (25.9 %)	0 (0.0 %)	7 (16.2 %)
0.5 ha and less	4 (14.8 %)	2 (12.5 %)	6 (13.9 %)
0.5 - ≤ 1 ha	3 (11.2 %)	2 (12.5 %)	6 (13.9 %)
1 - ≤ 2 ha	7 (25.9 %)	5 (31.2 %)	12 (27.8 %)
2 - ≤ 3 ha	3 (11.1 %)	3 (18.8 %)	6 (13.9 %)
3 - ≤ 4 ha	2 (7.5 %)	1 (6.2 %)	3 (6.9 %)
> 4 ha	1 (3.7 %)	3 (18.8 %)	4 (9.3 %)
Total	**27**	**16**	**43**

Table 1. *Land ownership in Sintuwu*

[8] A great deal of these land areas have recently been cleared after significant parts of local plots and secondary forests have been sold to the migrants. Because many of the newly cleared plots are located within the prohibited area of LLNP and because the researcher was too times forced to leave the village because people feared that quarrels over land may become public, we regret that we cannot present reliable data on the real extend of actual forest conversion and encroachment.

Through a combination of continuous land accumulation and surplus production in cacao cultivation, the Buginese migrants were able to establish domination over the village economy within less than 20 years. Fast land accumulation among the migrants is facilitated by the local predilection to sell land. As a consequence of the combined effects of immigration, the closing down of the forest margin as a land reserve by the establishment of LLNP and the sale of land, two separate processes have occurred. Whereas migrants could consolidate and expand their holdings, part of the locals have become land less (or almost land less) laborers, dependent on relatives and neighbors to provide them with work. Some Bugis on the other hand function as "middlemen" between small local plantation operators and the city-based cacao traders in Palu city (see Sitorus 2004: 113).

Table 2 shows the important role of purchase as the major means to obtain land in current Sintuwu. It is important to note however, that purchase is meanwhile an important means of land acquisitions among the local population as well. In March 2004, parcels purchased by locals from locals (n = 21) did already slightly outnumber the number of plots obtained by forest clearing (n = 20). There exists a systematic relationship between the year plots were acquired and how they have been obtained. Table 2 reflects a continuos decrease of land clearings as well as a decreasing role of grants in land allocation at the same time they indicate a situation in which access to land becomes regulated by the market.

Years of acquisitions (interval)	Types of Acquisition				
	First clearer	**Inheritance**	**Purchase**	**Grant**	**Total**
before 1970	11	1	0	0	12
1971 - 1975	1	0	1	0	2
1976 - 1980	1	1	4	2	8
1981 - 1985	3	0	8	1	12
1986 - 1990	2	2	9	1	14
1991 - 1995	2	0	14	0	16
1996 - 2001	6	1	14	0	21
2002 -	1	1	6	0	8
Total	**27**	**6**	**56**	**4**	**93**

Table 2. *Changes in types of land acquisitions in Sintuwu through time*

The transfer of land to migrants involved a large-scale conversion of seasonal dry fields and wet rice fields into perennial stands. For instance within the representative household sample, no Buginese farmer was involved in wet rice cultivation. Instead, rice was exclusively bought from local/regional producers at the market. In Indonesia, the logic of cacao dependence is intrinsically linked to the cacao/rice price ratio. From 1989-1995, one kg of cacao was enough to buy more than two kg of rice. In late 1997, when cacao prices rose up seven-fold, Indonesian farmers needed to sell one kg of cacao to obtain eight (!) kg of rice (Ruf and Yoddang 1999:250). Empirical evidence suggests further that – given a significantly higher return to labor – perennial farmers give up cacao only when the price ratio drops to one kg of rice to one kg of cacao (ibid). However, even then responses may be delayed due to "memories of boom times" (Li 2002: 431). According to official archives, during 1992-1998 nearly 72% of Sintuwu wet rice fields were converted into cacao groves with the wet rice area declining from 270 ha in 1992 to 75 ha in 1998 (Sitorus 2004: 109). The combined effect of immigration, land transfers, accumulation and conversion has reduced the average household availability of wet rice plots in Sintuwu to less than 0,2 ha. In an official letter issued in June 1998, the District head (*Camat*) of Palolo prohibited any conversion of wet rice fields into plantations by government regulations in order to safeguard the food security of Palolo valley.[9]

After local people became increasingly displaced from their land by piecemeal transactions, the key question is in how far the Central Sulawesi cacao boom has created new jobs or reduced job opportunities in Sintuwu. The process which transformed the migrants into the "new landed" and the locals into the "new land less" (Sitorus 2004: 111ff) implied not only changes in land distribution, but an increasing internal complexity in the distribution of labor opportunities. Ben White (White 1981: 141) made an important distinction in highlighting the difference between *resource scarcity* (the pressure of people on resources), which is a matter of demography on the one hand and differentials in the *access to resources* (the pressure of people on people) on the other, which is a matter of political economy.

---

[9] Decision Letter (*Surat Keputusan*) No. SK 525/0406/PMD/1998.

The combined effects of a migrant-dominated cacao sector and a severely diminished wet rice area indicate a scenario where significant parts of the local population have not only been displaced from their land, but have simultaneously been displaced from access to work. Given their low demand in labor, cacao enterprises are almost exclusively managed by household and family members. Further agricultural expansion being impossible,[10] local people find themselves increasingly in a situation where "one person's gain is always some other person's loss" (Harjono 1993: 110). Via almost irreversible changes in the agrarian structure, perennial trees, often perceived as custodians and ideal security providers which protect peasants from ecological and agrarian adversities and risks (Chambers 1993, Ruthenberg 1980), turned out to be the major means by which local people in Sintuwu became displaced from their land and livelihoods.[11]

The Past: Social Organisation, Land Use and Labour Recruitment

Following Kahn (1985: 82, cited in Schrauwers 2000: 103), one must distinguish between the commodification of land, products an labour. As mentioned above, the introduction of a commodified product (cocoa) involved a subsequent commodification of land, but not necessarily a commodification of labour. Instead, relations of production seem often to correspond with relations of kinship. In the Central Sulawesi uplands, the concept of the "household" as a unit of production, reproduction and consumption is analytically questionable (Schrauwers 2000: 104-110). The local term for household, *bantaya,* has its roots in the traditional longhouse organization where it referred to a compart-

---

[10] Until recently, encroachment into the National Park offered some relief for disinherited or land less people (see Sitorus 2004: 114). Much of the existing forest margin however is already located on slopes which are often less favourable for cultivation. In mid-2004, some plots had already been opened "behind" steep slopes in the forest interior making agrarian plots not only illegal, but also highly fragmented.

[11] This fact is also pointed out by Li (2002: 415) who deals with problems of displacement in Central Sulawesi in more general terms. In discussing changes in land control , Li does not address the question of work access and changes in employment opportunities.

ment hosting a married couple (*soe tambi*) or a nuclear family. The *bantaya* was also the place where the mobile property (e.g. bronze gongs, *keris*-dagger) of the couple was stored. The longhouse (*banua*) consisted of several *bantaya*, with each of them having its own "seating place" (*palongku*) in front. Surrounding the common hearth, these "seating places" formed the "inner gallery" which was used for common meals and for meetings held by the longhouse community (see figure 1). Usually, the *banua* members consisted of several *soe tambi* which belonged to the same *hantina*, that is a cognate descent group including third cousins.[12] Albeit there was never a *banua* in Sintuwu since longhouses were forbidden by the Dutch in 1908 (see Schrauwers 2000: 99), it is important to keep in mind that much aspects of the social and economic relationships to be found in Palolo today are derived from the former longhouse organization. Within the *banua*, each *soe tambi* owned its own plots. Couples worked their plots not only on their own responsibility, but also transferred them to their offspring by inheritance. As known from other Central Sulawesi settings (Schrauwers 1998), the social and economic core unit was the *soe tambi*, perceived first of all as the "couple". There is a tendency to transfer domestic responsibilities to the wider kin group as it is still reflected in the high numbers of fosterage[13] within the village. Thus, within our sample of 103 local households in four villages, 19,4 percent (n = 20) had fostered children (*anak tinggal*) of poorer relatives.[14] Forest resources surrounding a village were divided into tracts, each of them being claimed by a certain shifting cultivation group (*horobo*).

---

[12] As among most Malayo-Polynesian societies, the kinship terminology within the extended kinship group is of the Hawaiian type and stresses differences in generation rather than in lineality (Fox 1967: 256-259).

[13] For the sake of anthropological accuracy, it should be noted that in contradiction to 'adoption', 'fosterage' does not involve rights to inheritance (Schrauwers 2000: 119). Central Sulawesi villagers differentiate clearly between 'anak angkat' (adopted children) and 'anak tinggal' (fostered children). Numbers and accounts in this paper relate only to children who were reported as 'anak tinggal' during our survey.

[14] This figure is even higher than the 8 to 15 percent reported by Schrauwers in the Central Sulawesi district of Tentena (2000: 105).

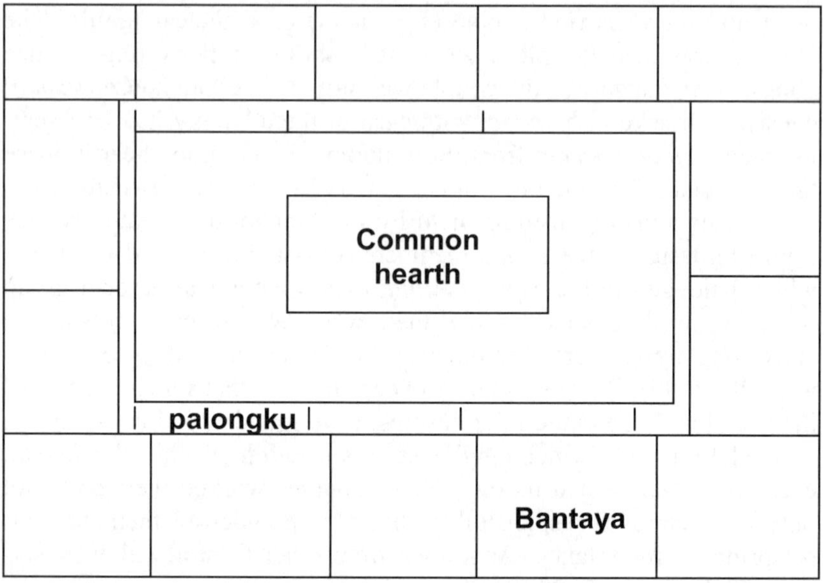

Figure 1. *The Banua longhouse community*

Whereas a rough correspondence between social (*horobo*) and physical (resource) boundaries did exist, the *horobo* were not very closed groups and the *soe tambi* could join and leave its *horobo* freely. Within the *horobo*, each couple lived from and consumed the products of its own fields. There was neither a system of communal labour, nor a pooling of harvests. Instead, the *soe tambi* cultivated its fields with the help of strict labour exchanges. The prototype of the current working groups in the villages was the *morambanga* (literally "walking together to perform a task"), which has meanwhile become extinct. These work-sharing groups worked on the basis of "exact reciprocity", with members moving from one field to the next during agricultural peak seasons. Within the working group, the labour input of each *soe tambi* was reckoned separately. The *morambanga* is meanwhile replaced by the so-called *palus*-groups which differ not only in terms of membership recruitment, but also with regard to their activities. The *palus* is not restricted to agrarian production, but includes also mutual assistance in the preparation and arrangements of life cycle rituals and religious festivities. Whereas the *palus* may also be rented by outsiders against cash; a principle of exact reciprocity however is still applied among its members.

Thus, rather than by membership in a certain *horobo* and the resulting "closeness of fields", (as in the case of *morambanga*), membership in the *palus*-groups is determined by residential neighbourhood and religious affiliation. According to Savitri (2006:59), *palus*-groups are characterised first of all by their gendered division of labour, a feature they preserved from former work groups. Based on his observations in the neighbouring district of Tentena, Schrauwers (2000: 105) points out to the fact that "unmarried individuals were unable to cultivate their own fields since they could not participate in the exchange of opposite-sex labour required for certain tasks". The finding that under past conditions access to labour was determined by marriage is confirmed by our respondents. Returning to the principle of strict exchange in agricultural labour, it follows that large land owners may not be able to reciprocate the labour provided by the *palus* and other *soe tambi*. Here, local practice offers two major strategies. A landowner may either lend out part of his plots to others (*pinjam garap*) or he may enlarge the number of household members (and his labour force) by fostering children of poorer kin. Thus, the rather blurred conception of household boundary and composition and a kinship ideology that delegates a great deal of domestic functions to the extended kin group (*hantinu*), provide local people with the flexibility to redistribute kin (and free kin labour) according to economic circumstances.[15] One must keep in mind however, that this rather flexible local social organisation and its peculiarities developed under conditions when land was freely available. In contrast to past conditions which forced people to marry in order to gain access to the external labour required to cultivate their fields – as a consequence of the processes described in the last chapter – the present situation is rather one in which married couples (*soe tambi*) may have neither access to sufficient land, nor to sufficient labour opportunities. Consequently, issues of kinship solidarity and the free use of kin labour are not at all exhausted with the phenomenon of fosterage but are played out first of all in the context of agrarian relations.

[15] It should be pointed out that local conditions reflect more or less a perfect reversion of the *Chayanov*-model of peasant economics.

"Bapetak": The Evolution of Exclusionary Work Arrangements

Irrespective of the changes in land ownership observed, an adequate analysis of agrarian change is not exhausted with a description of a dichotomy of "landed" and "landless". White and Wiradi (1989) remind us that agrarian differentiation is not so much about whether some farmers own more or become richer than others, but first of all about the changing relationship between them. Similarly, Hart points out that "a unidimensional measure like landownership is narrow and may produce a distorted picture of patterns of control over resources…. agrarian class structure cannot be understood simply in terms of distribution of resources, but must take account of relations among people" (Hart 1986: 102). These arguments are vital for the understanding of the following analysis.

Exclusionary work arrangements are well documented for the rice growing heartland of Java (Hart 1986, Harjono 1995, Naylor 1991); their appearance in the outer islands however is hardly noticed. Known as *kedokan* or *ceblokan* in Java,[16] exclusionary arrangements are based on the principle of deferred gratification: a tenant/worker plants a defined parcel of a wet rice fields (*sawah*) of a land owner under the promise of an exclusive harvesting right, receiving a defined share of the product. The worker receives no compensation for the period between planting and harvesting and harvest shares are in most cases lower than those obtained from share tenancy arrangements. In general, a system like this occurs only when access to land and outside employment opportunities are limited. The worker's right to harvest the parcel is exclusive, participation of others being strictly forbidden by the arrangement. The major feature of the arrangement in comparison to other, more open agreements is that it provides *job security* to a selected group of workers at the same time it is excluding others (Hart 1986). It is inherent in such arrangements that the more crucial is job security, the greater the effort workers make in order to renew their contact which in turn may once again enhance the value of job security (see Hart 1986: 180).

[16] In difference to the *ceblokan* system which is limited to planting and harvesting, *kedokan* usually includes other tasks such as weeding.

The local term for the arrangements, *bapetak*, stems from the term *petak* which refers to a dyked-off section of a wet rice field (*sawah*). The allotment of the harvest is done by using the *blek*, a local beaker equivalent to 10-12kg of milled rice. Usually, the worker may receive either one out of every five *blek*. The arrangement occurred in the early 90ties and is reported to have been first introduced by a landless Javanese migrant who offered his labor to local *sawah* owners before the system was embraced by the local population. At that point of time not only a significant part of *sawah*, but also the loin's share of annual dry fields (*ladang*) had already been converted into cacao plantations which – once established – could be easily managed by household labor. Thus, the tradwork sharing groups (*palus*) become increasingly ineffective as collective work actions were not anymore crucial in dry field cultivation. Owners of wet rice fields on the other hand could no longer rely on a stable work force formerly provided by the *palus*-groups as an increasing number of "free riders" withdraw their involvement in transplanting and harvesting, especially during rainfall. Given the high costs of delay, reliability of labor is of primary concern for *sawah* owners/operators,[17] especially in the case of transplanting. When rice seedlings reach the proper state of development, they must be transplanted immediately, otherwise they will be lost. Under the given circumstances, there was not sufficient labor available for exchange at the same time access to work for land-poor people was severely limited so that *bapetak* provided an appropriate means of recruiting and disciplining cheap labor. In Sintuwu, the value of job security was not only enhanced by large scale conversion of fields into cacao stands, but also by the dominance of *staged planting* in wet rice cultivation (in Sintuwu every stage in rice cultivation can be observed throughout the year) as opposed to the practice of *simultaneous planting*. Thus, cultivation tasks are spread out over longer periods and peaks in demand are considerably less sharp. This implies that at a certain point of time only a limited number of job opportunities in planting and harvesting can be provided.

[17] It should be noted that not all plots worked via *bapetak* are owned by those practising *bapetak out*. Within the sample, 119 plots (73.5%) have been owned by the contractor, 33 plots (20.3%) are plots owned by his/her parents and in 10 cases (6.2%) it was a tenant whose sharecropped-in land is planted and harvested by *bapetak*.

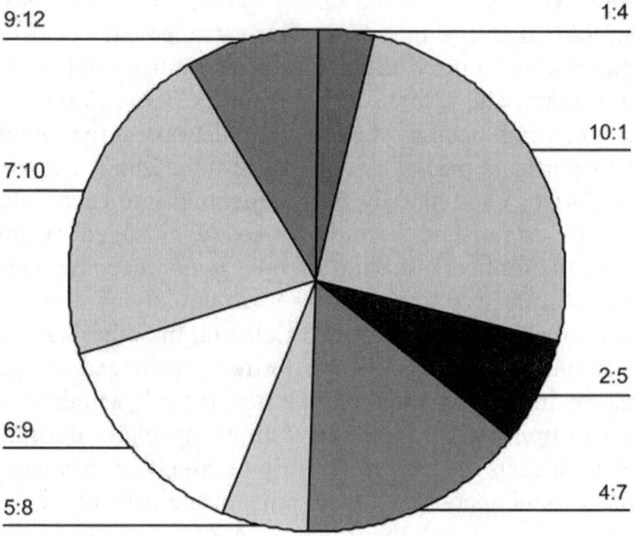

Figure 2. *Months of planting and harvesting in* ***bapetak****-arrangements (e. g. 7:10 = planting in July, harvesting in October).*

Out of the 162 arrangements in our sample,[18] 115 involve *sawah* plots cropped twice a year, whereas 47 contracts relate to plots where three crops of rice are cultivated per year. There is a strong tendency for sawah operators to limit contracts to members of their own ethnic group. Within the sample, 44.4% (n = 72) of all contracts are between Kaili operators and Kaili workers. In 37.0% (n = 60) of the contracts both parties involved are Kulawi whereas only 5.6% (n = 9) of all arrangements considered are between Kaili and Kulawi partners. Interestingly, Kaili land owners have more contracts (n = 17) with part-

[18] The unit of analysis was the *bapetak-contract*, not the individual farmers or plots involved in the arrangement.

ners of other ethnic groups (partly living in adjacent villages) than with Kulawi workers living in Sintuwu. Thus, a high degree of ethnic rivalry between the two groups is obvious. It would be misleading however to assume that only landless people or people with no access to sawah at all are entering into bapetak-contracts as workers. Out of the 71 respondents in our sample who are working the fields of others via bapetak, 11 (15.4%) are landless, 28 (39.4%) own/operate dry fields only, 29 (40.9%) own/operate dry fields plus sawah and three (4.3%) own/operate wet rice fields only.

It is important to note that 13 out of these 71 respondents who practise *bapetak-in* recruit labor for the planting and harvesting of their own *sawah* also via *bapetak-out*-arrangements with others. In eight cases there exists even reciprocity between households with members of household 'A' transplanting and harvesting the fields of household 'B' on the *bapetak*-premise and vice versa. This is of special value when planting and harvesting times differ, so that access to rice can be spread over longer periods. Of course, it is tempting to interpret *bapetak* in terms of *shared poverty* or as a *moral economy.* Sitorus, with special reference to the same village, speaks of *bapetak* as an "equalization mechanism" in order to guarantee the "fulfilment of subsistence" (Sitorus 2004: 114). The most obvious trait of *bapetak*-contracts, its tendency to appear within the confines of ethnic boundaries, indicates already that this "equalization mechanism" is not open for everybody. As Shrauwers (2000: 116) reminds us, "morality" cannot be separated from the relationships within which it is enacted. In contrast to an all-encompassing conception of *shared poverty* as a welfare principle working on an abstract level, our analysis revealed that there is not only differentiation between those with access to *bapetak*-arrangements and those without, but also differentiation between those who have stable contracts with fix partners over longer periods as opposed to those who have to rely on short-term contracts with changing employees. As will be exemplified below, this differentiation is enshrined within a locally dominant ideology of kinship. Despite the area devoted to wet rice has diminished continuously, since their appearance in the early-mid nineties, the number of *bapetak*-agreements has risen steadily. Figure 3. shows the year when individual *sawah* parcels have been worked by a *bapetak*-contract for the *first time* based on the information of the present or former owners of the plots. Because of the limited time span of many of the present contracts, the year a plot became involved in exclusionary arrangements was chosen as a measure instead of the year the present contracts were made. The most striking fact is the sharp drop in 1997/98 where almost no new contracts were made. This phenomenon is inseparably linked to the economic crisis and the sevenfold increase in cacao prices from 1997 to 1998 in Sulawesi (Ruf and Yoddang 1999: 248).

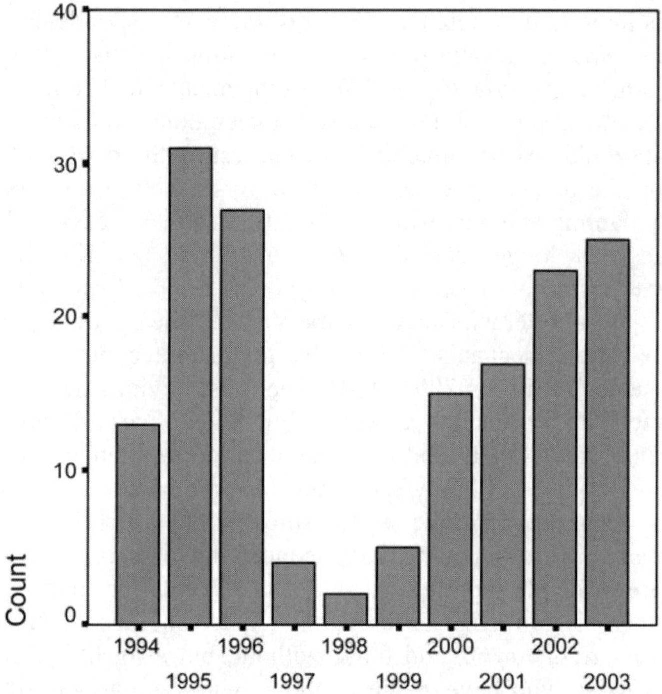

Figure 3. *First involvement of* ***sawah*** *plots into* ***bapetak****-arrangements according to year*

Given the fact that even the smallest landowners own at least some hundreds of cacao trees, the wind fall effect of cacao production led to higher incomes also for land-poor villagers, making them less dependent on *bapetak*-agreements for their living. Only after cacao prices fell again, from 1999/2000 onwards Sintuwu witnessed a steady rise in *bapetak*-arrangements again with now almost 90% of all *sawah* being planted/harvested by *bapetak*-workers. This significant decline in the number of new contracts during the crisis and the subsequent, not less significant re-emergence of new arrangements since 1999 conforms with findings in rural Java where "a notable feature of these exclusionary arrangements is their tendency to disappear and reappear, sometimes with astonishing speed" (Hart 1986: 15).

Out of 162 contracts effective in March/April 2004, in 82% (n = 133) of all cases, the partners involved belonged to the same kinship group. In 24.7% (n = 40) of the arrangements, relationships between partners have been defined as "close kin" (*keluarga dekat*), that is contracts between parents and children, between in-laws and between siblings. In these cases the partners may share the same hearth either as nuclear families (e.g. an unmarried daughter works on the fields of her parents who live in the same house) or as extended families (e.g. a daughter-in-law works the plots of her co-resident parents-in-law) or the partners may live in different houses or even different neighborhoods and hamlets. Because there is often a limited family pool left to equip only one heir, bapetak-contracts are first of all granted to younger siblings by their elder fellows. Thus, *bapetak* is first of all an arrangement between people, or between kin, not between households. That the agreements also occur between members of one and the same households shows that there are often overlapping units of production and accumulation within residential units and that functions of production, reproduction and consumption are not at all congruent with household boundaries. Despite that *bapetak* contracts are made between individuals, in practice the land owner gains implicitly access to the labor of the workers' family members. Schrauwers (2000, 2001) has pointed out to the central function of the extended kin group in upland Sulawesi in providing bride wealth because only through marriage one could assure access to the labor networks required to cultivate ones fields. Under conditions of shifting cultivation and free access to land, this conclusion is certainly true. Since land is not anymore freely available, nor are work exchange groups between households anymore vital in agrarian production, *bapetak* is not only a recruitment institution to secure quick and ready availability of labor, but also a mechanism through which the local kinship ideology of the caring extended family based on solidarity among its members (*shared poverty*) may be maintained. On first glance, statistic analysis seems to support the relevance of the locally dominant kinship ideology. The area of *sawah* operated correlates with the number of *bapetak*-out contracts (r = 0,81, statistically significant with 0,01). The preference among landowners for *bapetak*-contracts however is not as altruistic as they assert or as the quantitative data make us believe. Agrarian arrangements may well *formally* function as welfare mechanisms without being the outcome of welfare-oriented decisions and motivations. Besides providing quick and ready availability of labor, *bapetak*-arrangements contribute substantially to the reduction of production costs. As the changes in arrangement preferences during the economic crisis and its aftermath have shown, farmers are quite flexible in reducing or expanding the number of contracts according to outer circumstances.

Unfortunately for the adherents of neo-classic theory and moral economists alike, the use of free kin labor via *bapetak*-agreements – albeit enshrined in notions of kin-based charity – may well be a function of individual profit-maximisation. This can also be seen in the preference to contract other landowners at the cost of landless workers. The category of people most affected by this logic are regional immigrants from within Central Sulawesi, especially those from the poverty-striven Behoa and Bada valleys – located some 100-200 km to the south of Palolo – who lack both, capital and access to land as well as social capital in the form of kin relations and ethnic affiliation. Far from being pulled out of agriculture by new opportunities outside the agricultural sector (a model assuming that rural labor markets fulfil more or less the competitive norm), these people have definitively been pushed out of agriculture by the processes of change induced by market (cacao) penetration and large-scale conversion.

The rural institution of *bapetak* serves to maintain solidarity among the extended kin group while simultaneously it provides a profit-maximising (or at least cost reducing) tool for individual farmers. Differentiation is neither exhausted with a division into landowners and landless, nor with a division into workers with access to *bapetak*-arrangements and those without access. A more rigid analysis of the arrangements revealed that their individual conditions are far from homogenous. Individual contracts differed in terms of their duration and reliability, in terms of the size of the dyked-off *petak*-sections involved, in the number of harvests/year involved in the contact and in the shares rates included in the arrangement. The only stable element in the contracts seemed to be the principles of exclusion and deferred gratification. If we divide the relationship between the two parties involved in each contract into (1) related I / close family, (2) related II/ member of same extended (*hantina-*) kin group and (3) not related, 57.4% of the contacts (n = 93) are within the extended family, 24.7% (n = 40) within close family (*keluarga dekat*) and 17.9% (n = 29) between non-related partners. However, there is a statistically significant relationship between type of kinship/non-kinship and the area a landowner provides for his worker (s = 0.47, significant on the 0.01 level). If we compare the mean size of the each dyked-off *petak* in dependence from the type of relationship of the actors involved, there is only a minimal difference between the average area affected in contracts between non-kin (8.3 are, SD = 3.4) and between partners of the same *hantina* (9.4 are, SD = 4.3) on the one hand, but a significant difference of both in contrast to the mean area involved in contracts between close kin (19.9 are, SD = 8.5) on the other hand. Since, on average, sawah operators control only 0.49 ha of wet rice fields per household, the mean size of almost 0.2 ha involved in each contract which is granted to close kin is quite considerable. In terms of the total area affected, the area of 7.9 ha involved in only 40 contracts between close kin almost equals the area of 8.7 ha involved in 93 contracts between members of the extended kin group. There is a tendency of declining plot sizes within the contracts among all categories of workers. If we take the median of the year the contracts were made, that is the point in the distribution above and below there are an equal number of contracts, this is the year 2002, plot sizes involved in *bapetak*-arrangements dropped to 8.8 are among non-kin, to 8.2 are in contracts between extended kin and to 15.0 are in contracts between close kin. Because borders of dyked *sawah* plots (*petak*) are drawn anew every season, owners enjoy a certain degree of freedom in how much land to grant to a certain worker in order to fulfil their social obligations and secure own subsistence needs which have become more pronounced since saleable surpluses have become smaller.

Family relation	Actual number of *bapetak-in-*contracts with same partner			Seasonal extension	
	1	**2**	**3 or more**	**full**	**partly**
Not related	18	10	1	9	20
Ext. kin	40	38	2	44	49
Close rel.	7	7	24	29	11

Table 3. *Distribution of key defining security conditions of* ***bapetak****-arrangements*

On average, workers have 2,4 (SD = 2,1) contracts with 1,7 (SD = 1,3) *sawah* providers. The number of contracts per worker ranges from one to 11 plots, which are provided by 1 to maximal five *sawah* operators. Whereas some people may work a fix number of plots for the whole year (up to three harvests) and may thus count on an albeit small, but stable access to rice, others instead may have to look out for new contracts with changing providers every season. Albeit it is impossible to quantify, the fact that there are variations in the degree of contract security cannot be obscured. Looking at the number of contracts workers have with a fixed partner, it can easily be seen that a close family relationship with a *sawah* owner is the best guarantee for a stable access to *bapetak*-arrangements. In case of less close kin relations instead, as a rule owners seem to restrict their generosity to one or two plots. In the case of *bapetak*-relations between non-kin access to work arrangements becomes even more confined. This finding supports the complaints of some *sawah* owners, that nowadays too many kinsmen are requesting (*minta*) for *bapetak*-contracts. Whereas during the mid-nineties, when *bapetak* was adopted, it was first of all the landowners who searched for a *bapetak*-worker in order to secure their harvest labor, it is now usually the workers who are in search for arrangements. The same pattern applies to the "seasonal extension" of the agreements. As pointed out above, all *petak* involved in exclusionary arrangements can be harvested two or three times a year. Because one contract allowing access to one plot for three cropping periods may equal three limited arrangements that are valid for only one season, it seemed interesting to know in how far the contracts apply to all or only a part of the cropping periods. Once again, the findings show a clear privilege of close kin at the cost of competing workers. Besides the larger areas per *petak* involved, the higher number of contracts granted, and the longer seasonal extension, share rates are also higher when close kin is contracted. In minimal contracts, a worker transplants a *petak* and receives a share of one out of every five *blek* after harvest. In case of close kin however (and to a lesser extend in the case of extended kin), workers often receive two *blek* out of every five *blek*. Whereas among non-kin two *blek* are usually given out if additional tasks such as weeding are included, among kin the higher share rate is best described as a kind of bonus. There are no specified rules in regard to the tasks that are added or in regard to the manner the favour is returned. Instead, it is a way of sharing of one's own limited assets among the circle of one's closer kin. As indicated by table 4., the willingness of *sawah* operators to let their plots work by others who do not operate *sawah* themselves (landless, dry land farmers) is higher among close kin than in contracts with other partners. In most cases three contracts have been given to close, non-*sawah*-operating kin. On the other hand, having more than two contracts with a non-wet-rice-farmer of one's extended kin group is very seldom and does not at all occur if a non-kin worker is contracted. This indicates that one may speak of *bapetak* as a welfare mechanism providing one's fellows with a minimal access to rice only when close kin is affected. One effect of the large-scale conversion and the subsequent reduction of wet rice reserves is that inheritance is significantly delayed. Young couples (*soe tambi*) now often work the plots of their parents whom they have to sustain as well. Because there is often a severely limited family pool left, sufficient to equip often one heir only, *bapetak*-contracts are first of all granted to younger siblings by their elder fellows.

As pointed out above, the major reason for the emergence and fast acceptance of *bapetak* as the major labor recruitment institution was the crisis of the reciprocal work-sharing groups (*palus*) and the ensuing unpredictability of reliable labor. Given the dominance of perennials and the high returns to labor in cacao, there is little prospect for the *palus* to re-emerge in the dry land sector, or to be re-installed as a work organizing institution on the community level. All the more it is worth to note that meanwhile a new kind of *palus* (reciprocal work sharing group) had been introduced among the *bapetak*-workers. Far from equalising *bapetak*-workers however, this institution may well work in the direction of further differentiating the villagers. First, given the reciprocal nature of its organization, it is usually those who have more than one contract who join the groups. Those having most contracts are usually those who are in possession of the most important social capital in obtaining job opportunities: close family relations with *sawah* owners or *sawah* operators. Second, given the unpredictability of timing due to weather dependence and the high costs of delay in agricultural production, for *sawah* operators quick and ready availability of labor is of primary concern. In recruiting a *palus bapetak*-member, the *sawah* owner/operator gains access to the labor of additional group members which accelerates the process of planting and harvesting. Because of their reciprocal duty with a certain worker, *palus bapetak*-members show up automatically and do not have to be called or invited by the landowner. As Bardhan (1979) points out "the employer is usually keen on entering into some explicit or implicit contracts with workers about a dependable supply of labor at the right time or, at least, is aware of the significant hiring or recruitment costs to be incurred as and when such recruitment needs arise" (ibid: 488, cited in Hart 1986: 171). Thus, when in search for a new *bapetak*-worker, a land owner may always prefer someone who is already member of a working group instead of someone who is not, preferring those who are already in a privileged position anyway.

Given these findings, the processes of change in current Sintuwu have little to do with the stereotype of the timeless, socio-culturally homogenous upland community. Instead, they reflect some of the most pronounced features usually associated with the changing livelihoods in the lowlands. These features include an increasing polarisation in land ownership, the commodification of agrarian products, the prominent role of the market in access to land, and more rigid mechanisms in the allocation of work opportunities. The difference is thus less in terms of the patterns of agrarian change, but in their respective causes. Whereas in the lowlands, it is first of all the introduction of hybrid rice varieties, their dependence on external inputs (fertiliser, pesticides) and the concomitant technological changes (e.g. the substitution of the hand knife by the sickle in harvesting) as well as access to credits which work as the lubricants of polarisation (Collier 1979, Hansen 1981, Harjono 1995, Hart 1986, Naylor 1995, Stoler 1977), in

Sintuwu a similar reconfiguration of the agrarian structure was not caused by changes in the rice sector, but by the combined effects of immigration and the kind of changes of land use "usually associated with every cacao boom throughout history" (Ruf and Yoddang 1999: 248).

Family/kin relation	**Number of contracts with same partner**	**Agricultural farming type of *bapetak*-in**			total
		Landless	Dry land farmer/ no sawah	Sawah owner/ operator	
Non-kin	1	9	7	2	18
	2	2	4	5	11
	3	0	0	0	0
	total	**11**	**11**	**7**	**29**
Ext. kin (*hantina*)	1	11	12	17	40
	2	1	23	25	49
	3	0	3	1	4
	total	**12**	**38**	**43**	**93**
Close kin (*keluarga dekat*)	1	3	4	0	7
	2	0	1	8	9
	3	6	9	9	24
	total	**9**	**14**	**17**	**40**

Table 4. *Number of contracts according to family/kin relation and farming type*

Summary and Conclusion

The Sintuwu case challenges not only some established stereotypes in regard to upland-lowland dichotomies, but also some well known assumptions about peasant societies and socio-economic change in general. There are two major approaches in explaining peasant economic behaviour, that is (1) the neo-classical theory of farm production, (2) the so-called "risk aversion-approach" and (3) Neo-Marxian political economy. Neo-classical theory perceives of the peasant as an individual decision maker exclusively concerned with questions of individual profit maximisation. The major short-coming of this conception however is that profit maximisation is perceived as the one and only variable in peasant decision making. It overlooks that in peasant societies, land functions not only as a production factor, but is more often than not an important subsistence insurance. Besides, peasants often prefer non-market, reciprocal transactions which are not measured by market prices (Burkard 1999: 37). Such reciprocal arrangements often ensure survival for all community members, irrespective of their individual economic performance (Ellis 1988: 11). For an influential part of researchers, the principle of reciprocity is even the key distinctive feature of peasant societies in general. According to them, a high level of uncertainty in peasant livelihoods (natural hazards, market fluctuations, social uncertainty, tenure insecurity) implicates that peasant decisions aim first of all at survival, not at profit-maximisation. Thus, so the argument, under conditions of climatic instabilities and fluctuations in yields and prices, "survival in bad years" is always given priority over "maximum profit in average years" (Eijkemans 1995). Slogans like "Shared Poverty" (Geertz 1963), "Moral Economy" (Scott 1976) and "The Economy of Affection" (Hyden 1980) are some of the more prominent variations of this theme.

Although both, Neo-classical and Neo-Marxian political economy claim to analyse the role of market economy / capitalism in peasant development, their starting point is quite different. Whereas the former starts from the individual economic unit, the starting point of the latter is society as a whole. Of special relevance for our analysis is the Neo-Marxian concern with relations of production; especially the question of the persistence of peasant forms of agrarian relations within a dominant capitalist mode of production. The classic Marxist reasoning is that capitalist pressures will inevitably result in disintegration and social differentiation, dividing rural societies into capitalist farmers/landowners on the hand and an underprivileged class of wage workers on the other. Others instead believe that their reciprocal traditions, their flexibility in regard with cropping patterns and labor use as well as their tendency to self exploitation (e.g. an unlimited increase of labor in the production process) may help peasants to resist the pressures of capitalist relations of production and to sustain a distinctive peasant form of livelihood (Ellis 1988: 52).

The socio-economic processes of change underway in current Sintuwu have been labelled as the "cacao revolution" (*revolusi cokelat*, Sitorus 2004). Albeit a successive process and less abrupt than the term might suggest, the fundamental changes of the agrarian structure it implies may justify to describe the cacao boom in terms of a *revolution*. The term *involution* on the other hand, as popularised by Geertz (1963: 98), describes a process by which a high degree of social and economic homogeneity is maintained "by dividing the economic pie into a steadily increasing number of minute pieces" (ibid: 97). Given an average ownership of 0.2 ha *sawah* and 0.4 ha dry fields among *bapetak*-workers as against 0.5 ha *sawah* and 0.6 ha dry land among *sawah*-providers, it is tempting to interpret the innovation of *bapetak* in terms of an "equalisation mechanism" (Sitorus 2004: 114) informed by an ideology of "shared poverty". The notion of "shared poverty" however remains analytically problematic. First, in contrast to the Geerzian theory, the labor institution of *bapetak* is not a relict of a past, risk and benefit sharing "folk society", but is itself a product of a relatively recent market penetration and its concomitant processes of agrarian change.[19] Second, as pointed out above, any morality of sharing cannot be understood in isolation from the social relationships into which it is embedded. The previous discussion has made explicitly clear that far from spreading work equally within the community, *bapetak* instead is enshrined within a locally dominant ideology of kinship. Whereas this ideology is derived from a kinship system which pronounces solidarity within the extended kin group (*hantina*), in practice solidarity in form of providing access to work and rice is most significantly pronounced between close kin (*keluarga dekat*). The key question related to the emergence of exclusionary arrangements in the lowlands, if labor has been pushed or pulled out of agriculture (Hart 1986: 8), is easy to answer in the case of Sintuwu. Given an average population density of 79 persons per square km in Palolo sub-district (Sitorus 2004: 107), it is obvious that the problem faced

[19] In another Central Sulawesi setting, under quite different circumstances (no immigration) , a different pattern of land use (dominance of wet rice cultivation, no conversion), in focusing on different issues (development cycles of households, traditional exchange and fosterage), Schrauwers (2000: 103) comes to a similar conclusion.

by Sintuwu villagers is less a problem of "pressure of people on resources" as such, but a problem of "pressure of people on people" caused by differential access to those resources (see White 1981: 141). This means that a part of local rural society is pushed out of wet rice cultivation into other activities (forest clearing, rattan collection) with even lower returns to labor. In creating job security for a part of the local community at the same time it is creating insecurity for others, *bapetak* represents a rather exclusive form of "shared poverty".

It is within this web of kin obligations and responsibilities local farmers make their decisions. The mix of considerations – at the one hand one wants to minimise production costs by the use of free kin labor (profit-maximising farmer), on the other hand the number of *bapetak*-workers as well as the size of plots worked by them should not be unlimited in order to secure one's own subsistence needs (risk averse farmer) at the same time one has to share the limited assets one has with one's kin (moral economic farmer) – makes the abstract ideal types of neo-classical and moral economic theories of peasant decision become a farce. In regard to the (often) reductionist assumptions of Neo-Marxian political economy, it has to be pointed out that the effects of capitalist penetration are anything else than uniform. The most strongest capitalist pressure one can imagine, the Central Sulawesi cacao boom, had a totally diachronic effect on the commodification of the local village economy. Whereas agrarian products have been partly fully commodisised (cacao), and are partly produced for subsistence and sale (rice), the land market on the other hand has been commodified completely. In no case however, neither in the export crop sector (cacao production), nor in the domestic sector (rice production), have relations of production been commodified. Either based on household labor (cacao) or based on deliberate arrangements of access and exclusion (rice), in both agricultural spheres have relations of production been organized personally. Wage work instead is confined to renting the *palus*-groups for field preparation, land clearing or in paying individuals for ploughing the *sawah* according to local wages. This persistence of the peasant mode of production is not motivated by a concern for counterbalancing the impacts or the pressure of capitalism, but by a need to enhance social control, to discipline the labor force and to secure a ready availability of labor in transplanting and harvesting.

This paper is part of a reader on the stability of forest margins. The question is therefore how the processes of socio-economic change described above relate to the stability of the forest margin in Sintuwu. As I have pointed out elsewhere (Burkard 2002: 24ff), *sawah* cultivation can reduce the pressure on the forest cover. Given the fact that in comparison to rain fed agriculture, wetland rice production is characterized by a higher absorption of work force and lower returns to labor, it provides less time and surplus that could otherwise be invested in new land clearings. Given its extremely high returns to labor, cacao cultivation on the other hand is positively related to deforestation (see Burkard, this volume). These findings stem from aggregate data of several villages. Because the Sintuwu sample has the highest number of missing values and the most serious validity threats with regard to any quantitative data related to forest clearing, we have to combine the available overall qualitative information in order to obtain a more coherent picture of agrarian change and deforestation in Sintuwu. Thus, the large scale conversion of wet rice fields into cacao groves had for sure a twofold destabilizing effect on the forest margin: The first is a *capital-driven* form of encroachment by which rich Buginese cacao farmers pay local people for converting forest into agricultural land later to be planted with cacao (this practice is a well-known topic of daily village conversation and could be proved by several transects into the forest). Given relatively high export prices and a serious lack of enforcement and monitoring, successful cacao farmers are in a comfortable position to expand their holdings. The second form of encroachment is *poverty-driven* and is directly related to the emergence of exclusionary work arrangements. As described above in much detail, the question of "who obtains access to work and who not" is informed by a regionally dominant kinship ideology. Within this ideology, the former responsibility of the extended kin group (*hantina*) to provide its members with agricultural land, housing and access to labor exchange groups has been replaced by the responsibility to provide kin with the access to work on one's *sawah*. Thus, people (often 'regional migrants' from Central Sulawesi) neither sufficient access to land nor to labor have become severely marginalised by this process. Far from enjoying the benefits of the new world crop, encroachment by this kind of people is driven by despair. Besides opening agrarian plots for others, most of them are engaged in rattan collection by tied labor with town-based rattan bosses. People become thus pushed out of agriculture into activities with even lower returns to labor. Given the very limited role of formal and informal off-farm employment in upland Sulawesi, to be pushed out of agriculture means often to be pushed into the forest. However, there is a rather ironic aspect at work in the overall scenario. Albeit reliable quantitative data on encroachment are missing, the forest margin in Sintuwu seems still less disturbed as in several other communities around the National Park hosting high rates of migrants, such as Watumaeta (see Burkard, this volume). In most communities around the park there is a rule supported by customary law that forbids the selling of wet rice fields to outsiders. Thus, it was first of all secondary forests of different ages that was alienated to migrants resulting in rapid deforestation of the margin areas.

In Sintuwu, a significant part of the cacao plots are converted wet rice fields located inside the village territory. Thus in the first place, the non-existence of an indigenous security mechanism, the ban to sell *sawah*, gave a certain degree of relieve to the forest margin when wet rice fields did absorb part of the cacao plots which would have been otherwise established in the forest. Thus, cacao-induced deforestation was if not forestalled, so at least delayed. Recent anthropological and sociological research in rain forest margin areas has primarily focused on "community based resource management". With special reference to the "cacao boom" in Central Sulawesi, Li (2001: 88) has questioned this approach in pointing out that it is not "managerial interventions", but "agrarian struggles" and rapid class polarisation which counts for the ecological and livelihood crisis in this region. This is certainly true, but without more effective managerial interventions in the future, the forest margin of the Lore Lindu area cannot be maintained. An adequate regional analysis instead must consider *both* concerns simultaneously by (1) describing the mechanisms which shape agrarian processes at the local level and by (2) analysing the prominent role of local organizations and institutions in the processes of change currently underway. These processes are largely to be explained by issues of "socio-economic security" which lie at the interface of natural resource management on the one hand and class formation and agrarian struggles on the other.

References

Allen, B. (199e): The Problems of Upland Management. In: Brookfield, H. and Y. Byron (Eds): South-East Asia's Environmental Future: 225-237. Kuala Lumpur.

Barth, F. (1969): Ethnic Groups and Boundaries. The Social Organization of Culture Difference. London.

Burkard, G. (1999): Bauern, Kader und Migranten. Ländliche Modernisierungsprozesse am Beispiel einer Dorfgemeinschaft in der Sonderregion Yogyakarta. Pfaffenweiler 1999.

Burkard, G. (2002): Stability or Sustainability: Dimensions of Socio-Economic Security in a Rainforest Margin. Storma Discussion Paper No. 7. Bogor.

Chambers, R. (1997): Whose Reality Counts? Putting the First Last. London.

Collier, W. L. (1978): Rural Development and the Decline in Traditional Welfare Institutions in Java. Honolulu.

Dove, M. (1983): Theories of Swidden Agriculture and the Political Economy of Ignorance. *Agroforestry Systems* 1: 85-99. The Hague

Ellis, F. (1988): Peasant Economics. Farm Households and Agrarian Development. Cambridge.

Fremerey, M. (2005): Local Communities as Learning Organizations: The Case of the Village of Toro, Sulawesi. In: A. Neef (ed): Participatory Approaches for Sustainable Land Use in Southeast Asia: 253-2 Bangkok.

Fox, R. (1967): Kinship and Marriage. Victoria.

Geddes, W. R. (1976): Migrants of the Mountains. The Cultural Ecology of the Blue Miao (H'mong Njua) of Thailand. London.

Geertz, C. (1963): Agricultural Involution. The Process of Ecological Change in Indonesia. Berkeley.

Hansen, G. E. (1981): Agricultural and Rural Development in Indonesia. Colorado.

Harjono, J. (1990): Land, Labor and Livelihood in a West Java Village. Yogyakarta

Hart, G. (1986): Power, Labor and Livelihood. Processes of Change in Rural Java. Berkeley

Hart, G., Turton. A. and B. White (1989): Agarian Transformations: Local Processes and the State in Southeast Asia. Berkeley.

Hyden, G. (1980): Beyond Ujamaa in Tanzania: Underdevelopment and an Uncaptured Peasantry. London.

Kahn, J.(1985): Indonesia after the Demise of Involution: Critique of a Debate. *Critique of Anthropology,* 5 (1): 69-96.

Kahn, J. (1999): Culturising the Indonesian Uplands. In: Li, T. M. (ed.): Transforming the Indonesian Uplands: 79-104. Amsterdam.

Lehman, F. K. (1967): Ethnic Categories in Burma and the Theory of Social Systems. In: Kunstadter, P. (ed.): Southeast Asian Tribes, Minorities and Nations: 93-124. Princeton.

Li, T. M. (2001): Relational Histories and the Production of Difference on Sulawesi's Upland Frontier. *The Journal of Asian Studies*, No. 1: 41-66.

Li, T. M. (1999): Marginality, Power and Production: Analysing Upland Transformations. In: Li, T. M. (Ed.): Transforming the Indonesian Uplands: 1-44. Amsterdam.

Li, T. M. (2002): Local Histories, Global Markets: Cacao and Class in Upland Sulawesi. *Development and Change,* 33 (3): 415-437.

Naylor, R. (1991): The Rural Labor Market in Indonesia. In: Pearson, R. et al. (Eds): Rice Policies in Indonesia: 38-59. Berkeley.

Ruf, F. and Yoddang (1999): The Sulawesi Cacao Boom and its Crisis. *Plantations, Recherche, Development*, July-August: 248-53.

Ruthenberg, H. (1980): Farming Systems in the Tropics. Oxford

Savitri, L. A. (2004): Agrarian Change and Gendered Local Knowledge. Unpublished Master Thesis. Institut Pertanian Bogor (IPB). Bogor.

Schrauwers, F. (1998): 'Let's Party': State Intervention, Discursive Traditionalism and the Labor Process of Highland Rice Cultivators in

Central Sulawesi, Indonesia. *The Journal of Peasant Studies*, Vol. 25, No. 3, April 1998: 112-130.

Shrauwers, F. (1999): "Its Not Economical": The Market Roots of a Moral Economy in Highland Sulawesi. In: Li, M. T. (Ed.): Transforming the Indonesian Uplands. Amsterdam.

Schrauwers, F. (2000): Colonial 'Reformation' in the Highlands of Central Sulawesi, Indonesia, 1892-1995. London.

Scott, J. (1976): The Moral Economy of the Peasant. New Haven.

Sitorus, F. (2004): 'Revolusi Cokelat': Social Formation, Agrarian Structure and Forest Margins in Upland Sulawesi. In: Fremerey, M., Gerold, G. and E. Guharja (eds): Land Use, Nature Conservation and the Stability of Rainforest Margins in Southeast Asia: 105-118. Berlin.

Stoler, a. (1977): Rice Harvesting in Kali Loro: A Study of Class and Labor Relations in Rural Java. *American Ethnologist*, 4: 678-698.

Tsing, A. (1993): In the Realm of the Diamond Queen. Marginality in an Out-Of-The-Way-Place. Princeton

White, B. (1981): Population, Involution, and Employment in Rural Java. In: Hansen, G. E. (Ed.): Agricultural and Rural Development in Indonesia: 130-145. Colorado.

White, B. and G. Wiradi (1989): Agrarian and Nonagrarian Bases of Inequality in Nine Javanese Villages. In: Hart, G., Turton. A. and B. White (Eds): Agarian Transformations: Local Processes and the State in Southeast Asia: 266-302. Berkeley.

Chapter X

UNVEILING REGIONAL PREFERENCES FOR BIOLOGICAL DIVERSITY IN CENTRAL SULAWESI: A CHOICE EXPERIMENT APPROACH

Klaus Glenk, Jan Barkmann and Rainer Marggraf

Abstract

Knowledge on preferences for ecosystem services at the rainforest margin can facilitate the development of economically sound conservation strategies as mandated by the Convention on Biological Diversity. We applied an ecosystem service approach for the development of a choice experiment instrument that was employed in Central Sulawesi, Indonesia, around Lore Lindu National Park. Preferences were elicited for changes of the availability of rattan, water supply for irrigation, the population size of the dwarf buffalo anoa and types of cacao cultivation differing along a shade tree gradient. While, on average, a willingness to contribute to the environmental maintenance of the resource base was found for the first three attributes, respondents had preferences for more intensive ways of cocoa cultivation. Apart from estimating gross effects of changes of the attributes, we focus on in-depth insights into choice behaviour with regard to socio-economic and socio-demographic differences within the sample of respondents.

Introduction

An important issue in the biodiversity debate is the loss of species, for which land use change is an important driving force (e.g. Pearce and Moran 1994). Land use change alters ecosystems and thus influences the provision of ecosystem goods and services. The Central Sulawesi rainforests are part of the global Wallacea biodiversity hotspot (Myers et al. 2000), and Sulawesi's moist forests are among the world's most biologically valuable ecoregions (Olson and Dinerstein 1998). Lore Lindu National Park (LLNP) in Central Sulawesi is one of few large forest areas left on the island of Sulawesi (Waltert, Mardiastuti, and Mühlenberg 2004). Therefore, ensuring its integrity is an important contribution to global biodiversity conservation efforts as demanded by the Convention on Biological Diversity (CBD). The CBD recognizes that any conservation effort needs to consider the livelihood of people that depend on the use of natural resources potentially affected by conservation measures. This commandment has high importance for conservation efforts in the LLNP area because Central Sulawesi is one of the poorest provinces in Indonesia (Suryahadi and Sumarto 2001). Thus, residents are situated between (global) conservation objectives and urgent (local) development goals.

By eliciting preferences for biodiversity held by residents around LLNP, this study aims at (i) improving the knowledge base of decision makers for the development of economically informed conservation strategies, and at (ii) providing information regarding the economic behaviour of agents – mainly smallholder farmers impacted by potential conservation measures. Methodologically, the purpose of this study is to improve the *choice experiment* approach for an estimation of locally perceived values of biodiversity. We make explicit use of an ecosystem service approach to assess functional biodiversity benefits. In particular, combining the analysis of preferences for ecosystem services with socio-economic and socio-demographic characteristics of individuals provides further insights into the choice behaviour of the respondents.

The Research Area

The research region (figure 1) is located in the humid tropics about 1 degree south of the equator. It comprises seven administrative districts (*kecamatan*) in the province of Central Sulawesi, Indonesia. The districts of the research region mainly follow distinct valleys and their bordering mountain ranges. In more than 115 villages, the research area holds a population of about 130,000 – mainly smallholder farmers – on 7,220 km^2. LLNP is centred within the study region and covers some 2,200 km^2 of mountainous rainforest. A large number of species endemic to Sulawesi, including the mammals anoa (*Bubalus sp.*) and babirussa (*Babyrousa babirussa*) as well as many endemic bird species, can be found in the National Park area (Waltert, Mardiastuti, and Mühlenberg 2004). The geophysical conditions of the research region vary to a large extent. The altitude ranges from just above sea level up to 2,500 meters, and rainfall varies from 500 mm to 2,500 mm per year (Maertens 2004: 30). In combination with other heterogeneous physical features like relief and soil conditions, the prerequisites for agricultural activity are diverse among the 7 districts. Demographics and land use are characterized by strong dynamics, providing an interesting background in the potential area of conflict between development and conservation goals. In the period from 1980-2001, the population increased by 60% (2.4% p.a.). The spatial distribution of the growth is unequal. In one of the districts (Palolo), population size doubled, and in Lore Utara it almost tripled (ibid: 39). A large share of population growth results from migration – mainly from other parts of Sulawesi.[1] There are also migrants from other islands of the Indonesian archipelago. A main driving force for this migration processes was the availability of land for agricultural use (ibid: 39). The complex demographic history led to high ethnical diversity (Faust et al. 2003).

[1] Many 'cocoa migrants' were attracted by the availability of land, particularly from South Sulawesi (Faust et al. 2003, Schippers et al. 2007).

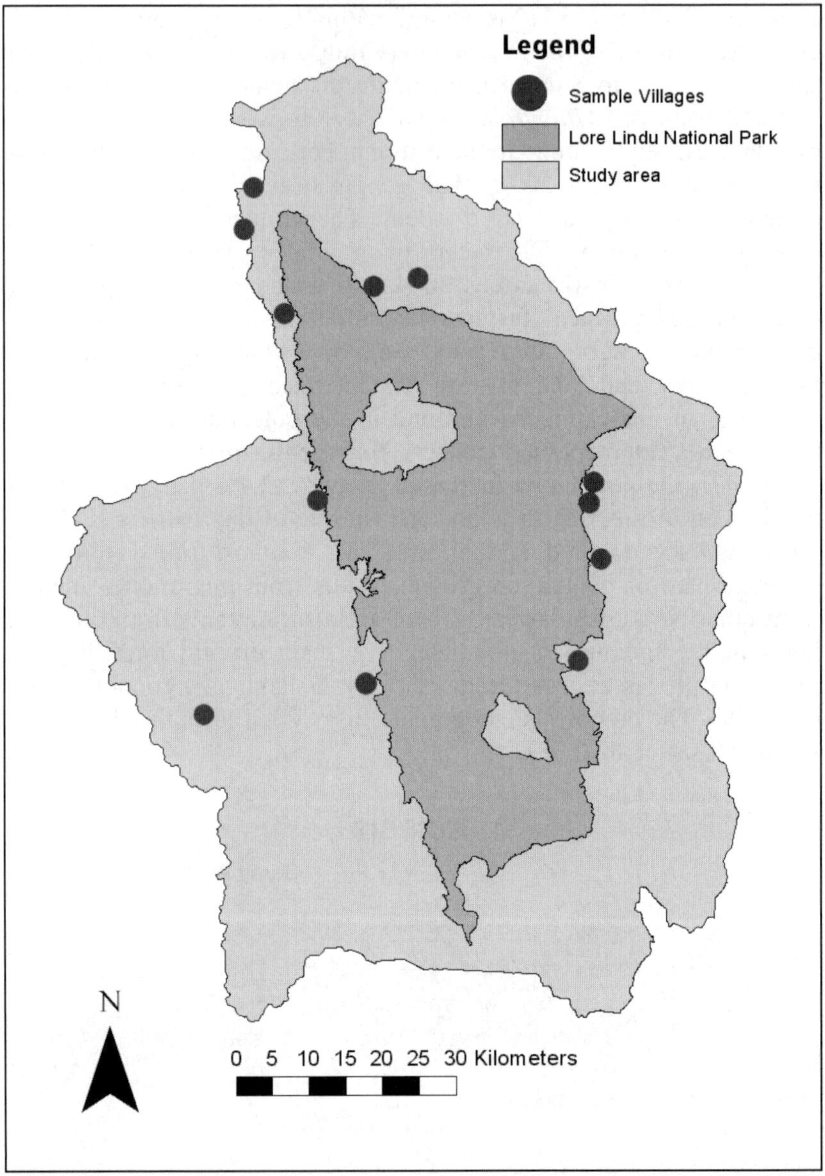

Figure 1. *Research region*

Furthermore, a large variation of land use patterns can be found (Schwarze 2004: 47ff). In the course of the 'cocoa boom' in Indonesia (Akiyama and Nishio 1996), cocoa became the dominant 'cash' crop in the research region being often cultivated in the uplands. Wetland rice remained the dominant 'food' crop being cultivated in the lowlands, resulting in a lowland-upland dichotomy throughout the research region. Based on data from household surveys, cocoa and wetland rice together account for 57% of the net crop income (Schwarze 2004: 47). Factors that facilitated the increasing cultivation of cocoa included, among others, the availability of suitable land, low production cost and the entrepreneurship of smallholders particularly of southern Sulawesi Bugis origin (Akiyama and Nishio 1996; Schippers et al. 2007). Thus, the agricultural areas in the Lore Lindu region itself increased to a large extent over the past two decades (Maertens 2004: 36). The related land use change (LUC) is closely related to the increasing production of (cash) crops (e.g. cocoa) (Steffan-Dewenter et al. 2007). LUC can be divided into conversion of (primary or secondary) forest into arable land, and conversion within land previously used for agriculture (e.g. wet rice fields to cocoa plantations, coffee to cocoa). Concerning forest products, collection of fuel wood is widespread for private consumption, while rattan is the most important marketed forest product (Schwarze 2004: 51). A broad array of factors can be identified as driving forces for LUC, ranging from market forces over changes in the natural environment (e.g. lack of water for irrigation) to social processes within local communities related to in-migration (e.g. Burkard 2006, Schippers et al. 2007).

Valuation of Functional Benefits of Biological Diversity – the Ecosystem Service Approach

The CBD regards ecosystems as functional units made up of a dynamic complex of plant, animal and micro-organism communities and their non-living environment. The ecosystem concept is an appropriate approach to compartmentalise nature into units that can be examined by researchers of both ecological and economic disciplines. Biodiversity is defined as the diversity within and between living organisms as well as between ecosystems. It provides flows of products and benefits to humans. Thus, biological diversity can be regarded as a natural asset. There are manifold links between biodiversity and ecosystems. "Diversity is a structural feature of ecosystems, and the variability among ecosystems is an element of biodiversity" (Alcamo et al. 2003: 9). Thus, changes in biodiversity and the related changes of ecosystems can affect the generation of ecosystem functions beneficial to humans.

In order to show the way these benefits influence human activity, environmental economics makes use of the *Total Economic Value* (TEV, Randall and Stoll 1983, Pearce and Moran 1994). On a first level, use and non-use values are distinguished. *Non-use values* are are most difficult to assess, and have been subject to extensive discussion (e.g. Kahneman and Knetsch 1992; Sagoff 1996), particularly concerning the so-called "existence value"[2] (Krutilla 1967). *Use values* are further divided in direct (active) and indirect (passive) use values. It is mainly the indirect/passive use values that capture functional benefits of ecosystems, such as flood control or the benefits from nutrient cycling. The TEV is based on an *economic* concept of value, and was introduced for improving *economic* valuation. The TEV does not pretend to be the only possible way of conceptualizing the values generated by nature.

The TEV does not explain which components or 'effects' of ecosystems beneficial to humankind belong to which value category. Particular difficulties arise when functional attributes of ecosystems are concerned. In particular, it is difficult to assign an economic value directly to ecosystem functions, such as nutrient cycling. 'Ecosystem function' itself is an analytic scientific term that has no direct economic meaning: neither do ecosystem functions directly enter economic production functions, nor are consumers sufficiently aware of their economic implications. Therefore, a concept that mediates between ecological processes and their impact on humans is required. What may be called ecosystem service approach is a promising com-

---

2 "Existence value" can be defined as the value arising merely from the information that a resource exists regardless of whether this asset now or in the future generates a productive, recreational or aesthetic use for the person holding the preferences (e.g. Bingham et al. 1995, Fromm 2000). In this sense, there is a notion that the resource or asset has a value of its own, which is of anthropocentric intrinsic nature.

plement to the TEV to fill this conceptual gap (Barkmann et al. *acc.*). As de Groot, Wilson, and Boumans (2002: 395) put it: "...observed ecosystem functions are re-conceptualized as 'ecosystem goods and services'[3] when human values are implied. The primary insight here is that the concept of ecosystem goods and services is inherently anthropocentric: it is the presence of human beings as valuing agents that enables the translation of basic ecological structures and processes into value-laden entities". The 'translation' process clarifies that the *effects* of ecosystem structures, functions and processes on human activity are the objects of valuation, not the ecosystem structure, function or process *themselves*. For the economic valuation of ecosystem services, a large variety of valuation techniques have been developed during the past decades. They are commonly divided into *revealed* and *stated preference* methods, the first using actual market data, the second being a survey-based technique and assessing values from hypothetical markets (Adamowicz 1998a). As the effects that are investigated by revealed preference (RP) methods are already captured in existing markets, an ecosystem service approach might be particularly beneficial for stated preference (SP) methods.

A prerequisite to perceive an ecosystem service benefit in the first place is the awareness of the specific interrelation between humans and nature. Awareness in turn is initiated by recognition, which needs prior identification. For many ecosystem services, such specific interrelations between humans and nature are well known through experience and tradition, especially those concerning the production of (material) goods. However, although „the science of ecology has matured, mankind's knowledge about the interconnectedness of ecosystem processes and structures has grown" leading to "...increasing concern about the effects of human actions on ecosystems" (Bingham et al. 1995: 76), much more remains unknown both for ecological experts and non-experts. Additionally, ecological non-experts face the problem to make sense of the often abstract insights of the ecological sci-

[3] There is no need to distinguish between ecosystem goods and services if the provision of goods by an ecosystem is classified as an ecosystem service itself. The distinction between production and consumption values (Marggraf and Birner 1998) as well as direct and indirect values within the TEV concept suggests that it may be appropriate to separate goods (production value, often direct use) and services (consumption value, often indirect use) from a pragmatic point of view.

ences with regard to their personal values and livelihoods. The remaining lack of scientific data on ecological relationships and functioning due to the complexity of nature, which can be termed "ignorance" (Fromm 2000), and, similarly, the lack of people's awareness of existing knowledge aggravates environmental and ecosystem valuation, because it limits the *array* of services that can be a proper object of a valuation exercise. Furthermore, these problems limit the *ability* of survey respondents to express their preferences for services of ecosystems already identified and recognized scientifically. Consequently, it is doubtful, whether ecological non-experts are able to express meaningful preferences regarding decisions on complex issues of ecosystem functioning (Nunes and v.d. Bergh 2001). The major conceptual improvement of the ecosystem service approach is to 'translate' complex aspects of ecological functions into concepts that non-experts recognize as important for their lives. In this form, they are potentially valuable by people. By using the ecosystem service approach in this case study for an assessment of functional benefits of ecosystems, its applicability and limitations can be evaluated. A more detailed account of the ecosystem service approach can be found in Barkmann et al. (*acc.*).

Choice Experiments – Theory and Models

In a CE, respondents state their preferences by (repeated) choices from different alternatives or 'goods'. Founded in Lancastarian consumer theory (Lancaster 1966; 1991), the goods are being transformed into objective characteristics (attributes) from which the consumer is assumed to derive utility. In environmental choice modelling, the alternatives are often described as different development or policy options (Bennett and Adamowicz 2001) that are associated with changes in the quality of the good in question. Another main pillar of choice modelling is random utility theory (RUT) (e.g, Thurstone 1927; McFadden 1973; Manski 1977).[4] Utility is partitioned into a determi-

[4] Further theoretical input was obtained from research of informational processing in judgement and decision-making in psychology (e.g. Luce 1959, Slovic and Liechtenstein 1971).

nistic, systematic component or 'representative utility' and a random part of utility "reflecting [the] unobserved individual idiosyncrasies of taste" (Louviere et al. 2001: 38):

$$U_{ij} = V_{ij}\left(X_{ij}, S_i\right) + \varepsilon_{ij} \quad \forall j \in C_i \tag{1}$$

where U_{ij} is the utility an individual i is assumed to obtain from of alternative j in choice set C_i. V_{ij} is the deterministic part described as a function of the attributes of alternatives X_{ij}, which is a vector of attributes as perceived by individual i for alternative j and characteristics of the individual S_i. ε_{ij} is the random term. Equation (1) describes a *Random Utility Model* (RUM). As ε_{ij} is unobservable to the analyst, s/he cannot determine *exactly* why an individual chooses an alternative j out of a set of competing options C_i $\forall$ $j,k \in C_i$ and $i = 1,...I$. The systematic component V_{ij} allows her/him to make only probabilistic statements about the choice. Assuming utility maximization, the probability that alternative j is chosen by individual i over any alternative k out of choice set C_i can be expressed as:

$$\begin{aligned} P\left(j \middle| C_j\right) = P\left(U_{ij} > U_{ik}\right) &= P\left[\left(V_{ij} + \varepsilon_{ij}\right) > \left(V_{ik} + \varepsilon_{ik}\right)\right] \\ &= P\left[\left(V_{ij} - V_{ik}\right) > \left(\varepsilon_{ij} - \varepsilon_{ik}\right)\right] \qquad \forall\, j,k \in C_i;\ j \neq k \neq 0 \end{aligned} \tag{2}$$

As an aside, equation (2) shows that the choice probability depends on the differences in the systematic components of utility V. Since V is characterised by the characteristics or attributes of an alternative, the probability depends on the differences in the attributes. In order to to estimate the probabilities of equations (2), assumptions have to be made about the nature of the random error term. The majority of discrete choice models assumes that the random error term is independently and identically distributed (IID), and related to the choice probability with a Type I extreme-value (Gumbel, Weibull, double-exponential) distribution. Given these assumptions, a closed-form expression can be derived for the probability of choosing an alternative from a choice set. The conditional probability that gives rise to the multinomial logit model (MNL, McFadden 1973)[5] is:

---

[5] The reader is referred to Alberini et al. (2007) for a more detailed derivation of the conditional logit model.

$$P(j \mid C_i) = \frac{\exp^{\mu V_{ji}}}{\sum_{k \in C_i} \exp^{\mu V_{ki}}} \quad \textbf{(3)}$$

where μ is a scale parameter usually set to 1 (constant error variances) and inversely proportional to the standard deviation of the error terms (Louviere, Hensher, and Swait 2001). V_{ij} is assumed to be linear and additive in parameters:

$$V_{ij} = \beta_{ASC} ASC_j + \sum_{n=1}^{N} \beta_n X_{nj} \quad \textbf{(4)}$$

where X_{nj} is the attribute level of attribute n of the jth alternative. β_n is the parameter value associated with attribute *n*. ASC_j is short for alternative specific constant that equals one for alternative *j* (otherwise: zero), and can be included for *j*-1 alternatives. If the alternatives are generic (unspecific, i.e. unlabelled), the specific nature (e.g. labelling) of these alternatives should not have an impact on choice behaviour. ASCs can therefore be set equal. "It is the role of the ASCs to take up any [observable] variation in choices that cannot be explained by either the attributes or the socio-economic variables" (Bennett and Adamowicz 2001: 60). Socio-economic, socio-demographic and attitude variables can be interacted with the ASC and/or the attributes. If a number *K* or *M* of individual-specific characteristics *Z* are interacted with the ASC or attributes, equation (4) becomes:

$$V_{ij} = \beta_{ASC} ASC_j + \sum_{k=1}^{K} \beta_{ASC_k} ASC_j \times Z_k + \sum_{n=1}^{N} \beta_n X_{nj} + \sum_{m=1}^{M} \beta_{n_m} X_{nj} \times Z_m \cdot \quad \textbf{(5)}$$

By using the maximum likelihood method as available in, e.g. NLOGIT 3.0 (Greene 2003), estimates for the attribute parameters in the MNL model can be obtained.

Utilities for attributes are expressed by parameter estimates of models of the logit family such as the multinomial logit (MNL) or the random parameter logit (RPL) or other models (for example, multinomial probit or heteroskedastic extreme value) (Louviere, Hensher, and Swait 2001). If one of the attributes reflects 'cost' (i.e., choosing an alternative is associated with a decrease of income), parameter estimates can be used to generate implicit prices (IP) and other policy-relevant welfare estimates. Implicit prices for an attribute are calculated using equation (1).

$$\text{Implicit price } (n) = -\left(\frac{\mu\beta_n}{\mu\beta_\$}\right) = -\left(\frac{\beta_n}{\beta_\$}\right) \quad \textbf{(6)}$$

where β_n is mean parameter of attribute n in a linear and additive utility function (cf. equation 4), $\beta_\$$ is the coefficient of the 'cost' attribute and μ denotes scale. The scale parameter μ is cancelled out for the calculation of implicit prices. Implicit prices reflect the maximum *marginal* willingness to pay (MWTP) for a marginal change in a single attribute on a *ceteris paribus* basis (Bennett and Adamowicz 2001).

The MNL is restricted by an assumption known as Independence from Irrelevant Alternatives (IIA), which is the behaviourally comparable 'counterpart' to the IID assumption (Hensher, Rose, and Greene 2005: 518f). IIA states that the ratio of probabilities of choosing alternative j over k out of a choice set C_i remains unaffected of the presence or absence of any other alternative. The hypothesis that IIA holds is often rejected in stated choice data. If IIA is violated, models such as the RPL (e.g., Train 1998) should be considered which relax these assumptions instead of MNL models (Louviere, Hensher, and Swait 2001). There is another limitation to MNL models: preferences or tastes are assumed to be homogenous across the sample. RPL estimation accounts for heterogeneity of preferences across the population by allowing the taste parameters β_i of the systematic part of the utility V_{ij} to vary for each individual i representing each individual's tastes. In RPL models, preferences vary continuously over decision makers in the population with density $g(\beta|\theta)$. The density is a function of parameters θ that characterise, for example, the mean and the covariance of a normal distribution of β. The researcher specifies the type of distribution with the normal and the lognormal being the most commonly used and estimates the parameters θ.[6]

---

6 If taste parameters βi for each individual were known by the researcher, the probability of individual i to choose alternative j from a choice set would become MNL,

The Choice Experiment Design

In the previous section, we have shown that choices made by individuals from choice sets can be used to obtain estimates of preferences for attributes of an environmental good. In this study, we are interested in preferences for ecosystem services of residents of the Lore Lindu area. Therefore, we first need to select attributes that are expressions of different levels of provision of ecosystem services. Because we are interested in the implicit prices for changes in the provision of ecosystem services, it is necessary to include a 'cost' attribute. Furthermore, a plausible context for the choices of local respondents and a definition of a base (reference) alternative must be developed. In order to ensure that estimated preferences for the attributes are not confounded with each other, the attribute levels have to vary across alternatives following an experimental plan that allows for statistical estimation of attribute coefficients.

Attribute Selection

Generally, only 'relevant' characteristics should be considered in a choice experiment. According to Lancaster (1991: 56), a characteristic is "relevant to the situation if ignoring its existence would change our conclusions about choice or ordering of the goods by the consumers". Blamey et al. (1997) distinguishes between demand-relevance and policy- or context-relevance. Demand relevance refers to the relative importance people assign to an attribute (demand-driven), whilst policy relevance refers to what the government or researchers perceive to be important (supply-driven). In order to be sure that people make sense from an attribute offered, any "... supply-driven attribute should always be screened from a demand-perspective" (Blamey et al. 1997:

---

since the error component εij of the utility is assumed to be IID extreme value. The probability of an individual's sequence of choices would then be the product of choice probabilities, where the probability for each choice within the sequence is as in equation (3). Because the $\beta$is are not revealed to the researcher, the probability of a sequence of choices involves integration over all possible values of β using the density $g(\beta|\theta)$. The integral of this probability cannot be estimated analytically because it does not have a closed form. Therefore, it is approximated numerically via simulation. See Revelt and Train (1999) for further details.

7). In other words, selection of an attribute can be motivated because the target population perceives it to be important within a certain context, and/or because knowledge on preferences is of relevance to policy makers or researchers. In the latter case, researchers should make sure that offering the attribute creates demand for it among the respondents. We followed this approach in the study presented here.

The choice of attributes and levels was based on information gathered in individual and peer-group interviews[7] in various villages of the research region, and was guided by an ecosystem service approach (Barkmann et al. *acc.*). Furthermore, information and data obtained by experts as well as from literature (e.g. Belsky and Siebert 2003; Siebert 2002; Keil 2004) were incorporated. We found that local farmers and residents showed a high degree of awareness regarding functional (indirect) benefits, particularly concerning water-related issues such as water supply or flood protection/erosion (figure 1). The attributes and the attribute levels are listed in table 1 and are described in detail below.

Because task complexity increases with the number of attributes offered (DeShazo and Fermo 2002), concerns arise about the maximum number of attributes. In pre-tests, we used 5 attributes including 'cost' in two choice scenarios and a status-quo alternative. First presented orally only, the bulk of information on attributes caused some fatigue and confusion among a number of respondents. Jae and Delvecchio (2004) found that the presence of a visual decision aid can improve choice by reducing task complexity and by facilitating the mediation of information to low-literacy consumers. Because the education of 53% of respondents did not exceed elementary school, paintings containing the main information were developed in cooperation with local farmers in order to meet their perception. The pictures and their respective informational background were simultaneously presented to

[7] Many CE case studies use 'focus groups' for that purpose (Kaplowitz and Hoehn 2001). During the pre-study phase of our research in the LLNP area, we found that it was hardly possible to conduct focus groups without the attendance of the most influential inhabitants of a village. The opinion of those opinion leaders (e.g the village head) often dominated the sessions. Understandably, village officials tend to cast a positive light on their village. Therefore, it seemed appropriate to focus on individual interviews and use the peer-groups to obtain further information mainly about prevailing policy issues.

the respondents during the explanation of the attributes. This way of presenting the context and attribute-specific information proved to be very successful in increasing the understanding of the choice task as well as raising interest and attention (see also Mathews, Freeman, and Desvousges 2007: 116). Visualizations were also included in the choice cards.[8] Using this tool, pre-tests suggested that respondents were able to cope with 5 attributes. For final refinement of the questionnaire, a pilot study was conducted (n = 96).

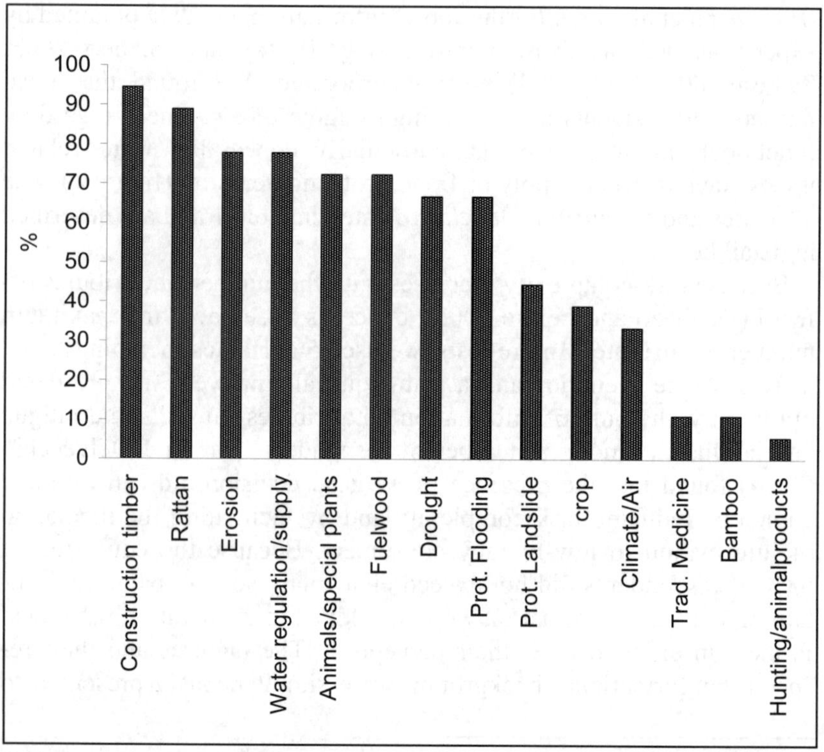

Figure 2. *Results of the pre-study. Benefits derived by the forest (% mentioned by n=18 respondents. Source: own calculations*

[8] The survey instrument is available at http://ufgb989.uni-forst.gwdg.de/DPS/pdf/SDP16b.pdf.

Cocoa pods are harvested from cacao trees (*Theobroma cacao*); cocoa beans are the dominant cash crop in the research region. In its natural environment, cacao trees grow under a dense forest canopy. There is indication throughout the research region that production is increasingly intensified, resulting in widespread monocultures with no or merely low levels of (planted) shade trees (e.g. *Gliricidia sepium*). Despite the generation of higher yields on average (cf. Steffan-Dewenter et al. 2007), intensification to 'sun-grown' cocoa leads to higher agronomic and socio-economic risks, e.g. soil degradation and negative impacts on local food security (Belsky and Siebert 2003). 'Shade-grown' cocoa farming can provide habitat for a wide range of native species, thus contributing to biodiversity conservation, while soil productivity may be retained (Siebert 2002, Steffan-Dewenter et al. 2007). Thus, this attribute contrasts (short-term) economic goals with (long-term) biodiversity conservation objectives. To the local farmers, it reflects trade-offs between a more or less risky way of cultivation along a shade tree gradient (5-35-65-9% under shade) for preponderance of cocoa plantations ranging from full-sun grown cocoa on one side, and cocoa cultivated beneath dense primary or secondary forest vegetation on the other side.

The production of wetland rice, the research region's main staple food ('paddy' rice), relies on the availability of sufficient water for irrigation purposes. Water shortages can increase the risk of reduced yield, or harvest failure in extreme cases. There is indication that deforestation on the hillsides leads to water shortages in the dry months of the year, particularly when the water originates from small watersheds in combination with prevailing simple irrigation techniques (own data[9], Burkard, this volume). Keil (2004: 84) found that perception and measured data on irrigation water availability and seasonal precipitation show the same patterns. Thus, an ecosystem level attribute on the provision of irrigation water was created. Effects of land conversion on water levels in rivers in streams were pointed out to respondents. The levels of this attribute, however, make no reference on this link and were described as the perceived months with water

---

[9] E.g., in one village (Sintuwu), one of the streams providing water for irrigation dried up, in another the water declined significantly so that irrigation is hardly possible any more (Sidondo II).

shortage for irrigation purposes.[10] More details regarding this attribute are reported in Barkmann et al. (*acc.*).

	Attribute	**Levels**	**Ecosystem service category**	**Value type (TEV)**
Rattan	availability of rattan (*Calamus spp.*) as expressed in distance from village	[*km*] 5, 10, 15, 20	provisioning service	direct Use/ option Value
Water	availability of irrigation water for wet rice cultivation as expressed in number of months with water scarcity	[*No of months*] 0, 1, 2, 3	regulating service	indirect Use Value
Cocoa	preponderance of cocoa plantations differing along a shade tree gradient	[*% under shade*] 5, 35, 65, 95	regulating services	indirect Use/ option Value
Anoa	populations of different sizes of the endemic dwarf buffalo anoa (*Bubalus depressicornis/quarlesi)*	[*No of animals*] 10, 180, 350*, 520	cultural/ provisioning service	existence/ direct Use Value
Cost	extra taxes or donation to village fund	[*1,000 IDR per year***] 18, 36, 54, 72	-	-

Table 1. *Attributes and levels (*present state; ** the amounts showing up on cards with a monthly payment scheme are 1/12 of the amounts shown in the table; 1US$ ~ 8,500 IDR at the time of the survey).*

[10] This description implicitly entails that perceived effects of irrigation technology on irrigation water supply cannot be separated from the effects of land use.

Rattan (*Calamus spp.*) is one of the most important non-wood forest products (NWFPs) in global trade. Indonesia is the biggest supplier of the resource (Sastry 2001). Most of the Indonesian rattan is harvested from wild stocks in forest areas and buffer zones, often under open access conditions. The sustainability of the current rates of harvesting, particularly regarding valuable large-diameter canes raised concerns (Siebert 2001, Clayton, Milner-Gulland, and Sarjono 2002). As a result of collection beyond sustainable thresholds, the distance from the forest edge to the harvesting locations increases (Siebert 2004). Rattan provides important socio-economic functions: rattan collection provides an income opportunity for poorest segment of the population (Vedeld et al. 2004), and can serve as an income alternative for others if economic disasters strike a household. Thus, rattan availability has an 'option' or insurance value component. Overexploitation of rattan stocks would result in a loss of these functions, and may pose a serious threat to the livelihoods of a particularly vulnerable part of the population in the future. In accordance with results from our pre-study investigations, previous research of Zeller and Birner (2003) showed that the encounter distance from the forest edge to harvesting locations increased from 4.4 km on average in 1990 to 14.5 km in 2001, indicating an overuse of rattan resources. Consequently, we operationalised the rattan attribute by the encounter distance to the nearest extracting location (5, 10, 15, 20 km).

Many endemic species, among them mammals and birds (Waltert, Mardiastuti, and Mühlenberg 2004, Whitten, Henderson, and Mustafa 1987), can be found within the boundaries of the LLNP. Large forest clearings inside the National Park (Adiwibowo 2005) show that the forest frontier in the research region is by no means secured. Because we were interested in how biodiversity conservation objectives are supported by the local population around the National Park, different population sizes of the dwarf-buffalo anoa (*Bupalus depressicornis, B. quarlesi)* were included as an exemplarily attribute dealing with an endemic species. Population sizes of anoa are in decline in the research region (Zeller and Birner 2003, Burton, Hedges, and Mustari 2005). Anoa populations provide a mixture of different TEV value categories. One is 'existence' value, e.g. the "... concern to protect [...] although he or she has never seen one and is never likely to ..." (Pearce and Moran 1994: 12). However, interactions between the living environment and the population are common, implying that other value components such as direct use value (e.g. hunting) or existence value (e.g. intergenerational aspects) have to be considered as well. Some respondents mentioned that they fear to be injured if they encounter one of these animals in the woods. Such perceptions may also influence choices.

We expected that respondents have preferences for fewer months with water scarcity and shorter encounter distance to the rattan harvesting locations. Results of a pilot study suggested that respondents prefer less shade in cocoa plantations, however only to a certain limit. For 'anoa', we assumed that direct use and bequest value aspects are overall more influential than aspects of fear, such that we expect to observe a preference for greater population size.

Framing

At the beginning of the questionnaire, respondents were asked to recall past changes in agriculture, infrastructure and living conditions. This way, we hoped to increase respondent awareness of potential *future* changes to be included as choice experiment attributes. Next, respondents were asked to choose those 3 out of 6 competing development issues, which they thought should have priority for spending public money. Besides contributing to establish a frame of reference for respondents, this question also served as a 'warm-up' exercise for the choice task (cf. Krupnick and Adamowicz 2007).

Framing involves the development of an appropriate context, in which the hypothetical scenarios are presented to the respondent. If the context is misleading or not credible, there is little incentive for respondents to take the choice task seriously. Contexts are credible and plausible if they are based on current policy processes, such as plans by the LLNP authority to create a buffer zone of restricted use within the Park. However, if respondents focus too strongly on contextual aspects such as their relationship with the LLNP authorities and current disputes about property rights, a survey may be perceived to be commissioned by an institution implementing the policy, or that local village authorities exert influence on individual responses. If respondents are strong proponents or opponents of a program or policy, they are more likely to fail to make trade-offs between the attributes. We decided to define the five choice experiment attributes as outcomes from a government development program on a village scale. Multi-level development programs have been implemented in the past in the research region. One example is the Central Sulawesi Integrated Area and Development Program (CSIADP, ANZDEC 1997).

The Cost Attribute

The 'cost' attribute was double split-sampled. One half of the respondents were offered alternatives with a rise in "house- and land" tax (*Pajak Bumi Bangunan*), the other half with a donation to a village fund (*iuran dana pembangunan desa*) affecting every household of the research region as payment vehicles. The second split-sample confronted respondents with either a monthly or a yearly payment scheme. The results of the split-sample tests will be reported elsewhere.

According to Whittington (1998), the highest price should be rejected by 90% - 95% of the respondents in CVM studies. The levels were derived following this rule of thumb by using different 'prices' in pre-tests in addition to information obtained by a payment-ladder approach (e.g. Bateman et al. 2002: 138f) asking for the willingness to pay for the subjectively *'best'* development program alternative. The income distribution in the study region is highly skewed: a few people live in concrete houses, have satellite television and sometimes even own a car, while others share a wooden hut without electricity. The few respondents of the first group did not hesitate to express a willingness to pay exceeding double the highest amount finally offered in the choice experiment (12,000 IDR per month or ~1.4 US$), whereas some members of the second group were virtually struck by the thought of paying the same amount. Offering the highest price to the poor could embarrass them, and could make "... the interviewers look insensitive and/or uninformed" (Whittington 1998: 24).

Experimental Design and Status Quo Alternative

An orthogonal main-effects design was derived using SPSS12.0 to reduce the number of possible combinations of attribute levels from a total of 4^5 to 16 profiles. The 16 profiles were blocked into four versions of the questionnaire, resulting in four choices of the experimental design to face for each respondent. Using a mix-and-match approach (Louviere 1988), a second set of profiles was created and subsequently combined into choice scenarios that consisted of two (generic) alternatives A and B. The last step was repeated several times to minimise level overlap and dominated pairs. The final design step ensured within-alternative orthogonality, which is the main concern when the alternatives are unlabeled or generic (Hensher, Rose, and Greene 2005: 152). Two additional choice sets not derived from the main effects design were included for further model testing to be reported elsewhere.

Bias amongst other sources may arise from the order in which the attributes appear on the choice cards. In order to minimize such effects, we created a second version of choice sets with an inverted order of the attributes. The two versions were randomly assigned to respondents in a way that approximately each half of the respondents received one of the versions. Results on this additional split-sample test are reported in Glenk (2007). Inclusion of a status quo option allows for economic welfare measures. The status quo ('do-nothing') option is the reference point from which the scenarios that the researcher offers to the respondents diverge. Blamey et al. (1997: 14) note that "[s]election of a base may have an important influence on CM results by affecting the utility of the base option relative to others, and by influencing the framing of outcomes, for example, as gains or losses".

We included a status quo as a third choice alternative that was described as the present environmental situation regarding the four ecosystem service attributes. Particularly due to the heterogeneity concerning environmental and socio-economic conditions among the villages, it would have been plainly unrealistic to define *one* common base for all villages. Therefore, respondents were directly asked which attribute level they perceived as most similar to their own present situation ('cocoa', 'water', 'rattan').[11] By this means, respondents created their 'individual-specific' status-quo or 'self-explicated' alternative (cf. Blamey et al. 2001: 137). There are several reasons that motivate this approach:

[11] Present population sizes of anoa cannot possibly be known. Hence, a common reference point was defined after consultation of locals and experts (350 individuals).

- A 'constant base reference' is likely to be incredible to respondents because of the substantial heterogeneity of environmental conditions in the research region. If respondents have doubts about the reference point, their choices may become meaningless. An 'individual-specific' status quo frames the attribute levels of the alternative options properly as gains and losses.
- Involving respondents in the preparation for the choice experiment demonstrates respondents that their perceptions are taken seriously. This is likely to reduce 'disbelief' about the hypothetical nature of the choice task and the survey. However, there might be some implications for welfare analysis if actual and perceived values diverge (Adamowicz et al. 1997).
- Prior to the choice task, respondents had to intensively engage themselves with the present state regarding the attributes. As a result, respondents may be more certain about their choices.

Individual-specific Characteristics and Model Specification

Individual-specific characteristics regarding the socio-economic background, perceptions and attitudes or information on the performance of the choice task can help to explain choice behaviour and preferences. For a useful overview on the role of supporting questions in stated choice studies, see Krupnick and Adamowicz (2007). We collected additional data on respondent and household characteristics including data related to the choice task (e.g. difficulty, confusion), background data concerning the attributes (e.g. past experience, present use) as well as several socio-economic characteristics (e.g. age, education, wealth status). The variables listed in table 2 are used in the succeeding analysis. Additionally, attitude items on value, risk and coping issues were included, which were analysed elsewhere (Barkmann et al. *acc,* 2007a). It is often assumed that attribute levels reflect an individual's utility for a characteristic in a linear fashion. In this case, attributes enter the models as continuous variables using the actual attribute levels. However, the utility curve of an attribute may actually not be linear but follow a different functional form. This can be assessed by using effects codes rather than actual values to derive point estimates of the utility for each attribute level. We did this with the results of the pilot study, which suggested that an (inverted) unimodal preference curve is likely for the cocoa attribute. While all attributes were coded as continuous and quantitative variables for the final analysis, we included a quadratic term for the cocoa attribute. The ASC is 1 for the development alternatives, and 0 for the status-quo option.

Variable	Description	Mean (std. dev.)	Interaction
Edu	Years in school (respondent)	7.68 (3.29)	ASC, Cost
Imppay	5 point Likert score for importance ascribed to paying tax / village fund on household well-being *[1: not important; 5: very important]*	3.03 (0.9)	ASC
Unds	5 point Likert score for overall understanding of the choice task as perceived by the interviewer (respondent comprehension rating) *[1: not at all; 5:very well]*	3.14 (0.85)	ASC
KL	Dummy variable showing whether respondent is from Lore or Kulawi districts	0.52 (0.5)	ASC, Cocoa
Colrat	Dummy variable showing if a respondent is collecting rattan	0.13 (0.34)	Rattan
Invwat	Dummy variable showing if a household is involved in wet rice cultivation	0.60 (0.49)	Water
Agecoc	Maximum age of cocoa plantations owned by respondent	4.46 (4.08)	Cocoa
Povind	Relative poverty index comprising a set of welfare indicators*	0.04 (0.76)	Cocoa
Prisec	Indicator for perceived discretionary income: share of total household income spend on primary needs (rather than secondary) *[1: 3/4 to everything; 3: ¼ up to ½]*	2.05 (0.69)	Cost
Depch	Dependency rate adults vs. children*	0.54 (0.48)	Cost

Table 2. *Socio-economic and socio-demographic variables. N = 288 for all variables*
**Data from subproject A4 (see Schwarze 2004); missing values (N=39) replaced by the mean*

We calculated both MNL and RPL models with and without interaction terms of attributes and individual-specific characteristics. For the RPL models, we gave all attribute parameters normal distributions except for 'anoa' and 'cost', which were fixed.[12]

Data Collection

For data collection, a stratified village sampling frame (Zeller, Schwarze, and van Rheenen 2002) was adopted. The strata for the village sample were ethnicity composition, vicinity to LLNP and population density of a village. Households were then randomly selected within each village. The choice experiment survey was administered to 301 households in 12 villages of the research region (December 2004 - March 2005). Face-to-face interviews were conducted by 6 well-trained local enumerators. To minimize potential interviewer effects, enumerators were randomly assigned to households.

Results and Discussion

Overall Model Results

All 301 households completed the choice task. Sixty-six respondents always chose the status quo. Eighty percent of them reported that they could not afford the payment (three respondents), they perceived the present situation to be "better" than the offered alternatives (36 respondents), or a combination of both (27 respondents). The first aspect

[12] The standard deviation of a fixed parameter is set to zero. All the behavioural information is therefore captured by the mean (Hensher, Rose, and Greene 2005:608, supra note 3). The coefficient of the standard deviation for a normally distributed 'anoa' coefficient was not significant with t-statistics between 0.19 and 0.35 for both models with and without interactions. Other distributions, such as the triangular or a uniform distribution did not perform any better in this respect. This does not mean that preferences for the anoa attribute are homogeneous across the sampled population. Indeed, earlier modelling (Glenk et al. *acc.*) has shown that preferences for anoa differ as a function of general attitude towards anoa, and perceived threat of its extinction. If heterogeneity exists, it may not take the distributional form of the distributions mentioned. See Revelt and Train (1999:15, supra note 10) for reasons to keep the 'cost' or price attribute fixed.

underlines that WTP measures have to be interpreted as a combination of people's *willingness* and their *ability* to pay. The remaining 13 respondents were classified as "... essentially not responding to the CE task" (Adamowicz et al. 1998b: 68) and were omitted.[13] Thus, data of 288 respondents was available for further analysis.

MNL and RPL model results are reported in table 3. Overall, all models were highly significant (P<0.001). As expected, likelihood ratio tests show that model fit improves significantly if interactions are included for both MNL and RPL models (P<0.001). Significant improvements (P<0.001) in model fit can be found for both RPL models compared to their MNL counterparts (χ^2 statistic for models without interactions: 49.33; with interactions: 49.33; 5 d.f.). Adjusted pseudo-R^2 (ρ^2) values are another measure of model fit. They can be compared with values of R^2 as in ordinary least square (OLS) regression models. A value of ρ^2 between 0.2 and 0.4 corresponds to OLS R^2 values of 0.7 to 0.9 and can be interpreted as a good fit (Hensher, Rose, and Greene 2005: 338f). Increases of ρ^2 values show the same pattern as described above. The value of 0.37 for the RPL model with interactions indicates a very good model fit.

The anoa attribute is significant at the 5% level. All other choice set attributes are significant at 1% or better. A positive sign shows that *more* of an attribute results in a higher probability of an alternative being chosen, while a negative parameter signifies that *more* of an attribute has a negative effect on the odds of an alternative being chosen. 'Water' and 'rattan' have – as expected – negative signs. The 'anoa' attribute is positive and significant, indicating that people do care for the maintenance of viable populations of this animal. The coefficient of the linear specification of the 'cocoa' attribute is positive and significant, while the coefficient of the quadratic term is negative and significant. We could therefore confirm the suspected unimodal relationship between shade and utility for the 'cocoa' attribute. Figure 3 shows the utility functions for different coding and speci-

[13] Reasons for choosing always the status-quo were 'protest' answers, payment aversion, and exceeding the cognitive capability to complete the choice task (one respondent).

fications of the 'cocoa' attribute[14] (MNL models, data not shown). The quadratic specification in figure 3 illustrates that a threshold exists for preferences *for* intensification. This may reflect a certain degree of risk aversion. Utility peaks at a level of shading of approximately 28%.

A considerable amount of heterogeneity of preferences exists for the 'rattan', 'water' and 'cocoa' attributes, since the standard deviation parameters for all three attributes are significant in both RPL models at 1% or better except for the quadratic terms of the 'cocoa' attribute. Based on results of model 3, the standard deviations around the mean of the ecosystem service attributes show that about 24% of respondents would prefer *greater* encounter distance to the rattan harvesting locations, and approximately 5% of the respondents would prefer *more* months with water scarcity. It is difficult to find a behaviourally meaningful explanation for this result. This phenomenon is well-known for RPL models that use a normal distribution for the random parameter specification because the unrestricted functional form of the normal distribution enforces the existence of a long "left tail" of any preference distribution.

For purely mathematical reasons, this left tail can *always* include behaviourally implausible negative values. To address this problem, constraints could be imposed on the distributions to avoid behaviourally unacceptable sign changes (Hensher and Greene 2003). While opting to refrain from any such restrictions, we interpret calculated negative preferences as a sign of indifference of some respondents for the 'rattan' and 'water' attributes. The significant standard deviation parameter for the linear component of the 'cocoa' utility suggests that *at least* 20% of the respondents prefer less shading over the whole range of shading (5% - 95%) offered.

[14] MNL models were used for the calculation. Models with effects-coding and a linear specification of the cocoa attribute are not reported here but are, of course, available from the authors upon request.

Variable	MNL model 1		RPL model 2		MNL interaction model 3		PRL interaction model 4	
'Rattan'	-0.0404 (-5.179)	***	-0.0669 (-4.762)	***	-0.031 (-3.484)	***	-0.4746 (-3.470)	***
'Water'	-0.8943 (-18.772)	***	-1.4107 (-11.128)	***	-0.4997 (-7.123)	***	-0.7344 (-5.688)	***
'Cocoa Shade'	0.0126 (2.067)	*	0.0228 (2.571)	*	0.0258 (3.669)	***	0.0366 (3.858)	***
'Cocoa Shade'² (quadratic)/100	-0.0247 (-3.913)	***	-0.0412 (-4.367)	***	-0.0247 (-3.482)	***	-0.0326 (-3.578)	***
'Anoa' Population Size	0.0009 (2.688)	**	0.0011 (2.437)	**	-0.0009 (-2.452)	*	-0.0009 (-2.074)	*
'Cost'	-0.0254 (-9.146)	***	-0.0343 (-7.876)	***	-0.0626 (-6.714)	***	-0.0775 (-6.214)	***
ASC (non-status quo choice)	0.4892 (3.486)	***	0.6173 (2.884)	**	1.401 (2.924)	**	1.6508 (2.468)	*
ASC**KL*					-0.3509 (-2.291)	*	-0.2979 (-1.343)	
ASC**edu* (years in school)					-0.1339 (-3.027)	**	-0.1761 (-2.864)	**
ASC**imppay* (importance of payment)					0.4344 (4.826)	***	0.5404 (4.104)	***
ASC**unds* (perceived comprehension rating)					-0.3417 (-3.620)	***	-0.4186 (-3.065)	**
Rattan**colrat* (dummy rattan collector)					-0.1194 (-4.103)	***	-0.1650 (-3.913)	***
Water**invwat* (dummy wet rice activity)					-0.7710 (-7.686)	***	-1.06 (-6.161)	***
Cocoa (linear)* *povind* (poverty index)					-0.0048 (-2.033)	*	-0.0066 (-1.891)	T
Cocoa (linear)* *KL* (Kulawi/ Lore)					-0.0152 (-4.133)	***	-0.0213 (-3.853)	***

Cocoa (quadr.)* *agecoc* (max. age of cocoa)					-0.0017 (-3.546)	***	-0.0028 (-3.613)	***
Cost**edu* (years in school)					0.0036 (3.941)	***	0.0047 (3.956)	***
Cost**prisec* (discretionary income)					0.0060 (2.602)	**	0.0072 (2.205)	*
Cost**depch* (child dependency rate)					-0.0092 (-2.652)	**	-0.0124 (-2.551)	*
Standard Deviation ASC			0.9142 (4.406)	***			0.717 (3.087)	***
Standard Deviation 'Rattan'			0.0992 (4.464)	***			0.0873 (3.856)	***
Standard Deviation 'Water'			0.8682 (6.680)	***			0.6997 (5.767)	***
Standard Deviation 'Cocoa Shade'			0.0257 (4.796)	***			0.0181 (3.123)	**
Standard Deviation 'Cocoa Shade' quadratic			0.0024 (0.250)				0.0021 (0.220)	
Log-likelihood	-857.34		-812.52		-757.69		-733.03	
Number of observations	1152		1152		1152		1152	
Adjusted ρ^2 (Pseudo-R^2)	0.26		0.30		0.35		0.37	

Table 3. *Maximim likelihood model estimation results; t-statistics in parentheses (coefficient/standard error); Source: own calculations*
**** = significant at $p \leq 0.001$;*
*** = significant at $p \leq 0.01$;*
** = significant at $p \leq 0.05$; ^{T} = tendency at $p \leq 0.1$*

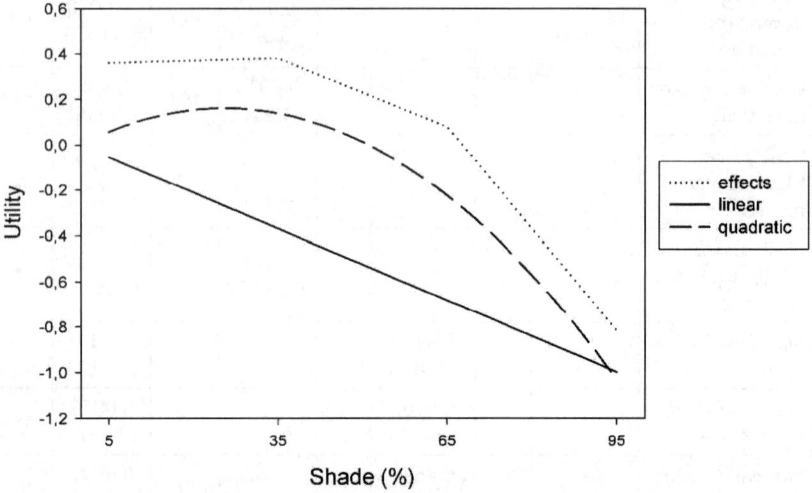

Figure 3. *Effects coding, linear and combined linear and quadratic effects for cocoa shading. Source: own calculations*

Status-Quo Effects

The coefficient of the ASC is positive and significant in all models. Overall, this suggests that there is no particular propensity to choose the status quo option relative to the alternatives as more commonly reported in literature (e.g. Adamowicz, Louviere, and Swait 1998a). One exception is the study by Mogas, Riera, and Bennett (2006), who report a positive ASC in a study on afforestation in Catalonia. A preference for the status quo, all else equal, is often referred to as 'status-quo bias' (Samuelson and Zeckhauser 1988). Status quo bias can arise if the status quo option is used as an 'easy way out', for example, in case of decision difficulty and/or limited cognitive capability (e.g. Luce 1998, Kontoleon and Yabe 2004). The positive ASC gives some evidence that this strategy was not particularly important in our study. Our respondents indicated on average – everything else held constant – more utility from departing from the present situation than from keeping it. This could be due to a number of reasons. For example, respondents include unobserved attributes associated with a governmental programme, or express a desire to comply with some non-existing task thought to be intended by the researcher. The high rate of status quo choices among all choices (53.2%) shows, however, that respondents did not feel generally 'forced' to choose among the alternatives of change. Furthermore, the RPL models show that there is significant heterogeneity around the mean estimate of the ASC coefficient. Based on results of model 3, about 25% of the respondents would exhibit a tendency to choose the status quo relative to alternatives A or B.

Unless specified explicitly, we use RPL model 4 (see table 3) for the subsequent analysis and discussion. The interaction terms of the ASC with individual-specific characteristics can shed some light on potential reasons for the observed overall tendency to move away from the status quo, and the heterogeneity of this tendency across the sampled population. The tendency to choose the management alternatives rather than the status quo decreases with increasing years in school (ASC**edu*), if a respondent is rated higher regarding the comprehension of the matter of the choice task (ASC**unds*). Model 3 indicates that it may also be relevant if a respondent is from Kulawi or Lore districts (ASC**KL*).

The education variable is correlated with the relative poverty index and with a dummy variable taking 1 if the respondent is involved in non-farm income activities, else 0 (Pearson's R^2: 0.256 respectively 0.362, both significant at $P<0.01$). Thus, respondents who are less poor and respondents who earn non-farm related income may be more satisfied with the current situation. Alternatively, they may perceive they are obtaining less benefit from changes in farming related attributes. On the other hand, better-educated respondents may make less use of information that was not included explicitly in the choice experiment. The same can be said for respondents with higher ratings of choice task comprehension (ASC**unds*). The likelihood of choosing the status-quo relative to the alternatives increases with higher ratings. This finding further supports the assumption that respondents did not have a tendency to use the status-quo as an 'easy way out' in case of difficulties associated with the choice task.

Agricultural land use in the Kulawi and Lore regions is – *on the whole* – less intensive compared to Palolo and Sigi Biromaru districts, and farther away from the urban capital Palu. Signs of environmental degradation are more obvious in Palolo and Sigi-Biromaru, resulting, for example, in water shortages for irrigation as well as household usage. In the district of Palolo, deforestation of the large forest area Dongi-Dongi was followed by a devastating flood event in December 2003 (see Adiwibowo 2005 for details). Personal experience of the potential consequences of environmental degradation could have resulted in a greater general demand for environmental improvements. The positive and significant interaction ASC**imppay* shows that respondents who believe that paying tax or donating to a village fund is of greater importance for well-being of their household have greater preferences for a change. This could be due to the fact that these respondents expect additional benefits from paying taxes or donating to the village fund that exceed those described in the choice task.

Attribute Interactions

Making sense intuitively, disutility for an impairment of rattan provision and water availability increases significantly if respondents are involved in either rattan extraction or cultivation of wet rice. Large parts of a utility gain from an improved provision of rattan and water services can be explained by the involvement of respondents in these activities. However, involvement alone does *not* explain the preferences regarding the rattan and water attributes. One may therefore assume that these two activities are not completely substitutable by other income generating activities because they fulfil purposes not necessarily related to the generation of income.

In this context, rattan availability provides livelihood security because it serves as an income alternative particularly for vulnerable households. In the course of our field work, the finding emerged that locals who were not involved in wet rice cultivation were mainly interested in sustained wet rice cultivation in order to maintain a local market for the most important staple food. The respondents expected that rice could then be obtained at rather low prices. Due to improved access to the urban market of Palu in most parts of the research region, this aspect may actually be less important than perceived. Besides, there may be an array of social and cultural factors associated with wet rice cultivation that have influence on the life within local village communities. Some aspects are pointed out in Burkard (security, this volume), who also vividly shows the complexity and heterogeneity of such factors in the research region.

The model results suggest that utility of shading in cocoa has a maximum at approximately 28%. The likelihood that the peak is reached at lower levels of shading with an increasing maximum age of cacao plantations and with increasing wealth of the respondents' household is assessed by a relative poverty index measure. Cacao treelets are usually planted under shade when young. Cacao becomes more tolerant to sunlight with increasing age. Thus, respondents owning older plantations are more confident in cultivation success under less shade than owners of very young cacao plantations.

The less poor people are, the more agronomic risk they may be willing to take, and the more disposable capital they may have to invest in fertilizer or pesticides. Additionally, respondents living in the Lore or Kulawi regions have greater preferences for more intensive cocoa cultivation compared to those living in Sigi-Biromaru and Palolo. This may at least partially be explained by a longer establishment of cocoa cultivation in Kulawi and Lore.

Respondents reported an average level of shading of about 45% as the status quo. The utility peak of about 28%, and the results of the interactions indicate that it is likely that (further) intensification of cocoa plantations will take place in the future. Thus, biodiversity conservation measures aiming at more sustainable ways of cocoa cultivation (measured here by a shade tree gradient) will probably not be successful without appropriate incentives for the cocoa farmers, including options such as price premiums (Steffan-Dewenter et al. 2007, Barkmann et al. 2007b). However, cocoa intensification can be highly beneficial from a developmental economics point of view: Keil (2004: 132) has shown that there is still much potential for increasing the technical efficiency and therefore the benefits obtained from cocoa cultivation, and Steffan-Dewenter et al. (2007) report much higher average annual net returns for less-shaded plantations. Despite the risks, the spread of cocoa and its intensification has improved welfare not only for migrants from South Sulawesi, who have a longer tradition in cocoa cultivation, but also for many "locals".[15] However, there is evidence that the arrival of the 'cocoa boom' may not necessarily lead to a genuine improvement of the living conditions for all local individuals (Sitorus 2002).

Weber (2005: 187ff) has discussed the risks associated with the rising dependency on (world market) prices for cocoa. He concluded that a severe price crash would mean the end of cocoa as a driving force for improved welfare of households, but that it is likely that welfare would not drop below the state before the beginning of the cocoa boom. It is a matter of dispute whether intensification can help to mitigate forest encroachment or whether it may actually enhance it (Keil 2007). Still, it can generally be assumed that the emergence of cocoa in the Lore Lindu region in the course of the cocoa boom has promoted forest encroachment (e.g. Maertens 2004: 36). Although relatively much is known about specific ecological and economic short term implications resulting from cocoa cultivation (Steffan-Dewenter et al. 2007), little information is available on long-term consequences. Thus, it is in the realm of speculation to judge whether the (private) net benefits of cocoa cultivation and its intensification exceed aggregated (social) costs, particularly in the long run. Yet, the demand for further intensification should seriously be considered by decision makers. While acknowledging economic development as a priority goal, informational and educational campaigns and extension services could point out the potential long term risks to cocoa farmers. A mix of economic incentives, education and regulation (property rights, encroachment) will be required to steer cocoa cultivation in the Lore Lindu area in a direction that accommodates biodiversity conservation objectives. Price premiums for "rainforest friendly cocoa" may be a market option to generate economic incentives, for example (Barkmann et al. 2007a).

As an aside, we could not find any significant differences of preferences regarding the cocoa attribute between "locals" and migrants as well as between households with an indigenous or non-indigenous head of household (data not shown). This finding gives rise to the

[15] We denote households as migrant households when the head of household or his/her father has moved to the village, and describe 'local' households when the household has been settling in a village for more than two generations.

hypothesis that (i) there are no differences in the preferred way of cocoa cultivation between these groups and (ii) that the key factor influencing the cultivation decision is profit maximization. This does not mean that there are no actual differences between cacao plots. However, these differences could be the result of, for example, different general cropping strategies, which are outlined by Burkard (security, this volume), or different (access to) knowledge about cultivation practices (Weber 2005: 192f). All included interactions of socio-economic and socio-demographic variables with the cost attribute were significant, and marginal WTP is a function of these variables. Individuals with higher education (Ed*u*) and a higher perceived discretionary (short-term) income (*prisec*) are less responsive to increases of the cost attribute resulting in positive effects on MWTP. In contrast, individuals with higher child dependency rates (*depch*) are more concerned about tax rise or village donation resulting in negative effects on MWTP. As an aside, we did not find a significant linear influence of relative poverty on the cost attribute (data not shown). In addition to the documented influence on cocoa shading, relative poverty may have a systematically non-linear influence on preferences for anoa, rattan and water (Glenk 2006). However, as relative poverty is a composite mid-term welfare measure, it does not directly reflect the current buying power or cash availability of respondents. These aspects are more likely to be determinants of sensitivity for changes in the cost attribute, and this is underscored by the significant interaction with the perceived discretionary income.

Implicit Prices

Using equation (6) and RPL model 4, implicit prices (marginal WTP) were calculated for the attributes. Confidence intervals were derived using a Krinsky and Robb (1986) procedure with 1,000 random draws (table 5). The estimates do only account for sampling variability, and cannot be interpreted as a heterogeneous distribution of preferences. High implicit prices can point out those attributes to decision makers that warrant more resources. However, direct comparison is complicated because the marginal units of the attributes differ. MWTP to avoid 1 month of water scarcity for irrigation is about 41,600 IDR

(~4.9 US $) per year, 100 more individuals of anoa is worth about 3,100 IDR (~ 0.4 US $) per year.

	Rattan [1 km]	Water [1 month]	Cocoa* [1%]	Anoa [1 individual]
Median MWTP	1,980	41,575	533	31
2.5% Lower bound	1,188	32,564	362	7
97.5% Upper bound	2,835	52,791	752	58

Table 4*. Implicit prices (MWTP) in IDR/year/household (US$) and 95% confidence interval; Source: own calculations*

* *Calculated as mean slope between 5% and 95%;*

What can be said about the absolute magnitude of the MWTP values? Are they in a reasonable range or particularly low or high? We collected some background data related to attributes that allow for a rough assessment, particularly for rattan and water availability. The following comparisons should not be conceived as a formal proof of external validity. Still, they are helpful in anchoring the MWTP values within a broader context. First, the average direct tax (*pajak bumi bangunan*) paid by households in the research region is about 15,000 IDR per year. Thus, MWTP for one month less with water scarcity for irrigation equals up to 200% of the average "house and land" tax. Typical irrigation fees are about 19,200 IDR per ha and year on average.[16] Therefore, MWTP for improved water availability reflect the order of magnitude of comparable payments in the region. Concerning rattan, we conducted a linear regression of the time needed to reach the rattan harvesting locations (h) on the distance (km) (figure 4).

[16] Mean value over four planting seasons (2003-2004) for households that paid irrigation fees in six of the sample villages. The authors like to thank Alwin Keil (University of Hohenheim, Germany) for the provision of these data.

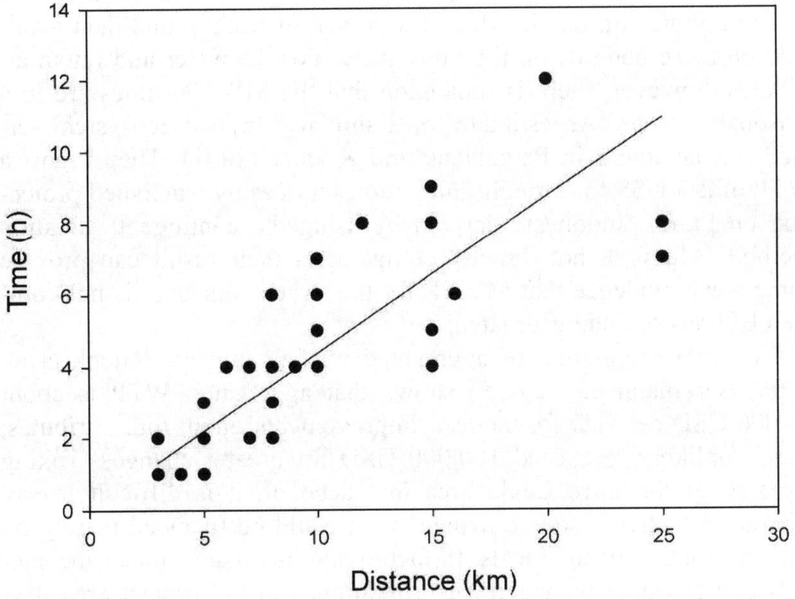

Figure 4. *Rattan collection: regression of time on distance; Time = 0.44 x Distance (R^2: 0.896); Source: own calculations*

One km less is associated with a time saving of about 0.4 hours (n = 37, significant at 99% level). To derive a monetary value of time savings, they may be related to the (local) wage labour market (Köhlin and Amacher 2005). Average income from unskilled wage labour is about 10,000 IDR per day (10h). On average, people went to collect rattan 18 times per year. One km less can therefore be associated with 18 x 0.4h = 7.2 hours saved per year. Assuming perfect substitutability of time and labour, one km less is equal to income forgone from wage labour of about 7,200 IDR/yr for the average rattan collector. Using results of model 4, MWTP for 1 km less is about 1,400 IDR/yr for non-collectors, while rattan collectors respond much stronger to changes in the rattan attribute and have a MWTP of about 6,400 IDR/year. The fraction of the sample stating to collect and/or sell rattan is 12.8%.

Comparisons of the above kind are more difficult for cocoa and particularly anoa due to the complexity of the benefits associated with these attributes on one hand, and to a lack of background data available on these benefits on the other hand. For the water and rattan attributes, however, there is indication that the MWTP values are in a reasonable range. An estimate for a similarly framed ecosystem service can be found in Pattanayak and Kramer (2001). They report a WTP of 2-3 US$ for drought mitigation services by watershed protection on Flores, Indonesia derived by using the contingent valuation method. Although not directly comparable, their result can provide some weak evidence that MWTP for the 'water' attribute is not completely over- or underestimated.

An initial calculation of aggregated welfare impacts (Glenk et al. 2006, Barkmann et al. *acc.*) shows that aggregated WTP is about 50,000 USD per year for modest 'improvements' in all four attributes, and is unlikely to exceed 100,000 USD for greater changes. Taking the size of the Lore Lindu area into account, it is difficult to say whether effective resource management could be financed merely by the contribution of the locals. In fact, numerous issues hinder the success and resource management programmes in the project area. Examples are unresolved property rights conflicts around and in LLNP, weak law enforcement, and continuously increasing pressure from population growth and agricultural intensification. Hence, the influence of our findings on the resource management in the Lore Lindu region should not be exaggerated. However, the importance ascribed to the attributes may encourage decision makers to find solutions that sufficiently consider the local demand for the provision of the ecosystem services analysed.

Concluding Remarks

With a carefully adjusted instrument, the choice experiment could be conducted successfully on one of Indonesia's outer islands. The model results provide some evidence that the choice experiment can be applied to complex ecological services in an ecologically and socially diverse rural area and in a developing country setting. The design strategies to adjust the status-quo scenario to the individual respondents and the translation of biodiversity values into ecosystem *services* contributed essentially to this result. Concerning the first, one of the major advantages of using individual-specific perceived levels for the status-quo is the availability to account for heterogeneous environmental conditions in an on-site study, therefore enhancing credibility among respondents as well as framing choices properly as gains and losses.

Using the ecosystem service approach and translating ecosystem functions into concepts that correspond to and interact with people's life may contribute to reduce potential problems arising from the divergence of perceived and actual values for welfare analysis. The ecosystem services approach proved to be a successful instrument to be used for the valuation of ecosystem services in general and for ecosystem services provided by complex ecosystem functions in particular. However, the link between ecosystem services as described in this study and the underlying ecosystem structures, functions and processes is not well established (also see Barkmann et al. *acc.*). It is for instance neither known, what management practices would be necessary to reduce the encounter distance to the harvesting locations for rattan by 5 km, nor is it clear, what kind of quantitative and qualitative land use change would result in a water shortage for irrigation of one month more. Thus, one of the major challenges for future applications of the ecosystem service approach is to improve the match of scientific ecosystem analysis with the data requirements of applied land use management studies. It remains the task for economists to identify the values of ecosystem services for consideration in decision-making processes. By combining the impacts on human-wellbeing with ecological background data and other societal aspects, a useful policy/management tool could arise. Hence, further efforts are needed to improve the communication between the two broad disciplines.

There is a willingness of local respondents to contribute actively to the maintenance of their resource base concerning water availability and rattan provision. This could also be beneficial from a biodiversity conservation point of view. On average, residents even have preferences for maintaining viable population sizes of the local endemic dwarf buffalo anoa. In the fast growing sector of cocoa agroforestry systems, respondents indicated a clear preference for more intensively managed plantations with fewer shade trees. Benefits due to increased private financial returns are contrasted by negative impacts on biodiversity. However, the precise relationship between private net benefits and social costs is neither clear for the short term nor for longer terms. Particularly for the cacao attribute, the conflict between locally predominant objectives of economic development and more globally defined conservation objectives is clearly reflected in people's preferences.

Plausible explanations for signs of significant interactions of attributes or the ASC with individual-specific variables in line with *a priori* expectations underscore the theoretical validity of the choice models. Beyond this technical aspect, they allow for a more detailed understanding of the choices made by respondents. Complemented by estimates of unobservable heterogeneity in RPL models, useful insights are provided that facilitate the interpretation of the stated choice behaviour and, most obvious for shading and cocoa cultivation, predictions about future behaviour. Measures of MWTP for an improved provision of ecosystem services ('water', 'rattan', 'anoa') were documented. We outlined that the magnitude of MWTP is quite substantial considering the living conditions of the inhabitants of the Lore Lindu region. Some indication is given that MWTP values calculated in this study are neither completely over- nor underestimated. The MWTP estimates reflect net benefits for a range of ecosystem services. Our results can serve as a benchmark for future research in which they can be used as measures of Hicksian compensating variation in a cost-benefit framework which addresses all relevant costs and benefits associated with a development program.

Acknowledgements

The authors thank Deutsche Forschungsgemeinschaft (DFG) for funding; the villagers of the Lore Lindu Region; the enumerators Sumarno Syamsuri, S.Pt., Muhammad Rifai Lasamana, S.P., Zulkifli Lasamana, S.H., Yonathan, SP., Muhammad Din and Harry Wibowo; Stefan Schwarze, Adithya Wardhono and STORMA sub-project A4 for providing the income data, and two of the variables in table 2 as well as Alwin Keil for further data.

References

Adamowicz, W. J. et al. (1997): Perceptions versus Objective Measures of Environmental Quality in Combined Revealed and Stated Preference Models of Environmental Valuation. *Journal of Environmental Economics and Management*, Vol.32: 65-84.

Adamowicz, W., Louviere, J. and J. Swait (1998a): Introduction to Attribute-Based Stated Choice Methods. Prepared by Advanis Inc. for the National Oceanic and Atmospheric Administration, US Departement of Commerce.

Adamowicz, W. et al. (1998b): Stated Preferences Approaches to Measuring Passive Use Values. *American Journal of Agricultural Economics*, Vol.80: 64-75.

Adiwibowo, S. (2005): Dongi-dongi – Culmination of a Multi-Dimensional Ecological Crisis: A Political Ecology Perspective. Ph.D. thesis, Kassel.

Akiyama, T. and Nishio, A. (1996): Indonesia's Cocoa Boom. Hands-off Policy Encourages Smallholder Dynamism. Policy Research Working Paper 1580. Washington, D.C.: World Bank.

Alberini, A., Longo, A. and M. Veronesi (2007): Basic Statistical Models for Stated Choice Studies. In: Kanninen, B. (Ed): Valuing Environmental Amenities Using Stated Choice Studies: 203-228. Dordrecht.

Alcamo, J. et al (2003): Millennium Ecosystem Assessment – Ecosystems and Human Well-Being: A Framework for Assessment. Washington.

Alpizar, F., Carlsson, F. and P. Martinsson (2001): Using Choice Experiments for Non-Market Valuation. *Economic Issues*, Vol.8: 83-110.

ANZDEC (1997): Report on the Central Sulawesi Integrated Area Development and Conservation Project. Palu.

Barkmann, J. et al. (*accepted*): Confronting Unfamiliarity with Ecosystem Functions: The Case for an Ecosystem Service Approach to Environmental Valuation with Stated Preference Methods. *Ecological Economics*.

Barkmann, J., Glenk, K. and R. Marggraf (2007a): Assessing Economic Preferences for Biological Diversity and Ecosystem Services at the Central Sulawesi Rainforest Margin – a Choice Experiment Approach. In: Tscharntke, T. et al. (Eds.): The Rainforest Margins: Linking Ecological, Economic and Social Constraints of Land Use and Conservation: 179-206. Heidelberg.

Barkmann, J., Schneider, E.-M. and S. Schwarze (2007b): Sweet as Chocolate: Stabilisation of Ecosystem Services by Production of Cocoa in High-Shade Agroforestry Systems in Central Sulawesi (Indonesia). *Book of Abstracts*, Tropentag 2005: 322. Witzenhausen

Bateman, I.J. et al (2002): Economic Valuation with Stated Preference Techniques – A Manual. Cheltenham.

Belsky, J.M. and F.S. Siebert (2003): Cultivating Cocoa: Implications of Ungrown Cocoa on Local Food Security and Environmental Sustainability. *Agriculture and Human Values*, Vol.20: 277-285.

Bennett, J. and W. Adamowicz (2001): Some Fundamentals of Environmental Choice Modelling. In: Bennett, J. and R. Blamey (Eds): The Choice Modelling Approach to Environmental Valuation: 37-69. Cheltenham.

Bingham, G. et al. (1995): Issues in Ecosystem Valuation: Improving Information for Decision-Making. *Ecological Economics*, Vol.14: 73-90.

Blamey, R., Louviere, J. and J. Bennett (2001): Choice Set Design. In Bennett, J. and R. Blamey (Eds): The Choice Modelling Approach to Environmental Valuation: 133-156. Cheltenham.

Blamey, R. et al (1997): Environmental Choice Modelling: Issues and Qualitative Insights. Research Report No. 4. School of Economics and Management, The University of New South Wales. Canberra.

Burkard, G. (2006): Community Based Resource Management in Central Sulawesi: Organizational, Legal and Structural Constraints. *Journal of Agriculture and Rural Development in the Tropics and Subtropics,* Supplement 84: 205-226.

Burton, J.A., Hedges, S. and A.H. Mustari (2005): Thetaxonomic Status, Distribution and Conservation of the Lowland Anoa *Bubalus Depressicornis* and Mountain Anoa *Bubalus Quarlesi. Mammal Review*, Vol.35(1): 25-50.

Clayton, L., Milner-Gulland, E.J. and A.P. Sarjono (2002): Sustainability of Rattan Harvesting in North Sulawesi, Indonesia. In: Maunder, M. et al. (Eds): Plant Conservation in the Tropics: Perspectives and Practice: 445-466. Kew.

De Groot, R., Wilson, M. and R. Boumans (2002): A Typology for the Classification, Description and Evaluation of Ecosystem Functions, Foods and Services. *Ecological Economics*, Vol.41: 393-408.

DeShazo, J.R. and G. Fermo (2002): Designing Choice Sets for Stated Preference Methods: The Effects of Complexity on Choice Consistency. *Journal of Environmental Economics and Management*, Vol.43(3): 360-385.

Faust, H. et al. (2003): Does Migration Lead to Destabilization of Forest Margins? Evidence from an Interdisciplinary Field Study in Central Sulawesi. STORMA Discussion Paper Series No. 11. Bogor.

Fromm, O. (2000): Ecological Structure and Functions of Biodiversity as Elements of Its Total Economic Value. *Environmental and Resource Economics*, Vol.16: 303-328.

Glenk, K., Barkmann, J. and R. Marggraf (2006): Locally Perceived Values of Biological Diversity in Indonesia – A Choice Experiment Approach. 8th Annual BIOECON Conference on Economic Analysis of Ecology and Biodiversity, Cambridge, UK.

Glenk, K. (2007): A Split Sample Experiment to Test for Effects of Attribute Order in Choice Experiments. In: Meyerhoff, J., Lienhoop, N. and P. Elsasser (Eds): Stated Preference Methods for Environmental Valuation: Applications from Austria and Germany: 81-104. Marburg.

Glenk, K. et al. (2006): Differential Influence of Relative Poverty on References for Ecosystem Services: Evidence from Rural Indonesia. 26th Conference of the International Association of Agricultural Economists 12-18 August 2006, Brisbane, Australia.

Glenk, K., Barkmann, J. and R. Marggraf (*accepted*): The Value of Ecosystem Services and Agrobiodiversity in Central Sulawesi. In: Kontoleon, A., Pascual, U. and M. Smale (Eds): Agrobiodiversity and Economic Development.

Greene, W.H. (2003). LIMDEP Version 8.0 NLogit Version 3.0. Plainview, N.Y.: Econometric Software, Inc.

Hensher, D. A. and W. Greene (2003): The Mixed Logit Model: The State of Practice. *Transportation*, Vol.30 (2), 133–176.

Hensher, D.A., Rose, J. M. and W.H. Greene (2005): Applied Choice Analysis: A Primer. Cambridge.

Jae, H. and D. del Vecchio (2004): Decision Making by Low-Literacy Consumers in the Presence of Point-of-Purchase Information. *The Journal of Consumer Affairs*, Vol.38 (2): 342-54.

Kahneman, D. and J.L. Knetsch (1992): Valuing Public Goods: The Purchase of Moral Satisfaction. *Journal of Environmental Economics and Management*, Vol.22: 57-70.

Keil, A. (2004): The Socio-Economic Impact of ENSO-related Drought on Farm Households in Central Sulawesi, Indonesia. Ph.D. thesis, Goettingen.

Keil, A., Birner, R. and M. Zeller (2007): Potentials to Reduce Deforestation by Enhancing the Technical Efficiency of Crop Production in Forest Margin Areas. In: Tscharntke, T. et al. (Eds): The Stability of Rainforest Margins: Linking Ecological, Economic and Social Constraints of Land Use and Conservation: 389-414. Heidelberg.

Kontoleon, A. and M. Yabe (2004): Assessing the Impacts of Alternative 'Opt-out' Formats in Choice Experiment Studies. *Journal of Agricultural Policy Research*, Vol.5: 1-32.

Krinsky, I. and A. L. Robb (1986): On Approximating the Statistic of Elasticities. *Review of Economics and Statistics*, Vol.68: 715-719.

Krupnick, A. and W.L. Adamowicz (2007): Supporting Questions in Stated Choice Studies. In: Kanninen, B. (Ed): Valuing Environmental Amenities Using Stated choice Studies: 43-66. Dodrecht.

Krutilla, J. V. (1967): Conservation Reconsidered. *American Economic Review*, Vol.57(4): 777-786.

Lancaster, K. (1966): A New Approach to Consumer Theory. *Journal of Political Economy*, Vol.74: 132-157.

Lancaster, K. (1991): Modern Consumer Theory. Brookfield.

Louviere, J. (1988): Analyzing Decision Making: Metric Conjoint Analysis. Sage University Paper Series on Quantitative Applications in the Social Sciences, 07-67. Beverly Hills.

Louviere, J., Hensher, D. and J. Swait (2001): Stated Choice Methods – Analysis and Application. Cambridge.

Luce, R.D. (1959): Individual Choice Behaviour. New York.

Luce, M.F. (1998): Choosing to Avoid: Coping with Negatively Emotion-Laden Consumer Decisions. *Journal of Consumer Research*, Vol. 24: 409-433.

Kaplowitz, M.D. and J.P. Hoehn (2001): Do Locus Groups and Individual Interviews Reveal the Same Information for Natural Resource Evaluation? *Ecological Economics*, Vol.36: 237-247.

Maertens, M. (2004): Economic Modelling of Agricultural Land-Use Patterns in Forest Frontier Areas. Ph.D. thesis, Goettingen.

Manski, C. (1977): The Structure of Random Utility Models. *Theory and Decision*, Vol.8: 229-254.

Marggraf, R. and R. Birner (1998): The Conservation of Biological Diversity from an Economic Point of View. *Theory in Biosciences*, Vol.117: 238-255.

Mathews, K.E., Freeman, M.L. and W.H. Desvousges (2007): How and How Much? The Role of Information in Stated Choice Questionnaires. In: Kanninen, B. (Ed): Valuing Environmental Amenities Using Stated choice Studies: 111-134. Dordrecht.

McFadden, D. (1973): Conditional Logit Analysis of Qualitative Choice Behavior. In: Zarembka, P. (Ed): Frontiers in Econometrics: 105-142. New York.

Mogas, J. Riera, P. and J. Bennett (2006): A Comparison of Contingent Valuation and Choice Modelling with Second Order Interactions. *Journal of Forest Economics*, Vol.12: 5-30.

Myers, N. et al. (2000): Biodiversity Hotspots for Conservation Priorities. *Nature*, Vol.403: 853-858.

Nunes, P. and J. v.d. Bergh (2001): Economic Evaluation of Biodiversity: Sense or Nonsense? *Ecological Economics*, Vol.39: 203-222.

Olson, D.M. and E. Dinerstein (1998): The Global 200: Representation Approach to Conserving the Earth's Most Biologically Valuable Ecoregions. *Conservation Biology*, Vol.12: 502-515.

Pattanayak, K. and R. Kramer (2001): Pricing Ecological Services: Willingness to Pay for Drought Mitigation from Watershed Protection in Eastern Indonesia. *Water Resources Research*, Vol.37(3): 771-778.

Pearce, D.W. and D. Moran (1994): The Economic Value of Biodiversity. IUCN – The World Conservation Union. London.

Randall, A. and J. Stoll (1983): Existence Values in a Total Valuation Framework. In: Row, R. D. and L.G. Chestnut (Eds): Managing Air Quality and Scenic Resources at National Parks and Wilderness Areas: 265-274. Colorado.

Revelt, D. and K.E. Train (1999): Customer-Specific Taste Parameters and Mixed Logit. Working Paper, Department of Economics, University of California, Berkeley.

Sagoff, M. (1996): On the Value of Endangered and Other Species. *Environmental Management*, Vol.20(6): 897-911.

Samuelson, W. and R. Zeckhauser (1988). Status Quo Bias in Decision Making. *Journal of Risk and Uncertainty*, Vol.1: 7-59.

Sastry, C.B. (2001): Rattan in the Twenty-First Century. *Unasylva*, Vol.52: 3-10.

Schippers, B., et al. (2007): Effects of the Interaction of Migration, Financial and Social Capital on the Adoption of Agricultural Innovation and on Social Stratification in Central Sulawesi (Indonesia). Book of Abstracts, Tropentag 2005 189. Witzenhausen.

Schwarze, S. (2004): Determinants of Income Generating Activities of Rural Households: A Quantitive Study in the Vicinity of the Lore-Lindu National Park in Central Sulawesi/Indonesia. Ph.D. thesis, Goettingen.

Siebert, F.S. (2001): Sustainable harvesting of wild rattan: Viable Concept or Ecological Oxymoron? *Unasylva*, Vol.52: 36-45.

Siebert, F.S. (2002): From Shade to Un-grown Perennial Crops in Sulawesi, Indonesia: Implications for Biodiversity Conservation and Soil Fertility. *Biodiversity and Conservation*, Vol.11: 1889-1902.

Sitorus, F. (2002): Land, Ethnicity and the Competing Power. Agrarian Dynamics in Forest Margin Communities in Central Celebes, Indonesia. STORMA Discussion Paper Series No. 6, Bogor.

Slovic, P. and S. Lichtenstein (1971): Comparison of Bayesian and Regression Approaches to the Study of Information Processing in Judgment. *Organizational Behavior and Human Performance*, Vol.6: 649-744.

Steffan-Dewenter et al. (2007): Tradeoffs Between Income, Biodiversity, and Ecosystem Functioning During Tropical Rainforest Conversion and Agroforestry Intensification. *Proceedings of the National Academy of Sciences*, Vol.104: 4973-4978.

Suryahadi, A. and S. Sumarto (2001): The Chronic Poor, the Transient Poor, and the Vulnerable in Indonesia Before and After the Crisis. SMERU working paper. Jakarta: SMERU Research Institute.

Thurstone, L.L. (1927): A Law of Comparative Judgment. *Psychological Review*, Vol.4: 273-286.

Train, K. (1998): Recreation Demand Models with Taste Differences over People. *Land Economics*, Vol.74(2): 230-240.

Vedeld, P, A. et al (2004): Counting on the Environment: Forest Incomes and the Rural Poor. Environment Department Papers No. 98. The World Bank. Washington.

Waltert, M., Mardiastuti,A. and M. Mühlenberg (2004): Effects of Land Use on Bird Species Richness in Sulawesi, Indonesia. *Conservation Biology*, Vol.18: 1339-1346.

Weber, R. (2005): Kulturlandschaftswandel während des 20. Jh. in Zentralsulawesi – eine Historisch-Geographische Analyse der Lore-Lindu-Bergregenwaldregion. Ph.D. thesis, Goettingen.

Whitten, A.J., G.S. Henderson, and M. Mustafa (1987): The Ecology of Sulawesi. Yogyakarta,

Whittington, D. (1998): Administering Contingent Evaluation Surveys in Developing Countries. *World Development*, Vol.26(1): 21-30.

Zeller, M., Schwarze, S. and T. van Rheenen (2002): Statistical Sampling Frame and Methods Used for the Selection of Villages and Households in the Scope of the Research Programme on Stability of Rainforest Margins in Indonesia (STORMA). STORMA Discussion Paper Series No. 1, Bogor.

Zeller, M. and R. Birner (2003): Teilprojekt A3. Der Einfluss von Märkten und Politikmaßnahmen auf die Ressourcennutzung. in: Sonderforschungsbereich 552 "Stability of Rainforest Margins in Indonesia" der Deutschen Froschungsgemeinschaft. Ergebnisbericht über die 1. Förderungsphase 2000/2-2003/1. Georg-August-University Goettingen and University of Kassel.

Chapter XI

CONCLUSIONS: EVOLVING SOCIAL ORGANIZATIONS AND FOREST SUSTAINABILITY

Günter Burkard

Abstract

This conclusion summarizes some of the major findings of this book and tries to link them to recent approaches in natural resource management. Evidence from Central Sulawesi suggests that successful decentralization in natural resource management must not only devolve authority to existing bureaucratic sub-systems, but must also consider "spontaneous orders of co-operation" as well as other "decomposable systems" with high rates of interaction. These subsystems in turn, must be ordered within a nested fashion of organizational hierarchies including spontaneous as well-designed structures in order to safeguard people's security concerns. Instead of making reductionist policy recommendations, researchers should provide integrated information to the subgroups and stakeholders concerned in order to facilitate learning processes, the contents of which cannot be known by the various subgroups and stakeholders themselves. Thus, in enhancing transparency, "integrative information" allows interests and perceptions to co-evolve. However, researchers must give decision-makers full freedom in searching for solutions and accepting decisions without suggesting certain preferences on scientific grounds.

The last decade revealed an increasing scepticism towards static approaches in the analysis of natural resource management systems. Standard economic models like *maximum sustainable yield*, *short term optimization*, and *individual profit-maximization* as well as reductionist preservation perspectives such as *single-species-*

management are basically derived from "images of closed and mechanic systems" (Rammel 2007: 26) characterized by an equilibrium of stable cause and effect relationships that allow for planned interventions with predictable outcomes (Wilson 2002: 334). These approaches are now challenged by more non-static, dynamic and process-oriented perspectives of local resource management as enshrined in concepts such as *co-evolution*, *adaptive collaborative management* and *complex adaptive systems*, which recognize explicitly the inherent complexity, variability and unpredictability of any natural resource management system. Thus, the concept of co-evolutionary potentials of human society and nature is increasingly adopted in environmental research. What does this mean? In biology, the term *co-evolution* denotes the mutual evolving influence between two species. The basic idea behind the concept in social science is that through technology, social organization and knowledge, society and the natural environment are not only interconnected, but also "co-evolve" through mutual processes of adaptation (Norgaard 1994). This mutual evolution however, is not a linear process. There is no "progress" in the sense of having reached or achieved a fixed stage of development on the road to sustainable management (Rammel et al. 2007: 30). In sustainable natural resource management we never "arrive". Further, embracing society and ecosystems alike, *complex adaptive systems* (CAS) may simultaneously display non-linear causal relationships at different levels. Thus, factors operating at different scales may impact particular events and cause-effect-relationships in varying degrees. Complex adaptive systems do not

> *"lend themselves to long-term prediction consistent with the needs of sustainability because of their changing, complex and usually non-linear causal relationships. We may be able to understand the structure and dynamics of these systems without being able to predict anything but broad patterns, or propensities" (Wilson 2002: 351).*

For researchers and policy makers alike, the recognition of natural resource management systems as CAS implicates a radical break with the long-established belief in our capacity to control and plan. Instead, it implies that sustainable development must not only take account of what we already know but also of what we do not (or not yet) know, so that the major goal is to make resource management systems *resilient* to those conditions of unpredictability and unexpected surprises that we have not been able to anticipate (Colfer 2005: 12).

What was said about complex adaptive systems in general is of special relevance to common pool resources (CPR).[1] Theories of common pool dilemmas however, are often based on generalizing models of abstract understandings of "rational action". Others (Ostrom 1990, Baland and Plateau 1996, Wade 1988) tried to identify the characteristics of user groups and institutional arrangements that may foster or impair sustainable resource use. Despite the various approaches used in addressing micro-institutional problems of resource degradation, there is a general consensus that "institutions matter" (Agrawal and Yadama 1997: 436) and that the main constraints on sustainable resource management are now organizational, not technical in nature (Grainger 1995: 3). The problem lies in the fact that it is often difficult to draw abstract conclusions on the basis of case studies, given the high degree of cultural variance and socio-economic differentiation to be found between local communities, even if they are located close to each other (as the contributions of this reader vividly demonstrate). In general, formal organizations are believed to be more robust than informal ones. Their preferred characteristics (the so-called "design principles") revolve mainly around specified user groups, clarity of resource boundaries, graduated sanctions and formalized institutions of, ideally, peaceful conflict resolution (Hobley and Shah 1996, Ostrom 1990). Despite the fact that most researchers agree that local institutions are crucial and that there must be some autonomy for local institutions in order to develop and to grow ("evolve"), few research focused on the question of how natural systems evolve. As Norgaard points out, in most cases "institutionalists have emphasized institutions apart from nature" (Norgaard 1984: 160).

The notion of rational action referred to above links up with approaches that explain common dilemmas first in terms of "market failure". The variations in problems, constellations, interests, situa-

[1] "Common pool resources" (CPR) are resources that are difficult to protect from potential abuse and that are threatened by exploitation and degradation in case deliberate rules of access are either missing or not enforced. They must be distinguished from "common property regimes" which are also abbreviated as CPR.

tions and events presented in the preceeding chapters show clearly that the observed "tragedy" of forest degradation in Central Sulawesi cannot be explained by the Neo-classical model of "market failure".[2] Instead, we agree that the uncontrolled use and abuse of forest resources may stem also from "community failure" (McCay and Jentoft 1998: 25). Sunito (this volume) raises the question of what actually is "community" in community-based natural resource management (CBNRM). One may answer that - as in any context - "community" in CBNRM is more than the sum of its individual parts (in this case its resource users), but a symbolic construction, a socially integrated *we* (Etzioni 1988, cited in McCay and Jentoft 1998: 23). This *we* however, does not necessarily match with the confines of a certain village. The Indonesian notion of "village" (*desa*) denotes first of all an administrative and geographical social entity, not a symbolic construction. In our view, "community" is best described in terms of an "adaptive system", that is a "process of communication and adjustment that constantly modifies the relationships among the various components, while at the same time preserving certain process mechanics" (Bennett 1985: 41). The often static perception of "community", and the focus on "institutionalization" in conservation and development projects correlates with a similar static perception of "sustainability" as a fixed objective of conservation efforts.[3] From a co-evolutionary point of view, "sustainable development is an open evolutionary process of improving the management of social-ecological systems through better understanding and knowledge" (Rammel et al.2007: 9). This perception seems more appropriate in that it grants decision-makers and resource managers more flexibility in dealing with chaos, uncertainty, and unpredictability as well as with different and changing temporal,

---

2 As Norgaard (1984: 161) remarks correctly, Neo-classical theory perceives of environmental problems in economics first of all as problems of "market failure" rather than as evidence for the applicable limits of the "market model" itself.

3 To a great extend, the Storma-research project, in which all contributors in this book have participated and in the context of which our data and conclusions are derived, shares this shortcoming. The project title STORMA (Stability of Rain Forest Margins) suggests a conception of forests as static entities separated from man rather than as evolving systems which adapt to and interact with society.

spatial and organizational hierarchies.[4] It further allows taking account for historical, cultural and social contexts, and for unexpected, non-linear events, chances and situations inherent in the dynamics of local communities.

The impending tragedy of the commons encountered in the villages of Wuasa (Sunito, this volume), Watumaeta and Sintuwu (Burkard, this volume) and in Dongi-dongi (Adiwibowo, this volume) are the product of complex situations, configurations, and interactions that led to irreversible changes in the agrarian structure and disruptions of long-established security conditions. These intricacies escape the explanation capacity of economic models. Referring to the notions of "thin" in contrast to "thick" description and analysis as conceptualized by American anthropologist Clifford Geertz (Geertz 1971), McCay and Jentoft (1998) lament the dominance of "thin" modes of community and commons analysis at the cost of "thick" ones and call for a stronger awareness of the situated "embeddedness" of local organizations and resources within encompassing systems (and sub-systems) at different layers (ibid: 22). Most contributions in this book are of the "thick description" type, indicating a more ethnographic, micro-detailed and qualitative approach, taking into consideration economic and social structures and relations of dominance within the communities. In contrast to generalizing models and grand theory, my contention is that micro-analysis in local resource management (or community based natural resource management) has not been micro enough.

In general, co-evolutionary adaptations between society and the natural environment are most likely to occur in times of instability, conflict, disturbance and social or ecological crisis. Apart from an unstable national institutional environment related to the transition from autocratic rule to regional autonomy/decentralization, village heads, park rangers, county officers and other decision makers in the research area find themselves in an uncertain situation characterized by market pressure, unprecedented changes in land use, communal

[4] Following Simon (1974, cited in Rammel et al. 2007: 18), our understanding of "hierarchies" exceeds top down structures of authority and control, and includes also semi-autonomous, transitory levels and intersections that are created by interactions of several sub-systems. Such intersections operate as "moving targets" emerging at multiple scales.

violence, uncoordinated in-migration, low institutional reliability, vaguely defined authority, legal pluralism and ecological vulnerability at the same time certain "process mechanics" (Bennett) or "structural configurations" (Rammel et al. 2007) such as co-operation in routine activities in agriculture, kinship exchanges, ritual activities and a minimum of neighborhood solidarity remain intact. Such situations of mixed conditions of stability and instability are usually referred to as "bounded instability" (Stacey 2003, cited in Wollenberg et al. 2007). The cases of Dongi-dongi (Adiwibowo), Watumaeta (Burkard), Wuasa (Sunito), and to a lesser extend Sedoa (Döbel) and Bolapapu (Savitri and Fremerey) presented in the preceding chapters may well be read in terms of a continuum of "bounded instability". In such circumstances, co-operation between individuals and groups may often be spontaneous, self-organized, and flexible and may not match the calm and stable image usually accredited to local communities. This may present tremendous problems for development and conservation practitioners/facilitators looking out for "grass-roots-organization" in order to build up on existing "indigenous" structures. A hallmark of such spontaneous co-operation is that it is not necessarily congruent with a clearly defined organizational, administrative or social hierarchy. This includes spontaneous co-operation and decision-making that may operate not only at levels and scales that are different from formally designed ones, but also scales that differ from pretended "indigenous structures".[5] In times of crisis, rather than relying on formal designs of co-operation or long-established grass-roots-organizations (which are often de facto imagined entities), facilitating efforts are probably better advised to focus on emerging "spontaneous orders of co-operation" (Wollenberg et al.2007: 3).

With regard to uncertain and unpredictable conditions, McCay (2002: 375) has argued for the adoption of a so-called "muddling-through"-approach, characterized by small steps with low initial costs in order to give communities the necessary room for manoeuvre, adjustment and experiment. The "muddling through" metaphor was first popularized by Lindblom (1959) before it gained wider attention since

[5] From another perspective, this may forestall a situation where grass-roots organizations "become the human software through which investments can be made with least local opposition" (Hildyard et al.2001: 59).

the late 80ties as scientists became increasingly confronted with extreme complexities (e.g. Bendor 1995, Bernstein and Fortun 1998, McCay 1988, Wollenberg et al. 2007). This small-step, incremental attitude is of special importance when the risks of far-reaching change are very high (Wollenberg et a. 2007: 5). Following McCay's reasoning (ibid: 374), such steps should not be guided by larger values and goals in the first place. In practice however, this means that "facilitating" must develop a high degree of patience, creativity and flexibility to respond to emerging opportunities. This however, might be difficult to achieve in Indonesia, where significant numbers of extension officers, forestry staff and other bureaucrats are engrained with clichés of upland people as more or less uneducated forest destroyers heavily in need of hillside development. They may not accept that the obvious disorder of those they are trained to "develop" may be meaningful to their facilitating and ordering efforts. As Murray Li (2001: 651) reminds us and as the articles in this book have demonstrated in much detail, many of the tasks usually associated with central governments are now performed by "non-governmental organizations" (NGOs). Investigating the Dongi-dongi case, Adiwibowo shows convincingly how – guided by their respective "moral visions" – advocacy, development and conservation NGOs engage in serious hostilities when the reform and improvement of local communities (as informed by their specific agendas) is at stake. The "muddling through" approach is intrinsically linked to the questions of program implementation and participation, irrespective of whether the implementing agencies are government institutions or NGOs. The question arising then is: If facilitating objectives (as the approach outlined above implies) must be defined by the results of (often small step) negotiations with various groups at different layers, are there any fundamental issues or principles that must be "preserved at all costs"? (Brosius et al.: 5).[6] We must admit that there is no generally valid answer to this question.

The recent history of the community of Toro (Shohibuddin, this volume; Burkard, this volume) showed that in this special case, "facilitation" by far exceeded the official objective of "empowerment".

---

[6] Such fundamental issues may be the realization of a national development goal, the strict conservation of a certain bio-diversity hotspot etc.

With the successful presentation of Toro as an environmentally sound customary community at the UNDP (United Nations Development Program), it was not only the community, but also the facilitating NGO that won international acclaim. This, in turn, may open up further sources to be tapped for additional funds. In his analysis of local institutions in Toro, Shohibuddin found a somewhat diachronic picture in which increased "democratization" coincided with the re-installation of noble privileges. Burkard's findings, on the other hand, suggest that given a high degree of participatory exclusion, it is not the common community member who has been "empowered" by the facilitation process. In the Napu valley, (two villages located in Napu, namely Wuasa and Watumaeta are discussed in this book by Sunito and Burkard respectively) on the other hand, the facilitating NGO (*The Nature Conservancy*) is an official partner of the government and was actively involved in the management of the park from the beginning.

Despite site-specific differences, both cases question how far NGO-knowledge in general (and conservation knowledge disseminated via NGOs in particular) represent a "knowledge of resistance" or a "knowledge of domination". Linking up with this issue in their book *Communities and Conservation*, Brosius et al (2005: 6) ask the question if NGOs are better understood as parts of the "imperial regime" or if they represent – via the support of grass-roots movements – a counter-culture/counter-force in opposition to the "empire". The author of this concluding remarks is inclined to presume the existence of only *one, single empire* as laid out by Hardt and Negri (2000, cited in ibid: 6). Following this viewpoint, NGOs are absorbed into the bio-political contexts of the constituting "empire" itself (Hardt and Negri 2000: 50f). Thus, NGOs anticipate the power of the "pacifying intervention", which represents one of the most significant hallmarks of the emerging transnational "empire". Both, the mapping of "indigenous" forest areas by a local advocacy NGO in Toro (Shohibuddin, this volume), and the mapping of the national zoning scheme by an international conservation NGO in Watumaeta (Burkard, this volume), represent well known forms of "domestication" and "simplification" that are usually associated with the "high modernism" of state control

(Scott 1998: 184).[7] The same holds true for the codification of customary law by the advocacy and the codification of village agreements by the conservation NGO respectively. "Steeling the master's tools" (Brosius et al 2005) has meanwhile become a cornerstone of the implementing strategies of many NGOs. Similar to their official counterparts, NGOs rely on standardized routines and work often under similar efficiency criteria. As Adiwibowo illustrates in the case of the zoning delineation of the National Park conducted by TNC, through their dependence on external agencies in order to raise funds, NGOs may be severely pushed by time pressure. Thus, it is questionable if governmental or non-governmental agencies engaged in organization building and institutionalization are sufficiently equipped and flexible enough to respond to spontaneous orders of co-operation and the arising opportunities they may imply.

From the various case studies presented in this book, the cases of Dongi-dongi and Watumaeta have not only been the most dramatic ones, threatening to turn into a real "tragedy of the commons" at the beginning of our research; in retrospect, both also showed most clearly the characteristics of evolving systems. The problem, however, is how to maintain spontaneous co-operation, as it emerges in often unpredictable manners in Dongi-dongi and Watumaeta. Dongi-dongi is without doubt the place where complexity, events, spontaneous surprises, and the degree of incalculability have been most extreme given the political significance and the operations of the various outside "forces of power" which surrounded the occupation. If spontaneous co-operation emerges under such conditions, it requires continuos efforts and inputs as well as flexibility and patience to deal with notorious unpredictability in order to maintain it. Thus, when spontaneous orders of co-operation between the hostile conservation and advocacy NGOs of Central Sulawesi emerged in June 2001 (Adiwibowo, this volume), no efforts were made to maintain or manage it and no further opportunities for co-operation were identified Thus, a promising initiative ended within two weeks. In Watumaeta co-operation has been

7 The case of Toro as seen from different angles by Shohibuddin and by Burkard raises not only the question of "whose idea of community is mapped?" (Brosius et al. 2005 : 20), but also the empirically related question of "whose perception of property is mapped"?

better-maintained due to clearer lines of conflict. Maintenance of co-operation seemed vital given the threat of communal violence that may be imported by the bloodshed of the Poso-crisis (of course, the factors that fostered co-operation in Watumaeta in the first place are not necessarily the same which may lead to sustainable resource use in the long run). The situation in Toro is another matter again. Here, a process of a facilitated organizational consolidation as it was informed by national statutes and an institutionalization of noble privileges, lost momentum and legitimacy as it disclosed a serious inflexibility in resource allocation and massive shortcomings with popular participation. As Stacey (2003: 226, cited in Wollenberg et al. 2007: 3) points out, a spontaneous co-operation requires "great effort to retain its structure and relatively little to change it," as opposed to an equilibrium structure (or a consolidated institution) that "requires no effort to retain its structure but great effort to change it".

As Berkes (2002) argues correctly, neither pure local-level nor pure state-level control of resources have proven to be adequate units of natural resource management. This leads us to the problem of nested hierarchies and cross-scale interactions: Discussing the village of Watumaeta, Burkard (this volume) pointed out to the necessity of village co-operation across horizontal scales in order to strengthen the bargaining power of rural communities vis-à-vis the central and provincial government. It was further argued that vertical integration in the form of "nested hierarchies" is of the same importance, because in times of crisis community leaders may need support from upper government agencies when dealing with communal violence and threatened security. On first glance it might look contradictory that one researcher argues that one village is probably a unit too small to deal with the given complexities. At the same time, a co-researcher in a neighboring village (Sunito, this volume), argues for the consideration of social units below the village level as targets of community-based resource management programs. This, however, is not a contradiction in the real sense. As pointed out above, the Indonesian "village" is not congruent with a symbolically constructed "community" imbued with meaning and values (Cohen 1985). Of course, there can (and must) be horizontal alliances between sub-systems other than the "village" if those are more meaningful in regard to a specific problem. This also has the advantage that persons with close day-to-day relationships or people confronted with similar challenges can find a compromise more quickly. With sub-systems, which also include semi-autonomous transitory levels and intersections, horizontal alliances and vertical inter-linkages are even more important. As Berkes (2002: 316) points out, if natural resource management becomes too decentralized, the necessary feedback between user groups of adjacent areas may be lost.

Wilson (2002: 342) further develops this idea by referring to Simon's conception of the "decomposable sub-system" (Simon 1962), the boundaries of which are defined in terms of the highest rates of interaction. Within decomposable (or almost decomposable) sub-systems, information is generated, shared and disseminated by that part of the system with the most profound and pertinent first hand knowledge of a given problem or issue. Further, the problem of accountability can be addressed more effectively (ibid: 342). This organization principle matches well with the perquisites of a "learning organization" (Senge 1990). Because of the complexity of the given systems (heterogeneity of groups as well as of endowments and forest conditions), no one individual, sub-group or category of people may be able to identify all relevant problems or to "know" the appropriate learning content for their solution. The idea is thus to bundle activities which are characterized by stable interactions and to link them to the rest of the system only when the sub-system's activities need to be coordinated with the activities of other parts of the system. The advantage is that "the scale of operation of each component of the organization can always be chosen" (Wilson 2002: 343). It goes without saying that spontaneous orders of co-operation derived from real life problems are crucial in the formation of meaningful decomposable sub-systems.

Sunito (this volume) argues that the attitude of Wuasa villagers towards the establishment of the National Park is far from homogenous, but is heavily informed by, for example, the location of one's home lot in the village and the location of one's plots in relation to the park borders. In facilitating "village conservation agreements" for 60 villages surrounding the park (Mappatoba 2004; Burkard, this volume), in 2002 TNC started to introduce a separate formal organization on the village level, the *Village Conservation Council* (Lembaga Konservasi Desa, abbr. LKD) which is in charge of the implementation of the agreements (including monitoring tasks) and of facilitating communication between the village society on one hand, and the Park Authority and TNC on the other. Conflicts between villages, when several sites in the forest are simultaneously claimed by different communities, or problems that cannot be solved by the LKD alone, are to be handled by a corresponding *District Buffer Zone Forum* (Forum Wilayah Penyangga, abbr.: FWP). Whereas the importance of vertical "cross-scale inter-linkages" (Berkes 2002) was acknowledged via the integration of intermediate levels of organization (*kecamatan*), the organizational layout conforms strictly to the national bureaucratic structure. The lowest subsystem within the organizational structure remains the village (*desa*). No other possible orders of sub-systems have been taken into account. In mid 2004 serious problems arose in Watumaeta when the LKD started to plant the border of the National Park with candle nut (*kemiri*). Candle nut trees have been planted in four rows with a planting distance of 5 x 5 m in such a way that two rows are located inside and two rows outside the National Park border (figure 1). The plan followed a suggestion made by the head of the National Park subdistrict office in neighboring Wuasa. The former (inner rows) are to cover the operational costs of the monitoring organization (LKD); the latter (outer rows) are to be harvested by the owners of adjacent plots as an incentive to comply with the border demarcation.[8] Unfortunately, the affected plots near the forest margin are relatively far away from the core village area and function also as home lots inhabited almost exclusively by Buginese migrants who belonged to the most recent immigration stream that entered the village since the economic crisis in 1997/98 (Burkard, this volume). Because the villagers concerned were not involved in the procedure, owners of plots at the forest margin thought that with the help of the candle nut markers the border itself should be moved outwards so that their plots became located inside the protected area.

---

[8] At the time of planting, prices for candle nut have been Rp. 2000 per kg.

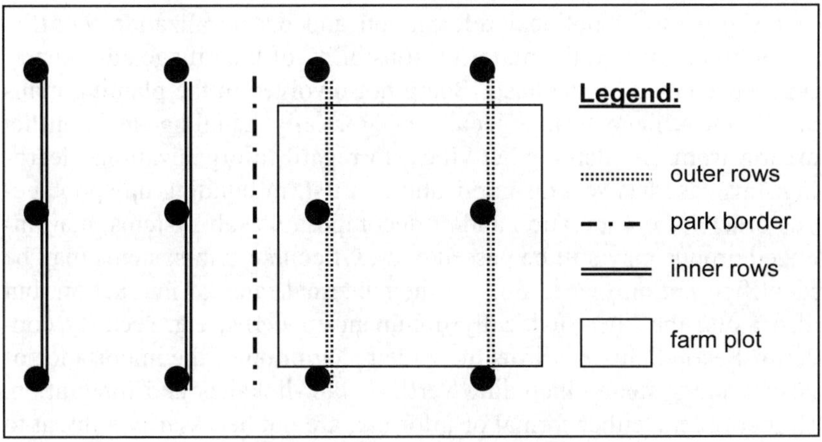

Figure 1. *Planting pattern of border demarcation (inner and outer candle nut rows)*

Being Muslims in a Christian diaspora where accumulated resentments had previously resulted in communal violence (Aragon 2001, Lieb 2003), most migrant farmers try to keep a low profile and refrain from involving themselves in village organizations and politics. The population concerned had neither been involved in the planning activities, nor been informed before the implementation started. The village-level LKD was the smallest sub-system within the management structure, albeit only a part of the villagers were directly affected by the implementing steps of the agreement. Within the planned framework of the conservation agreement spatially and socio-politically marginalized subgroups had neither been identified nor considered. Discussing this issue, the head of the National Park branch office in Wuasa pointed out that Watumaeta has always been a village of "troublemakers". The concrete problems faced by a sub-system (migrants living at the forest margin) was perceived as a problem or aggregate phenomenon of the higher hierarchical sub-system "village" that already fulfilled a certain image in his mind-set.

When the LKD (supported by several park rangers) started the planting of candle nut at the forest margin, its equivalent on the next higher administrative level, the FWP, was well informed. However, the village head had been neither involved nor consulted. Although there has been a revival of the customary law council (*lembaga adat*) in the process of political reformation and decentralization, conflict resolution remained the major responsibility of the village administration represented by the head. Being not involved in the planting plans at all, the village head rejected categorically handling any conflict arising from the planting activities. In regard to organizational learning, this case has several implications. First, in building up spontaneous orders of co-operation and/or decomposable sub-systems, marginalized groups may still be passed over. Of course, sub-systems may be identified not only according to their internal rates of interaction, but also along the lines of locally dominant concerns, e.g. security concerns. Second, this confirms the wide recognition of the importance of nested sub-systems, including vertical inter-linkages and integration. Higher orders, either formal or informal, are not necessarily a threat to sub-systems, but may play a crucial role as their security provider. Third, identifying meaningful orders of co-operation and interest does not mean denying the importance of formal structures. Instead, a nested fashion of hierarchical vertical and horizontal inter-linkages must include both formal as well as informal structures. This kind of poly-centric organization may well run counter to the convictions of government officials, often ingrained with centralist or even autocratic ideas, despite the official decentralization policy. Whereas it will probably be more in line with theories on collective learning and organizational performance, it is certainly not congruent with the ideas of "species-centred approaches" that are adopted by many conservationist NGOs.

Many problems discussed in this book are linked to issues of *security* or *socio-economic security*, which is a central determining factor in local resource utilization and man-forest interactions. However, rather than representing a separate "context variable" on the same taxonomic level with and clearly distinguished from other contexts such as "legal reliability" or "autonomy of organization", as primarily supposed when the research project was designed (Fremerey and Amini 1999), *security* is better perceived as an underlying, "residual category" influencing people's situated choices and decisions. According to circumstances, perceived insecurity may be related to land tenure, issues of social organization, legal reliability, gender relations, ethno-religious tensions etc. In times of massive market penetration and subsequent changes in land use ("cacao revolution"), the major dimension which defines security/insecurity may well shift from social organisation and leadership structures during the phase of land alienation (large tracts of local reserve land have been transferred to migrants by the village elite) to relations of production after land alienation resulted in significant land use changes (exclusionary work arrangements). However, further developments are almost impossible to predict. The case of Watumaeta (Burkard, this volume) reveals a pattern where – despite remaining problems and shortcomings like the one discussed above – a situation of mutual vulnerability and broadened organizational autonomy provided by decentralization policies resulted in meaningful self-organization, including a more democratic form of leadership. This, however, is linked to the specific constellations in this village and specific encounters, events and situations as they emerged at a certain point of time. It would therefore be misleading to suggest a general trend towards village democratization as a result of the decentralization policy. In the neighbouring village of Sedoa a relatively democratic style of leadership was replaced by a more autocratic one (Döbel, this volume) and the well-honed revitalization of customary law in Toro implied first of all a revitalization of noble privileges and hampered rather than fostered popular participation (Burkard, this volume and Döbel, this volume). This unpredictability of co-evolving systems at several levels urges facilitators and researchers alike to adopt a "go-slow", incremental approach as it is coined in the "muddling-through-metaphor" referred to above.

A second red thread running through the chapters of this book is the notion of *local knowledge*. Following an actor-oriented approach focusing on "knowledge encounters" and "knowledge interfaces" as laid out by sociologist Norman Long (1992), it is especially the intersections of nested sub-systems where "agents" interact with other individuals, groups and the ecosystem (e.g. when leaders are tapping outside sources of knowledge) and where knowledge is acquired, transformed, and created. As the case of Toro suggests (Burkard, this volume and Shohibuddin, this volume), it is not only important that villagers adapt external knowledge to their own world. Instead, the more important question is how far they regard themselves as "owners" of that knowledge. Whether in the case of community empowerment (or "leadership empowerment") in Toro or in the context of gendered local knowledge (Savitri and Fremerey, this volume), acquiring relevant knowledge succeeded in both cases in the creation of relations of dominance and power or an unequal relational structure. As pointed out above, no one individual or sub-group of people may be able to identify all relevant problems or to "know" the appropriate learning content for their solution. Although not explicitly dealt with in this volume, here the idea of complementary sources of knowledge is of primary importance. There is no doubt that methods of local knowledge generation (as long-term observations and trial-and-error-procedures) have produced exact and reliable results. However, facing a tremendous acceleration of change in complex natural as well as socio-economic and political systems, it may need the recourse to complementary sources and modes of knowledge in order to put local resource management on a sounder base. In the future, scientists must pay attention to the question of how different spheres of knowledge converge and in how far "knowledge gaps" in one sphere can be complemented by knowledge from other spheres. In this sense, as the previous papers have demonstrated in much detail, local knowledge should not be treated as an isolated entity, but be conceived relationally, that is, as a product of social practices rather than in terms of a "fixed commodity" (Mosse 2001: 17).

However, one must keep in mind that when local people integrate external knowledge into their existing life worlds, certain knowledge contents may be imbued with meanings different from those of the original knowledge provider (Fremerey 2005: 256). This statement is of special relevance in regard to ecosystems knowledge. As Glenk skilfully shows (this volume), it is of eminent importance to "translate" the complex aspects of ecological functions into concepts that community members may perceive as valuable and significant for their lives. Of course, heterogeneity excludes any attempts to define one common basis of ecosystem service valuation for all communities or villages.

This leads to the question of the role of the researcher working within complex adaptive systems. Meanwhile, it is common to add a "policy recommendation" at the end of a scientific paper. The findings presented in this book may be read as a kind of warning against conclusions and policy recommendations made too quickly. Given the complexity of ecological and social systems, interventions may not only have outcomes, but may have effects on variables and sub-systems at different layers which may not be anticipated by the recommending researcher. Thus, "outcomes" of interventions may themselves become independent variables with non-linear consequences. "We learn best from our experience, but we never directly experience the consequences of many of our most important decisions" (Senge 1990: 23). Or in the words of French philosopher Michael Foucault: "People know what they do; they frequently know why they do what they do; but what they don't know is what what they do does" (Michel Foucault, quoted in Dreyfus and Rabinow 1982: 187). Instead of a short-hand "policy recommendation", researchers should provide what Rammel et al. (2007: 25) called "integrative information". That is, trying to facilitate a mutual understanding of ecosystems on the one hand, and stakeholders, sub-groups, and their specific conditions and life worlds on the other. Here, the social scientist is especially challenged to create awareness for the "multiple realities" of the actors involved. In enhancing transparency, "integrative information" allows interests and perceptions to co-evolve and supports a shared understanding to develop by adaptive learning. However, researchers must give decision-makers full freedom in searching for solutions and taking decisions without suggesting certain preferences on scientific grounds.

References

Agrawal, A. and N. Yadama (1997): How Do Local Institutions Mediate Market and Population Pressures on Resources? Forest *panchayats* in Kumaon, India. *Development and Change*, Vol.28: 435-65.

Aragon, L. V. (2001): Communal Violence in Poso, Central Sulawesi: Where People Eat Fish and Fish Eat People. *Indonesia* 72: 45-79.

Baland, J. and J. Platteau (1996): Halting Degradation of Natural Resources: Is There a Role for Rural Communities? Oxford.

Bendor, J. (1995): A Model of Muddling Through. *American Political Science Review* 89(4): 819-840.

Bennett, J. W. (1985): The Micro-Macro Nexus: Typology, Process, and System. In: DeWalt, B. R. and P.J. Pelto (Eds): Micro and Macro Levels of Analysis in Anthropology: 29-53. London.

Berkes, F. (2002): Cross-scale Institutional Linkages: Perspectives from Bottom Up. In: Ostrom, E. et al. (Eds): The Drama of the Commons: 293-317. Washington.

Bernstein, H. J. and M. Fortun (1998): Muddling Through: Pursuing Science and Truth in the Twenty-first Century. New York.

Brosius, P., Tsing Lowenhaupt, A. and C. Zerner (2005): Introduction. Raising Questions about Communities and Conservation. In: Brosius, P., Tsing, Lowenhaupt A. and C. Zerner (Eds): Communities and Conservation. Histories and Politics of Community-Based Natural Resource Management: 1-36. New York, Oxford.

Cohen, A. P. (1985): The Symbolic Construction of Community. London.

Colfer, P. (2005): The Complex Forest. Communities, Uncertainty, and Adaptive Collaborative Management. Bogor.

Dreyfus, H. L. and P. Rabinow (1982): Michel Foucault: Beyond Structuralism and Hermeneutics. Chicago

Fremerey, M. (2005): Local Communities as Learning Organizations: The Case of the Village of Toro, Sulawesi. In: A. Neef (Ed): Participatory Approaches for Sustainable Land Use in Southeast Asia: 253-2 Bangkok.

Fremerey, M. and S. Amini (1999): Social Organization and Processes of Ecological Stabilization and Destabilization. Research Proposal forwarded to the German Research Association (DFG), unpubl.

Geertz, C. (1971): The Interpretation of Cultures. New York.

Grainger, A. (1995): Changes in Land Use and Forest Management in Southeast Asia: An Evolutionary Perspective. In: Sandbukt, O. (Ed): Management of Tropical Forests: Towards an Integrated Perspective: 3-17. Oslo.

Hardt, M. and A. Negri (2000): Empire. Die Neue Weltordnung. Frankfurt.

Hildyard, N. et al. (2001): Pluralism, Participation and Power: Joint Forest Management in India. In: Cooke, B. and U. Kothari (Eds): Participation: The New Tyranny?: 56-71. London.

Hobley, M. and K. Shah (1996): What Makes a Local Organization Robust? Evidence from India and Nepal. *Natural Resource Perspectives* 11: 1-9.

Li, M. T. (2001): Masyarakat Adat, Difference, and the Limits of Recognition in Indonesia's Forest Zone. *Modern Asian Studies* 35: 645-676.

Long, N. (1992): From Paradigm Lost to Paradigm Regained. In: Long, N. and Long, A. (Eds): Battlefields of Knowledge. The Interlocking of Theory and Practice in Social Research and Development: 16-46. London and New York.

Lieb, R. (2003): The Adaptation of Internally Displaced Persons in a Rain Forest Margin area. The Case of Kecamatan Lore Utara in Central Sulawesi, Indonesia. Diploma thesis Faculty of Organic Agricultural Sciences, University of Kassel.

Lindblom, C. E. (1959): The Science of Muddling Through. *Public Administration Review* 19: 79-88.

Mappatoba, M. (2004): Co-Management of Protected Areas. The Case of Community Agreements in the Lore Lindu National Park, Central Sulawesi – Indonesia. Doctoral Dissertation, Goettingen.

McCay, B. (1988): Muddling Through the Clam Beds: Cooperative Management of New Jersey's Hard Clam Spawner Sanctuaries. *Journal of Shellfish Research* 7(2): 327-340.

McCay, B. (2002): Emergence of Institutions for the Commons: Contexts, Situations and Events. In: Ostrom, E. et al. (Eds): The Drama of the Commons: 361-402. Washington.

McCay, B. and S. Jentoft (1998): Market or Community Failure? Critical Perspectives on Common Property Research. *Human Organization*, Vol. 57, No. 1: 21-29.

Mosse, D. (2001): "People's Knowledge", Participation and Patronage: Operations and Representations in Rural Development. In: Cooke, B. and U. Kothari (Eds): Participation: The New Tyranny? 16-35. London.

Norgaard, R. B. (1984): Coevolutionary Development Potential. *Land Economics*, Vol. 60, No. 2: 160-177.

Norgaard, R. B. (1994): Progress Betrayed: The Demise of Development and a Coevolutionary Revisioning of the Future. London.

Ostrom, E. (1990): Governing the Commons. The Evolution of Institutions for Collective Action. Cambridge.

Rammel, C., Stagl, S. and H. Wilfing (2007): Managing Complex Adaptive Systems – A Co-evolutionary Perspective on Natural Resource Management. *Ecological Economics*, Vol. 63, 1: 9-21.

Scott, J. C. (1998): Seeing Like a State. How Certain Schemes to Improve the Human Condition Have Failed. Yale.

Senge, P. M. (1990): The Fifth Discipline: The Art and Practice of the Learning Organization. New York.

Simon, H. (1962): The Architecture of Complexity. *Proceedings of the American Philosophical Society* 106: 467-482.

Wade, R. (1988): Village Republics : Economic Conditions for Collective Action in South India. San Francisco.

Wilson, J. (2002): Scientific Uncertainty, Complex Systems, and the Design of Common-Pool Institutions. In: Ostrom et al. (Eds): The Drama of the Commons: 327-360.

Wollenberg, E. et al. (2007): Facilitating Cooperation During Times of Chaos: Spontaneous Orders And Muddling Through in Malinau District, Indonesia. Ecology and Society, Vol. 12, No. 1 (3): 1-9. Paper available online via: http//www.ecologyandsociety.org/vol12/iss1/art3/

List of Contributors

Adiwibowo, Soeryo

Senior Lecturer
Head of Demographic, Agrarian and Political Ecology Division
Department of Communication and Community Development,
Faculty of Human Ecology, Bogor Agricultural University (IPB)
Kampus IPB Darmaga
Bogor 16680
Indonesia
Email: bowo@indonet.id

Barkman, Jan

Post-Doc Researcher
Department of Agricultural Economics and Rural Development
Georg-August-Universität Goettingen
Platz der Goettinger Sieben 5-7
37073 Goettingen
Germany
Email: jbarkma@gwdg.de

Burkard, Günter

Lecturer, Candidate for Habilitation
Faculty for Organic Agricultural Sciences
University of Kassel, Section: Rural Sociology
Steinstr. 19
37213 Witzenhausen
Germany
Email: storma-guenz@gmx.net

Döbel, Reinald

Senior Lecturer
Institute for Sociology
Westfälische Wilhelms-Universität Münster
Scharnhorststr. 121
48151 Münster
Germany
Email: dobel@uni-muenster.de

Fremerey, Michael

Professor for Sociology
Director of the Institute for Socio-cultural Studies (ISOS)
University of Kassel, Section: Rural Sociology
Steinstr. 19
37213 Witzenhausen, Germany
Email: fremerey@wiz.uni-kassel.de; mfremerey@t-online.de

Glenk, Klaus

Researcher
Socio-Economics Group
The Macaulay Land Use Research Institute
Craigiebuckler
Aberdeen AB15 8QH
United Kingdom
Email: k.glenk@macaulay.ac.uk

Hunowu, Momy

Lecturer of Sociology
Institut Agama Islam Negeri (IAIN)
Universitas Sultan Amai Gorontalo
Gorontalo
Indonesia
Email: momy_hunowu@yahoo.com

Marggraf, Rainer

Research Group Leader
Environmental and Resource Economics
Department of Agricultural Economics and Rural Development
Georg-August-Universität Goettingen
Platz der Goettinger Sieben 5-7
37073 Goettingen
Germany
Email: rmarggr@gwdg.de

Savitri, Laksmi Adriani
Researcher
Center for Strategic Studies
University of Indonesia
Kampus UI
Depok 16424
Indonesia
Email: savitri_la@yahoo.com

Shohibuddin, Mohamad
Lecturer, Director of Sajogyo Institute
Bogor Agricultural University
Kampus IPB Darmaga
Bogor 16680
Indonesia
Email: shohib@gmail.com; shohib@brighten.or.id

Sunito, Satyawan
Center for Agrarian Studies
Bogor Agricultural University
Kampus IPB Darmaga
Bogor 16680
Indonesia
Email: awan@indo.net.id, pka_ipb@indo.net.id

Southeast Asian Modernities

Aurel Croissant, Beate Martin (Eds.)

Between Consolidation and Crisis

Elections and Democracy in Five Nations in Southeast Asia

LIT

Southeast Asian Modernities

Aurel Croissant, Beate Martin, Sascha Kneip (Eds.)

The Politics of Death

Political Violence in Southeast Asia

LIT

Aurel Croissant; Beate Martin (Eds.)
Between Consolidation and Crisis
Elections and Democracy in Five Nations in Southeast Asia

Between Consolidation and Crisis focuses on five countries in Southeast Asia to examine how their elections have been conducted in the past two years, their domestic implications, and how the elections have differed from one another and from elections in other parts of Asia. Case studies on Cambodia, Indonesia, the Philippines, Malaysia and Thailand provide an overall understanding of the impact of elections on the consolidation or crisis of new democratic and semi-democratic polities in the region of Southeast Asia.

Bd. 3, 2006, 392 S., 29,90 €, br., ISBN 3-8258-8859-2

Aurel Croissant; Beate Martin; Sascha Kneip (Eds.)
The Politics of Death
Political Violence in Southeast Asia

This volume analyzes four aspects of political violence in Southeast Asia: elections and violence; intra-ethnic conflict, communist insurgency; terrorism and religious extremism and lethal crime and politics. Together, the ten case studies on Cambodia, Indonesia, Malaysia, the Philippines and Thailand challenge the idea that democratic governance will bring an end to internal violent conflict. As some examples in the region suggest, semi-democratic polities in Southeast Asia even may be more successful in reducing levels of internal violence, compared to new democracies in their neighbourhood and other types of political regime they have tried in the past.

Bd. 4, 2006, 384 S., 29,90 €, br., ISBN 3-8258-8860-6

LIT Verlag Berlin – Hamburg – London – Münster – Wien – Zürich
Fresnostr. 2 48159 Münster
Tel.: 0251 / 620 32 22 – Fax: 0251 / 922 60 99
e-Mail: vertrieb@lit-verlag.de – http://www.lit-verlag.de